IET TRANSPORTATION SERIES 6

Low Carbon Mobility for Future Cities

Other related titles:

Volume 1	**Clean Mobility and Intelligent Transport Systems** M. Fiorini and J.-C. Lin (Editors)
Volume 2	**Energy Systems for Electric and Hybrid Vehicles** K.T. Chau (Editor)
Volume 7	**Evaluation of Intelligent Road Transportation Systems: Methods and Results** M. Lu (Editor)
Volume 38	**The Electric Car** M.H. Westbrook
Volume 45	**Propulsion Systems for Hybrid Vehicles** J. Miller
Volume 79	**Vehicle-to-Grid: Linking Electric Vehicles to the Smart Grid** J. Lu and J. Hossain (Editors)

Low Carbon Mobility for Future Cities

Principles and applications

Edited by
Hussein Dia

The Institution of Engineering and Technology

Published by The Institution of Engineering and Technology, London, United Kingdom

The Institution of Engineering and Technology is registered as a Charity in England & Wales (no. 211014) and Scotland (no. SC038698).

First published 2017

The Institution of Engineering and Technology
Michael Faraday House
Six Hills Way, Stevenage
Herts, SG1 2AY, United Kingdom

www.theiet.org

British Library Cataloguing in Publication Data
A catalogue record for this product is available from the British Library

ISBN 978-1-78561-197-1 (hardback)
ISBN 978-1-78561-198-8 (PDF)

Typeset in India by MPS Limited
Printed in the UK by CPI Group (UK) Ltd, Croydon

Contents

Preface

The reform of urban mobility remains one of the major challenges confronting policy makers. Today, and despite decades of investment in transport infrastructure, mobility and access to economic opportunity is still hindered by high levels of congestion, long travel distances and unreliable travel times. These issues will become more pressing in the future as our world becomes more urbanised. While this urban growth will be largely driven by economic development and the search for a better quality of life, it will dramatically change the scale and nature of our communities, and put a tremendous strain on the built environment and infrastructure that delivers vital services like transport.

One of the key themes running throughout this book is a recognition that past (and still current) practices in urban and transport planning are fundamental causes of the transport problems we face today. The policies and practices that were adopted in the past are now having widespread undesirable effects on urban form, liveability, health and economic productivity. During the last half-century, cities worldwide have experienced rapid urban growth which led to urban sprawl, high demand for motorised transport and environmental problems, including noise, air quality and carbon emissions. The traffic gridlocks experienced on urban roads and motorways were the catalyst for most urban transport strategies and policy responses during that time period. The solution that was prescribed for most cities was to build out of congestion by providing more infrastructures for private vehicles, without equal attention to managing the demand for travel or improving alternative modes of transport in a sustainable manner. Equating mobility with building more roads has only nurtured a tendency towards increased motorisation and an ever-increasing inclination to expand the road network. This resulted in a range of unintended adverse environmental, social and economic consequences most of which are rooted in the high priority given to private vehicles in land-use and transport planning practices.

This volume was compiled on a central premise: urban transport policies and practices can be transformed in a sustainable and socially equitable direction for the benefit of future generations. To achieve this, the book calls for a conceptual leap and renewed thinking of how we address the contemporary challenges facing urban mobility and accessibility in our cities. The book is released at a time when these challenges are greater than ever, and brings together valuable contributions which analyse the complexity of urban transport and the interactions between transport, land-use, infrastructure, mobility, accessibility, emissions, health and well-being.

The book represents a call for action to overcome the unsustainable practices of the past. Rather than focusing on the infrastructure required to facilitate the physical movement of people and goods, the emphasis should shift to providing the mobility required for access to employment, goods and services. The book points out that considerable investments are still required in urban transport infrastructure in most cities, particularly in developing countries. It emphasises, though, that such investments should be guided by low carbon mobility priorities when and where they are most needed, and that such investments should be commensurate with community values and expectations.

The need, urgency and benefits of steering our cities towards a path of low carbon mobility are unmistakable. This has been emphasised throughout the book. This need has been recognised before but progress has been slow. Yet, the changing context for how we build future cities – smart, healthy and low carbon – presents new opportunities for progressing innovative policies and programmes for sustainable mobility.

The policy instruments identified in this book prioritise pathways and solutions with the likelihood of greatest impact in achieving reductions in transport energy use. These pathways focus on integrated land-use and transport policies; investments in dense and human scale cities; transit-oriented and pedestrian-oriented developments; optimised road network use; public transport and active travel options. They also recognise the role of digital innovations and disruptive mobility in addressing the modern-day demands of urban living and travel. If well planned and implemented, the proposed interventions will collectively improve accessibility to jobs and opportunities which are preconditions for sound economic development and achieving the emission reduction targets in cities around the world.

The Editor

Hussein Dia
Centre for Sustainable Infrastructure
Department of Civil and Construction Engineering
Swinburne University of Technology
Melbourne, Australia

Biographies

Hussein Dia is Associate Professor at the Centre of Sustainable Infrastructure at Swinburne University of Technology, Australia. He is a Civil Engineer with three decades of experience in Intelligent Transport Systems and transport modelling. He has worked extensively in collaboration with both the public and private sectors, including strong engagement with industry. His research interests are in smart infrastructure systems and the convergence of technology, infrastructure and human elements in urban environments. His current work is focused on disruptive mobility and harnessing digital innovations to unlock low carbon mobility opportunities. He is a Chartered Professional Engineer, Fellow of the American Society of Civil Engineers, Fellow of Engineers Australia and Fellow of the Institute of Transportation Engineers.

Peter Newman is the Professor of Sustainability at Curtin University in Perth, Australia. He has written 19 books and over 300 papers on sustainable cities. His books include 'The End of Automobile Dependence' (2015), 'Green Urbanism in Asia' (2013) and 'Sustainability and Cities: Overcoming Automobile Dependence' which was launched in the White House in 1999. Peter was a Fulbright Senior Scholar at the University of Virginia Charlottesville and was on the IPCC for their fifth Assessment Report. In 2014 he was awarded an Order of Australia for his contributions to urban design and sustainable transport. He is a Fellow of the Academy of Technological and Engineering Sciences Australia. Peter has worked in local government as an elected councillor, in state government as an advisor to three Premiers and in the Australian Government on the Board of Infrastructure Australia.

Michael Taylor is Emeritus Professor of Transport Planning in the School of Natural and Built Environments at the University of South Australia. He is an internationally acknowledged expert in travel demand modelling and demand forecasting, traffic systems modelling, the environmental impacts of transport and transport network reliability. His current research is in four areas: climate change adaptation for transport infrastructure, energy consumption and emissions modelling for road traffic, low carbon mobility, and transport network vulnerability. He has over 45 years professional experience, as a traffic engineer, a CSIRO research scientist, and in academia. Professor Taylor gained his PhD at Monash University. He is a Chartered Professional Engineer and a Fellow of Engineers Australia, the Chartered Institute of Logistics and Transport, and the Institute of Transportation Engineers. In 2014, he was awarded the Engineers Australia Transport Medal for outstanding individual contributions to transport in Australia.

Peter Newton is a Research Professor in Sustainable Urbanism at Swinburne University of Technology in Melbourne where he leads research on sustainable built environments. He is actively involved in three national Co-operative Research Centres: CRC for Low Carbon Living (as Research Program Leader), CRC for Spatial Information (as Research Project Leader) and CRC for Water Sensitive Cities (as a member of its Research Advisory Committee), is a partner in the Sustainable Built Environment National Research Network and is on the Board of the Australian Urban Research Infrastructure Network. He is a Fellow of the Academy of Social Sciences Australia and obtained his PhD at the University of Canterbury. Prior to joining Swinburne University in 2007 he held the position of Chief Research Scientist at the Commonwealth Scientific and Industrial Research Organisation (CSIRO). His most recent books include *Resilient Sustainable Cities* (2014), *Urban Consumption* (2011), *Technology, Design and Process Innovation in the Built Environment* (2009) and *Transitions. Pathways Towards More Sustainable Urban Development* (2008).

John Stanley joined the Institute of Transport and Logistics Studies (ITLS) at University of Sydney in July 2008 as Adjunct Professor and Bus Industry Confederation Senior Research Fellow in Sustainable Land Transport. Prior to taking on this role, he spent 9 years as Executive Director of Bus Association Victoria, after 8 years as Deputy Chairman of the National Road Transport Commission. He is a member of the Ministerial Advisory Committee advising Victoria's Minister for Planning on Melbourne's Metropolitan Planning Strategy. John has published widely on transport policy and planning and is co-author of the book, *An Introduction to Transport Policy*. He was awarded a Centenary Medal for services to public transport and conservation.

Chris Rissel is Director of the NSW Office of Preventive Health and Professor of Public Health in the School of Public Health, University of Sydney. The NSW Office of Preventive Health focuses on childhood and adult obesity prevention. His current research interests focus on obesity prevention and active transport, cycling advocacy, tobacco control and sexual health. He has a strong record of published research in the area of active travel and cycling and is one of the authors of the national report *Cycling: Getting Australia Moving*. He recently completed a 3-year grant to promote increased use of cycle paths in south-west Sydney – *Cycling Connecting Communities*. He is an Investigator with the Safer Cycling Study, and he is now working on evaluating new cycling infrastructure in the City of Sydney with an ARC Linkage grant.

Billie Giles-Corti is Director of the McCaughey VicHealth Community Wellbeing Unit, Centre for Health Equity, University of Melbourne. She is an NHMRC Senior Principal Research Fellow, an Honorary Fellow of both the Planning Institute of Australia and the Public Health Association, and a Fulbright Senior Scholar. For two decades, she and a multi-disciplinary research team have been studying the impact of the built environment on health and well-being. She leads an NHMRC Centre of Research Excellence in Healthy Liveable Communities established in 2014. She has published over 300 articles, book chapters and reports, and by citations, is ranked in the top 1% of researchers in her field globally.

Rocco Zito is Head of Civil Engineering at the Flinders University of South Australia. As a part of this leadership role he manages numerous research projects and teaching programmes at the undergraduate and postgraduate level. His research interests include the use of Global Positioning Systems (GPS) for transportation applications involving real-time modelling using Geographical Information Systems (GIS). More recent work has involved an investigation of the environmental impacts of various transport policy initiatives including low emission and autonomous vehicles.

John Stone is a Senior Lecturer in Transport Planning in the Faculty of Architecture, Building and Planning at the University of Melbourne. His research seeks to improve public transport performance in Australian cities through a greater understanding of the professional practice and the political and institutional context for public transport planning in similar cities in Canada and in German-speaking Europe. His research has identified many opportunities for more effective and efficient public transport in Australian cities through better service coordination and more efficient network design. Since completing his PhD in 2008, his work has been funded through grants including an ARC Discovery Project and a Commonwealth Endeavour Research Fellowship. This work follows over 20 years' engagement with public transport management in Melbourne that has included work in local government and the community sector.

Yvonne Kirk is a transport planning and policy researcher in the Melbourne School of Design at the University of Melbourne. Her published work has examined the structure of the Melbourne public transport franchises, Australian sustainable transport futures and history of urban planning in Melbourne. Her Master's thesis examined the structures of a reformed public transport authority established in Melbourne in 2011 to assess whether the elements required for success were in place. She is interested in many aspects of urban planning including transport, housing and community engagement and actively participates in community campaigns to improve public transport outcomes in Melbourne.

Ghassan Abu-Lebdeh is Associate Professor of Civil Engineering/Transportation at the American University of Sharjah, UAE. His current research is on the interactions between urban traffic operations and public health. His research interests/teaching is in the areas of urban traffic operations and control with focus on congested networks and sustainability in transport systems. He completed his studies at the University of Illinois at Urbana-Champaign (Ph.D.), USA. His academic and industry experience spans 27 years in transportation systems operations and planning with periods at Michigan State University, University of Kentucky, and the Metropolitan Planning Organizations in Worcester, Massachusetts, and in Urban-Champaign, Illinois.

Susan Thompson is Professor of Planning and Head of the City Wellbeing Program in the City Futures Research Centre, The University of New South Wales, Australia. Susan's academic career encompasses research and teaching in areas ranging from cultural diversity in urban planning, to migrant women's meanings of home, and the use of qualitative research methodologies in the built environment disciplines. For the past 12 years, Susan has focussed on the interdisciplinary area of healthy urban planning in her research, education and advocacy work. She directed the Healthy Built Environments Program at UNSW (2010–14) and in 2013 was the first urban planner to be appointed by the NSW Minister for Health to her Advisory Committee on Preventive Health. Susan's longstanding contribution to urban planning was recognised in 2012 when she was elected Fellow of the Planning Institute of Australia and in 2015, with the awarding of the prestigious Sidney Luker Memorial Medal.

Susan Shaheen is an Adjunct Professor in the Department of Civil and Environmental Engineering and a research engineer with the Institute of Transportation Studies at the University of California, Berkeley. She is also co-director of the Transportation Sustainability Research Center at UC Berkeley. She has authored 59 journal articles, over 100 reports and proceedings articles, four book chapters, and co-edited one book. Her research projects on carsharing, smart parking, and older mobility have received national awards. In May 2016, she was named one of the top 10 academic thought leaders in transportation by the Eno Transportation Foundation. She is a desk editor for Transport Policy and is a member of the editorial boards of the International Journal of Sustainable Transportation and the International Journal of Transportation Science and Technology. She is a member of the ITS Program Advisory Committee to the U.S. DOT and chair of the subcommittee for Shared-Use Vehicle Public Transport Systems of the Transportation Research Board.

Apaar Bansal is currently pursuing a Master of Science degree in Transportation from the Massachusetts Institute of Technology. Prior to this, he was at the Transportation Sustainability Research Center for two years, where he worked on research relating to one-way carsharing, roundtrip carsharing, bikesharing, on-demand mobility, and various conferences and papers pertaining to shared mobility. He is interested in the integration of shared mobility and public transportation operations and planning, as well as transportation in the developing world. He holds a Bachelor's degree in Civil Engineering from the University of California, Berkeley.

Nelson Chan is a survey researcher at the Transportation Sustainability Research Center. He has been researching the rise of shared mobility services, from traditional travel alternatives such as carpooling and vanpooling, newer carsharing and bikesharing systems, to the recent growth of ridesourcing services. He is particularly interested in understanding the impacts of these shared mobility services on reducing vehicle miles traveled, greenhouse gas emissions, and automobile ownership and how they can improve urban mobility. Nelson holds a Masters of Engineering degree in Civil and Environmental Engineering from the University of California, Berkeley.

Adam Cohen is a shared mobility researcher at the Transportation Sustainability Research Center at the University of California, Berkeley. Since joining the group in 2004, his research has focused on shared mobility, smartphone applications, and emerging technologies. He has co-authored numerous articles and reports on shared mobility in peer-reviewed journals and conference proceedings. His academic background is in city and regional planning and international affairs.

Eleni I. Vlahogianni is an Assistant Professor at the Department of Transportation Planning and Engineering of the National Technical University of Athens. She holds a diploma in Civil Engineering and a Ph.D. from the National Technical University of Athens, Greece (NTUA) (specializing in Traffic Operations).

Her primary research field is traffic flow analysis and forecasting. Other research fields include mobility modeling, driving analytics, ICT applications to transportation, Intelligent Transportation Systems, traffic management, advanced technologies for monitoring transportation infrastructures. She has a strong algorithmic background, which includes nonlinear dynamics, applied statistical modeling, data mining and machine learning techniques, and computational intelligence.

Her professional and research experience includes projects and consultancies, in a national and European level focusing on urban mobility and traffic flow management, public transport, and traffic safety.

She is an author of more than 50 articles in in international peer reviewed journals and more than 60 publications in international conference proceedings with over 1,000 citations.

Emmanouil N. Barmpounakis is a Ph.D. Candidate in the field of Traffic Operations, at the Department of Transportation Planning and Engineering at the School of Civil Engineering of the National Technical University of Athens since July 2013. He received a diploma in Civil Engineering (Transportation Engineering Cycle) in November 2012.

He has published two papers in International Journals, seven conference papers, and one diploma thesis. He is a scholar of the NTUA for Doctoral Studies, and during his undergraduate studies he was a scholar of Ergazakeio Foundation.

He has been awarded the Thomaideio Award 2014 and 2015 for scientific publication in International Conference and he holds awards and distinctions from the Hellenic Mathematician Society as a student.

Chapter 1
Introduction

Hussein Dia[1]

This volume examines the state of urban transport and explores the linkages and interactions between urban form, infrastructure, technology and human elements. It also presents the policy instruments that can be used to promote low carbon mobility and accessibility in our cities, with a view to determining the essential conditions for promoting sustainable transport in urban settings.

The chapters that comprise this book are principally the product of the intellectual efforts of the contributing authors who have dedicated their careers to researching the complex systems of urban mobility. The authors provide detailed analyses and compelling arguments to support this notion: for our cities to thrive; we must quickly begin to rearrange the priorities and incentives that guide the way we plan and design the built environment. Providing access to jobs, places and services using safe and efficient transport and mobility systems plays a major role in determining how that occurs, and how successful it will be.

The actions we take today will affect the liveability of our cities in the future. The authors fully appreciate the importance and urgent need for a paradigm shift to transform how we provide mobility in our cities. In their contributions, they share with the reader much valuable knowledge and insights about how we might get urban transport to correctly align with the challenges of providing the necessary access to our increasingly urbanised planet.

This introductory chapter provides a framework for the remainder of the book. It also describes the contents of each chapter and draws linkages to assist the reader in navigating the volume.

1.1 Chapter 2: Urban mobility: challenges and opportunities (Hussein Dia)

This chapter describes the major transport challenges facing out cities. This includes rapid urbanisation, road crashes and injuries, traffic congestion, environmental emissions, ageing transport infrastructure and limited budgets to maintain and upgrade the urban infrastructure necessary for resilient modern transport systems.

[1]Centre for Sustainable Infrastructure, Swinburne University of Technology, Melbourne, Australia

Another major challenge highlighted in the chapter is related to the vulnerability of a city's transport system and its limited ability to resist, absorb, adjust to and recover from the effects of acute stresses or 'shocks' (such as natural disasters or human-induced stresses) in a timely and efficient manner. Through discussion of these challenges, which are faced by cities in both developed and emerging economies, the chapter sets the scene for the complexity of urban transport.

The chapter then describes how the traditional approaches to addressing mobility, that is building additional road capacity, has met with limited success. It outlines the limitations of the 'predict-and-provide' approach and highlights how the dominance of motorised transport has resulted in high levels of mobility and urban sprawl, which translated into high volumes and long distances needed for travel to access economic opportunities and services, leading to high carbon mobility [1].

Next, the chapter proposes that addressing the challenges facing cities in the twenty-first century requires a conceptual leap, new approaches and a holistic vision which incorporates different strategies that emphasise land-use transport interactions and prioritise sustainable modes of mobility. Finally, the chapter introduces a framework for achieving this vision, which includes strategies for promoting integrated transport planning, public and active transport, and the role of digital innovations and disruptive technologies in improving connectivity, mobility and access to jobs and activities in urban centres [2].

1.2 Chapter 3: Policy principles for low carbon mobility (Hussein Dia)

Achieving energy efficiency improvements in urban transport is a challenging task. To deliver substantial reductions in emissions and promote low carbon transport, major policy, behavioural and technological changes would be required to achieve fuel security and climate change targets.

This chapter presents the strategies available to decision makers that can be used to steer cities on a sustainable course of urban mobility. These strategies can be grouped into four broad categories: those that allow travel to be 'avoided'; those that 'shift' travel to more efficient modes; those that promote 'sharing' of vehicles and rides; and those that 'improve' the efficiency of vehicles and supporting infrastructure.

The chapter then examines the strategic pathways to low carbon mobility and presents the policy principles that can help achieve efficiency improvements and emissions abatement, while also addressing urban mobility challenges such as congestion and poor air quality. These include planning dense cities; developing transit-oriented and pedestrian-oriented cities; optimising the road network and its use; public transport improvements; and planning for walking and cycling. A broad range of measures that could be used to influence behaviour are also examined to explore what could be achieved with respect to reduction of emissions.

Using best evidence from the literature on successful responses to policy interventions, the chapter then presents some case studies which demonstrate the benefits that accrue to communities, and how these can help to enhance the fabric of urban environments and make cities more liveable. Through these case studies, the chapter identifies the barriers and success factors for policies that have worked best elsewhere, and the most promising combinations of measures to encourage a reduction in the use of motorised transport.

The chapter also argues that well-planned mobility solutions for tomorrow's cities should improve accessibility to jobs and opportunities which are preconditions for sound economic development and for mitigating emissions and pollution in urban centres. Finally, the chapter includes policy pathways and steps for practitioners to support them in the development, financing, implementation and evaluation of low carbon mobility policies based on lessons learnt from successful implementations from cities around the world.

1.3 Chapter 4: Low carbon mobility and reducing automobile dependence (Peter Newman)

In this chapter, Newman explores the development of cities and the transport planning processes that have guided them, from walking cities to transit cities to the modernist era of automobile cities. He discusses how automobile dependence has been the main paradigm driving transport planning for the period from the 1940s until recently. The resulting urban fabric was quite distinct and required different approaches to recognise, respect and rejuvenate them, but the majority of transport planning guidance did not enable such area-specific solutions. The manuals of modernism have been built around automobile-based approaches with little room to enable walking and transit-oriented approaches and urban fabric.

Today, the demand is to reduce automobile dependence so it will need new manuals and new approaches. Newman suggests that low carbon mobility in our cities will need to be a combination of structural change that reduces automobile dependence as well as technological change. To determine the potential mechanisms for accelerating this transition, he outlines the emerging trends in reduced automobile dependence in the world's cities together with emerging technological trends. He highlights that these alternatives require substantial commitment to change in both how we live and the infrastructure priorities we have in our cities and regions. Newman also discusses that the transition will take time, but already it is clear that the early adopters in cities around the world are establishing economic competitiveness. The first signs of change towards emerging technologies can now be seen: the re-urbanisation of our cities; the dramatic growth in electric transit; the rapid move towards electric vehicles and solar energy with a 30%–40% per annum growth in global renewables; the emerging use of natural gas and biofuels; and new technologies like Skype and telepresence.

Newman provides some case studies and shows that the application of these interventions into large-scale urban demonstrations is now underway in places like Kronsberg and Vauban in Germany, Masdar in UAE and the Low Carbon Cities of China. Newman offers that the potential for creating oil-free cities is here and suggests that we can be oil-free by 2050 and renewably based oil-free by 2100.

1.4 Chapter 5: Integrated land-use and transport planning for future cities: and the importance of active transport (Michael A.P. Taylor)

In this chapter, Taylor discusses land use-transport integration (LUTI) which addresses the development, management and operation of urban transport systems that provide sustainable outcomes for our cities. LUTI is primarily concerned with the optimal distribution of facilities and services in an urban area so that these are available to all of its inhabitants in ways that minimise adverse environmental impacts whilst maximising opportunities for sustainable economic development and social interaction.

Taylor proposes LUTI as a key consideration in low carbon mobility. The means for achieving this would largely be through the encouragement and use of active transport modes (walking and cycling) and public transport – because access to public transport generally requires substantial use of the pedestrian and cycle modes. The key concept behind LUTI is in fact the basic driver of urban development: accessibility. In the planning context, accessibility is about the effort required for people and enterprises to engage in everyday life. Integrated planning of transport and land use can offer high levels of accessibility based on the use of active transport. Amongst the many issues facing planners are environmental impacts, both local (e.g. noise and air quality) and global (climate change and greenhouse gas emissions), economic productivity, conservation of resources (especially the use of land), traffic congestion, social severance and public health. Taylor poses that following the established principles and practices of LUTI offers a systematic path to long-term resolution of these issues. This chapter explains why and how this can be accomplished.

Through this chapter, Taylor offers that attaining sustainable transport and low carbon mobility is vital for the future success of our cities. For transport systems, this means focusing on moving people and goods instead of cars and trucks. Taylor also suggests that land use planning and transport planning have key roles in shaping our cities and in delivering social, economic and environmental sustainability. The successful delivery of these outcomes requires strong integration of land use and transport planning. Only then can we achieve a sustainable urban land use and transport system featuring low carbon mobility. LUTI can facilitate a sustainable city by (1) reducing the need to travel, (2) reducing the length of trips, (3) making it easier and safer for people to access services and facilities, (4) reducing transport impacts on communities, (5) providing for the efficient distribution of goods and services to businesses and the community, (6) providing choices in travel modes, and (7) ensuring flexibility now and into the future to address changes in demand in the economy and in society.

To conclude, Taylor offers that the most important contribution that LUTI could make to achieving low carbon mobility would be the reorganisation of facilities and services in an urban area so that trip destinations are generally closer to trip origins, reducing travel distances and making the use of active transport modes more feasible and attractive. Arranging residential areas clustered around activity nodes and with good access – again by the active modes – to line haul public transport is also important for longer distance travel, but the proximity of destinations remains a key feature of the integrated land use and transport system.

1.5 Chapter 6: Decarbonising suburban mobility (Peter Newton, Michael A.P. Taylor, Peter Newman, John Stanley, Chris Rissel, Billie Giles-Corti and Rocco Zito)

In this chapter, the authors propose that greening suburban mobility in low-density cities will require a step change in both the demand and supply sides of urban travel. Supply-side innovations are critical in order that the opportunities and alternatives are established for car-dependent suburban residents to embrace more effective forms of active, low carbon travel.

The chapter provides a high-level roadmap of the innovations and planning strategies required for a low carbon transition in each of the key supply-side arenas of urban transport. The authors offer that achieving this vision requires bi-partisan political leadership to deliver city plans, rather than government plans. They cite some examples from Australia where a 'planning deficit' has been evident for most of the past 50 years in relation to long-term planning of Australia's major cities – an increasing catalogue of strategic plans that have either been abandoned with a change of government or at best inadequately implemented.

A major casualty of these failures is the mobility available within cities, and especially the outer suburbs. The authors see a continuing need to halt automobile-dependent urban sprawl, a process that is compounding now well-known negative externalities such as loss of agricultural land in addition to those impacts catalogued in Chapter 2.

A combination of settlement strategies is required that can harness centripetal as well as centrifugal forces of population geography: disperse population growth to regional cities as well as redirect population inwards to the established suburbs via infill and urban intensification. In both cases, the authors find that sound strategic planning by governments that can activate market investment is critical.

The failed decentralisation policies of an earlier era need to be replaced by those that recognise the agglomeration economy benefits that accrue to large cities and to ensure regional cities can become part of a functional mega-metropolitan region – a 'system of cities' – via fast rail. For example, the authors cite high-speed rail where a 350-km/h high-speed service between Melbourne and provincial cities such as Geelong, Ballarat, Bendigo and Warragul would transform these centres into the equivalent of present day middle ring suburbs of Melbourne where 30-min commutes to work are typically the norm. The authors also discuss how urban regeneration in the established suburbs of cities is also required to deliver more sustainable outcomes for future populations. Significant transformation in transport and land use would be required to achieve this.

1.6 Chapter 7: The 'disruption' we really need: public transport for the urban millennium (John Stone and Yvonne Kirk)

In this chapter, Stone and Kirk discuss how mass transit is the only form of motorised transport that can move large numbers of people to the same destination at the same time without either alienating the space needed for social and economic interaction, or allowing cities to encroach further upon vital natural environments or agricultural land.

Despite the rapid development and deployment of 'disruptive' technologies in urban transport, the authors find that mass transit will still have a vital role to play in the transport systems of the world's great urban regions in coming decades. This is largely because cities of the future will face increasing competition over space. No urban region has been entirely successful in creating mass transit networks that offer speed and convenience approaching that of the private car, but some have done much better than others. The authors present some case studies of relative success in the creation of space-efficient transport systems in the urban regions of Vienna, Zurich and Vancouver. They provide an overview of transport system performance including operating costs, infrastructure investments, and mode share, together with contextual demographic data. In each case, this is accompanied by a short outline of the political and institutional processes that have enabled these outcomes to be achieved.

Common to features in all three cities is the consistent and skilful engagement in local political processes by transit advocates and planners; and coherent use of transport planning practices that give primacy to meeting the needs of mass transit users at the lowest possible cost. The authors suggest that the case studies of Vienna, Zurich and Vancouver demonstrate that transport planners have been able to achieve something that has often been considered impossible. The public transport systems in these urban regions have consistently increased their mode share at the expense of the private car in their inner and middle suburbs. And, this has been done at a cost that was affordable to the wider urban economy.

These results have also been achieved in cities in both North America and Europe, but the element that is common to all is a particular approach to transport planning which includes (1) a coordinated commitment to giving priority for public and active transport over the private car in investment and in the allocation of road space; (2) integrated land-use and transport planning to encourage the development of destinations (employment, services, schools and colleges, etc.) at nodes that can be serviced by multidirectional transit networks; (3) planning and development of effective conditions for intermodal transfers, including fare structures that do not penalise passengers making transfers and which allow for cross-subsidies from highly patronised lines to support necessary frequencies on feeder and circumferential lines; and (4) creation of institutions for the planning and operation of effective public transport networks (these are typically public agencies that are given some measure of budgetary independence).

1.7 Chapter 8: Urban transport and impacts on public health (Ghassan Abu-Lebdeh)

The focus of this chapter is the impacts of urban transport on health: physical, mental, well-being, and liveability. Specifically, and besides the traditionally obvious and well-studied subjects of transport-induced crash injuries and pollution health impacts, Abu-Lebdeh presents the modern day non-contagious ailments as stress, cardiovascular diseases, physical inactivity and obesity and highlights their links and interactions with urban transport systems.

These health impacts are mostly significant in the long term. They are subtle, slow and pervasive. They have traditionally not been addressed or even identified as transport related. Only recently and just qualitatively, these impacts came to the fore of the interrelated subjects of urban form, transport and modern day health problems. Abu-Lebdeh also addresses the question of how to objectively measure and quantify these health impacts with a particular view to evaluating the benefits that can be realised through low carbon mobility solutions.

He also offers that not all aspects/impacts on public health, though, are well understood. Abu-Lebdeh identifies the needs for research to both further the understanding of the dynamics between transport and urban form on one hand, and the identified health impacts on the other. He also poses that the definition of a 'smart city' as practised today needs to expand to also mean "health/healthier". He suggests that even though health and healthy is sometimes implied, it is not enough. Health needs to be addressed explicitly. Abu-Lebdeh sees that incorporating consideration of health in transportation projects, planning design, construction, operation and maintenance is necessary. Significant negative impacts can be avoided; others can be minimised.

He also emphasises the points made in earlier chapters that past (and still current) transport planning paradigms, which are grounded in the 'predict and provide' model are not sustainable nor are they efficient when issues beyond economic efficiency, as public health, are incorporated and accounted for. He also suggests that although transport demand management and sustainable transport have now become part of the renewed transport planning thinking, progress has been slow, and the current transport system is still carbon-intensive with continued emphasis and/or valuing of the private car.

Finally, Abu-Lebdeh suggests that a paradigm shift to low carbon mobility is necessary and that a systems approach is a must to avoid unintended consequences. All activities in an urban environment are components of one interconnected system. For it to function properly and optimally, all subcomponents (transport, land use, economy, politics, health system etc.) must be considered as one whole. Devising solutions to any of these components must account for potential unintended side effects on other components.

1.8 Chapter 9: Active transport: policy directions for creating built environments that support health and well-being (Susan Thompson and Michael A.P. Taylor)

This contribution builds on and extends Chapter 8 which examined the impacts of motorised transport on health. Specifically, this chapter focuses on how planning policy, practice and design can facilitate health and well-being through low carbon active transport interventions such as walking and cycling. Thompson and Taylor discuss the potential co-benefits emerging from the adoption of such mobility policies. These co-benefits are largely found in the areas of environmental impact, public health, social cohesion and economic productivity. A framework for identifying and assessing potential co-benefits is provided, as are case studies of planning and policy interventions to encourage low carbon active transport forms.

The authors point out that health and well-being are affected by the way people travel, which itself is significantly influenced by the shape and form of the built environment. In turn, this largely determines the accessibility and ease-of-access of different travel modes. In their contribution, the authors identify a set of three domains relating to specific behaviours that support the achievement of good health. The 'getting people active' domain has direct connections with travel and transport systems in urban areas. Walking and cycling, as active transport modes, make major contributions to individuals' health outcomes, both through the physical activity involved in daily life and recreational physical activity undertaken for fun, fitness and enjoyment.

The chapter also includes sets of basic questions and corresponding actions which planners can use to help create built environments that encourage physical activity to bring about positive community health outcomes, which also have environmental and productivity benefits, particularly around the reduction of car-based congestion. Thompson and Taylor emphasise that adopting a co-benefits approach to professional policy, practice and community action provides a strong platform to consolidate and extend effective and efficient approaches to addressing these issues. It can also facilitate cross-disciplinary exchange, mutual understanding and a better appreciation of the synergies between the disciplines of health, transport and environmental sustainability. Furthermore, adopting co-benefits can help professionals to link different policy environments, facilitating an ability to see what they can contribute to their own policy and research endeavours. The authors find this critical to support policy and practice initiatives for low carbon travel which supports health for humans and the planet upon which they reside.

In their case studies, Thompson and Taylor provide examples of current initiatives that are improving community well-being and health outcomes, along with reductions in carbon emissions. Finally, the authors emphasise that recognition of the roles of the active transport modes in meeting people's accessibility needs should be a vital component of urban policy formulation and implementation.

1.9 Chapter 10: Mobility and the sharing economy: industry developments and early understanding of impacts (Susan A. Shaheen, Apaar Bansal, Nelson Chan and Adam Cohen)

Shared mobility – the shared use of a vehicle, bicycle or other mode – is an innovative transportation strategy that enables users to gain short-term access to transportation modes on an 'as-needed' basis. Shared mobility includes various forms of carsharing, bikesharing, ridesharing, for-hire driver services and microtransit.

In this chapter, the authors describe the different models that have emerged in the shared mobility space and review research that has quantified the environmental, social and transportation-related impacts of these services. The authors also project future trends as automated vehicles begin to emerge.

The case studies included in this chapter demonstrate a number of environmental, social and transportation-related benefits of shared mobility services, such as a reduction of vehicle miles/kilometres travelled, vehicle use and vehicle ownership. Users have often cited cost savings and convenience over private vehicle ownership as reasons of using a shared mode. The authors suggest that although this has been demonstrated for roundtrip and one-way car-sharing and public bike-sharing, more research is needed to better understand the impacts of for-hire driver services, microtransit and courier network services.

The authors also point out that despite shared mobility's impacts and opportunities, challenges remain particularly in the area of public policy. Cities and states have faced difficulties in enacting and/or adapting policies to keep pace with shared mobility innovations while addressing issues of public safety, insurance and liability, and fair labour practices. All too often, equity considerations remain unanswered – early anecdotal evidence is emerging which suggests that lower income populations may lack sufficient access to certain shared mobility services. In addition, labour issues concerning for-hire vehicle drivers raise questions regarding whether drivers should be considered independent contractors or employees. Public–private partnerships, such as the Ride KC Bridj pilot, which employs drivers from the Kansas City Area Transportation Authority, may offer guidance on how public agencies can address labour issues.

Finally, the authors provide a glimpse of what the future might look like when automated vehicles arrive. This technology is poised to dramatically transform urban transportation, whereas shared autonomous vehicles may become a major public transportation mode, filling gaps in the existing fixed-route network and allowing first- and last-mile connectivity. Although research and development into autonomous vehicles has increased, the authors find that their wider impacts on travel and emissions remain unknown. The next chapter looks into this in some more detail.

1.10 Chapter 11: Autonomous shared mobility: shaping the future of urban transport (Hussein Dia)

Autonomous vehicles may be on our roads within 5–10 years. Although still a few years away, they have already started to shape some visions for a mobility transformation driven by a number of converging forces: self-driving technologies, mobile computing, on-demand car and ride sharing, and vehicle electrification. The coming together of these powerful trends is shaping an urban mobility future inspired by a vision of zero road injuries and low carbon living.

The introduction of autonomous vehicles will have major socio-economic impacts on our society – some good and some bad. Although they are expected to pave the way to drastic improvements in road safety and urban mobility, they can potentially have adverse impacts on the liveability of our cities if they are not included as part of a holistic approach to promote low carbon mobility. The fast pace of developments in the autonomous vehicles and collaborative mobility marketplaces is increasingly pointing to a wave of disruptions and business models that will have much wider impacts than the technology itself and might even impact on-road public transport services in the long run.

This chapter provides an environmental scan and an objective review and analysis of the technological, social and economic impacts and trends surrounding autonomous vehicles, and how they are likely to influence the transport and mobility industries and marketplace in the future. The chapter also outlines the technological, societal and regulatory challenges that must be overcome before widespread deployment. It gathers and collates information from a wide body of literature to help city planners, infrastructure engineers and other stakeholders in responding to the likely needs and impacts of this disruptive technology, and to inform their policy positions in shaping the future of urban mobility.

Observations of the current developments in this area suggest that there is a huge momentum from the private sector and also similar developments, although at a slower pace, in the area of regulations and public policy. Given the fast pace of developments in this space, all stakeholders impacted by these developments need to stay abreast and identify potential threats and opportunities to their businesses and industries.

1.11 Chapter 12: Gamification and sustainable mobility: challenges and opportunities in a changing transportation landscape (Eleni I. Vlahogianni and Emmanouil Barmpounakis)

Motivating people to change their mobility behaviour towards more sustainable ways of travel is one of the major challenges in the reform of urban transportation. Nowadays, with information communication and technology (ICT) solutions and especially smartphones, it is possible to deliver accurate mobility-related information to users to help them make smart travel decisions with the aim of reducing congestion and air pollution, and improving safety. However, although delivering mobility-related information is important, it is not a sufficient condition for initiating behavioural changes. The last can only be achieved, if the communicated information is targeted to influence the attitudes and motivators of road users.

In this chapter, the authors discuss how best to influence road users to change their behaviour. Specifically, the chapter investigates gamified applications as tools to trigger behavioural changes in urban mobility. The gamification tools are analysed and critically discussed, and example applications are reviewed with emphasis on their impacts on traffic operations, urban mobility and environmental conditions in cities. The opportunities stemming from the advances in technology and communications, and the potential of gamified mobility solutions for engaging citizens in sustainable mobility, are also presented.

The case studies provided in this chapter show that integrating gamification concepts into current and emerging ICT solutions can enrich the users' motivation and engagement, which may lead to influencing their mobility profile in a sustainable manner. However, the authors find limited evidence and quantified results regarding the impacts and suggest that future research should focus on the proper selection of the factors that are likely to trigger behavioural change and make such applications efficient and sustainable.

1.12 Chapter 13: Digital innovations and disruptive mobility: mapping the value beyond the hype (Hussein Dia)

The fast pace of breakthroughs in new technologies and scientific advances is relentless and continues to unfold on many fronts. As the list of 'next big things' grows longer, it becomes challenging to identify which new or emerging technologies may unleash disruptive changes and impact the growth and performance of transport systems in our cities.

The continuous waves in digital innovations over the past few years, and the sweeping changes anticipated by disruptive mobility, have inspired visions of a very different future, as well as a good deal of hype. To distinguish between the hype and reality, this chapter discusses the role of disruptive technologies within low carbon mobility networks and describes how digital innovations are changing the urban mobility landscape, enabling real-time measurement and analysis of urban mobility, and promoting sustainable smart cities.

The chapter then maps the value of data-driven innovations in mobility and draws on practical applications, case studies and modelling results to demonstrate the benefits. It also describes the major behavioural and technological changes, and the new mobility business models that would be required to achieve the desired targets and outcomes. The chapter also provides a critical reflection of the value derived from typical smart mobility 'use cases' and identifies the policy principles which are central to the success of smart mobility.

The chapter provides strong evidence that the benefits of investing in smart systems are compelling, particularly given the improvements that could be made in terms of providing innovative solutions to lift economic efficiency and living standards. To spur change programmes and capture potential savings, it is suggested that cities must move beyond a project-by-project approach and upgrade systems for planning, operating and delivering the low carbon smart mobility solutions. This sort of investment will give our cities an opportunity to address the transport challenges of the twenty-first century and meet consumers' needs and demands for more integrated and reliable mobility options.

1.13 Chapter 14: Summary and future directions (Hussein Dia)

This closing chapter provides a summary of the key themes arising from the individual chapters and outlines directions for future research, evaluation and monitoring frameworks. It also provides a research agenda for future work guided by the recommendations and findings from individual chapters. It also presents a framework for rethinking urban mobility and points to three key dimensions for how this can happen: (1) emphasis on a systems approach to take into account the compounding influence and interaction between the different elements; (2) consideration of transport not only as a 'derived demand' but also as a 'valued activity' which can no longer be considered 'wasted time'; and (3) shifting the emphasis from mobility to accessibility [1].

The chapter also summarises the policies and strategies available to decision makers and reframes these in six categories to demonstrate their strategic linkage and their cumulative potential for triggering policy and operational change [2]. Finally, the chapter provides a research agenda and some practical research routes that can be followed to inform low carbon mobility policies [3–5].

References

[1] Banister, D., 2007. The sustainable mobility paradigm. *Transport Policy*, vol. 15, pp. 73–80.

[2] UN-Habitat, 2012. *Planning and Design for Sustainable Urban Mobility: Global Report on Human Settlements 2013*. Available at: http://unhabitat.org/planning-and-design-for-sustainable-urban-mobility-global-report-on-human-settlements-2013/ [Accessed September 17, 2016].

[3] Philip, M. & Taylor, M., 2014. *Research Synthesis Report: A research agenda for low carbon mobility*. CRC for Low Carbon Living. Available at: http://www.lowcarbonlivingcrc.com.au/resources/crc-publications/crclcl-project-reports/rp2009-research-synthesis-report-research-agenda [Accessed August 17, 2016].

[4] Dia, H., 2014. *Workshop on Greening Suburban Travel*. Swinburne University of Technology. Available at: http://www.lowcarbonlivingcrc.com.au/research/program-2-low-carbon-precincts/rp2021-greening-suburban-transport [Accessed August 17, 2016].

[5] Bruun, E. & Givoni, M., 2015. Six research routes to steer transport policy. *Nature*, vol. 523, no. 7558, pp. 2–4.

Chapter 2

Urban mobility: challenges and opportunities

Hussein Dia[1]

Abstract

This chapter describes the major transport challenges facing our cities. This includes rising urban populations, road crashes and injuries, congestion, environmental emissions, ageing transport infrastructure and limited budgets to maintain and upgrade the urban infrastructure necessary for resilient modern transport systems. The chapter also describes how the traditional approaches to meeting the demand for travel, for example building additional road capacity, have met with limited success. Finally, the chapter introduces the elements of a framework for low carbon mobility which includes strategies for promoting integrated transport planning, public and active transport and the role of digital innovations and disruptive technologies in improving connectivity and efficiency of services.

Keywords

Low carbon mobility; rapid urbanisation; road safety; traffic congestions; emissions; resilient infrastructure; asset optimisation; sustainable mobility

2.1 Introduction

As part of a more interconnected world, our cities are playing an increasingly active role in the global economy. According to the McKinsey Global Institute [1], just 100 cities currently account for 30% of the world's economy. New York City and London, together, represent 40% of the global market capitalisation. In 2025, 600 cities are projected to generate 58% of the global gross domestic product (GDP) and accommodate 25% of the world's population. The McKinsey Global Institute also expects that 136 new cities, driven by faster growth in GDP per capita, will make it into the top 600 by 2025, all from the developing world, 100 of them from China alone [1]. The twenty-first century appears more likely to be dominated by these global cities, which will become the magnets of economy and engines of globalisation.

[1]Centre for Sustainable Infrastructure, Swinburne University of Technology, Melbourne, Australia

Although the forecast urban growth will be largely driven by economic development and the search for a better quality of life, the resulting success will dramatically change the scale and nature of our communities and put a tremendous strain on the infrastructure that delivers vital services like transport, electricity, water and communications [1]. Today, more than half the world's population lives in towns and cities and the percentage is growing. By 2050, two-thirds of the world population will be living in cities and urban areas [2]. Already, ageing infrastructures in many cities around the world are at a breaking point with governments' budgets for major infrastructure projects under increasing pressure [3].

The reform of urban mobility remains one of the biggest challenges confronting policy makers around the globe [3,4]. According to the World Health Organisation [5], it is estimated that 1.2 million people are killed on the world's roads each year. If left unchecked, this number is predicted to reach 1.9 million fatalities worldwide by 2020. The human cost is profound – unimaginable suffering and grief. The economic cost is also staggering, totalling $100 billion a year in developing countries alone [6]. The World Health Organisation [5] has described road casualty figures as being of 'epidemic' proportions, with road-related trauma being the biggest single killer of those aged between 15 and 29.

In addition, it has also been estimated that the social, economic and environmental costs of avoidable congestion account for a large percentage of a country's GDP. At the international level, the avoidable cost of congestion is estimated to account for more than 1% of the GDP across the European Union, and currently cost the United States more than $115 billion each year [7]. In addition, road traffic continues to account for around 80% of all transport carbon dioxide (CO_2) emissions and is expected to reach 9,000 Mt per year by 2030 if the current mobility trends are not curbed [8].

This chapter is divided into two parts. The first part covers the challenges faced by cities in providing adequate transport to meet current and growing travel demands. The second part explores the opportunities available for cities to develop in order to meet such challenges.

2.2 Rapid urbanisation

The global urban population is increasing rapidly. Today, more people live in urban areas than in rural areas. Around 200 years ago, only 3% of the world's population lived in urban centres [9]. The urban population of the world grew from 746 million people in 1950 to 3.9 billion in 2014 [10]. Despite its lower level of urbanisation, Asia is estimated to accommodate around 53% of the world's urban population, followed by Europe with 14% and Latin America and the Caribbean with 13% [10].

In 2007, for the first time in history, the global urban population exceeded the global rural population (Figure 2.1). Since then, the world population has predominantly remained urban [10]. In 2014, 54% of the world's population was

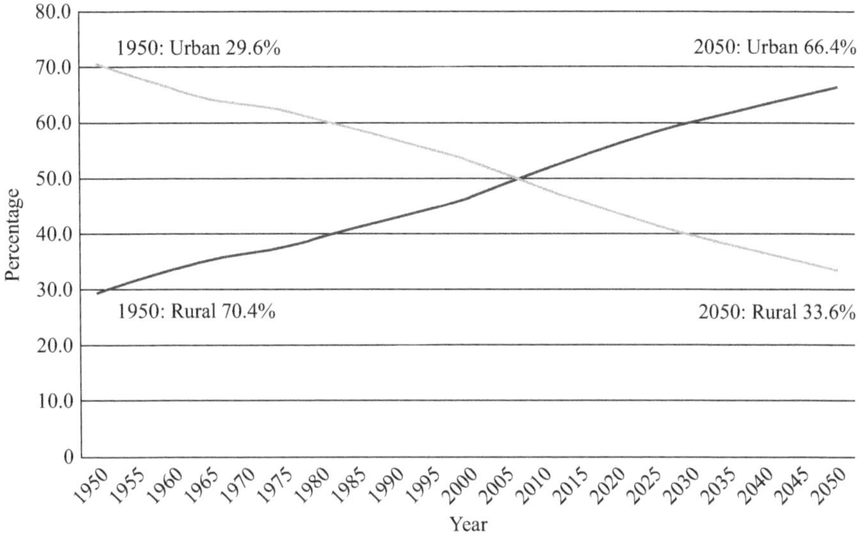

Figure 2.1 Percentage of world population living in urban and rural areas 1950–2050. Data source: [10]

urban, and the trend is expected to continue so that by 2050, the world will be 66% urban and roughly one-third rural [10].

The levels of urbanisation vary greatly across the world's geographies. Although urbanisation has occurred in all major regions, Asia and Africa are still mostly rural but are urbanising more quickly than other geographies. The proportion of urban population in Asia is increasing by 1.5% per annum, compared to 1.1% per annum in Africa [10]. The UN projections show that urbanisation combined with the growth of the world's population could add another 2.5 billion people to urban populations by 2050, with close to 90% of the increase concentrated in Asia and Africa (particularly in India, China and Nigeria) [10]. These three countries will account for 37% of the projected growth of the world's urban population between 2014 and 2050. If the current trends continue, by 2050, India is expected to add 404 million urban dwellers, China 292 million and Nigeria 212 million.

The world's urban population is expected to exceed 6 billion people by 2045, with much of the growth taking place in the developing world. The countries in these emerging regions already face massive problems, and it will become more challenging for them to meet the needs of their growing urban populations including housing, infrastructure, transportation, energy, employment and access to opportunities and basic services such as education and healthcare [10].

2.2.1 Increasing number of mega-cities

In 1990, there were ten 'mega-cities' with 10 million people or more. These cities accommodated around 153 million people or slightly less than 7% of the global urban population at that time [10]. In 2014, the number of mega-cities increased to 28 worldwide and accommodated around 453 million people or about 12% of the world's urban population. Of these 28 mega-cities, 16 were located in Asia, 4 in Latin America, 3 each in Africa and Europe, and 2 in Northern America [10]. By 2030, the world is expected to have 41 mega-cities each with 10 million inhabitants or more. A number of selected current and future mega-cities are shown in Figure 2.2.

As of 2014, Tokyo was the world's most populous city (38 million people), followed by Delhi (25 million), Shanghai (23 million) and Mexico City, Mumbai and São Paulo, each with around 21 million people. Osaka had just over 20 million, followed by Beijing with slightly less than 20 million. The New York–Newark area and Cairo complete the top ten most populous urban areas with around 18.5 million people each [10].

In the future, Tokyo's population is expected to decline, but it is projected to remain the world's largest city in 2030 with 37 million people, followed by Delhi (36 million in 2030). Osaka and New York–Newark were the world's second and third largest urban areas in 1990. By 2030, however, they are projected to fall in rank to the 13th and 14th positions, respectively, as mega-cities in developing nations become more prominent [10].

2.2.2 Urban growth expected to concentrate in small- and medium-sized cities

With the rapid rate of growth, the number of urban dwellers is expected to increase by three billion by the end of the century [12]. Most of this urban growth is expected to take place in small- and medium-sized cities each with a million or fewer inhabitants. In 2014, nearly half of the world's 3.9 billion urban dwellers lived in relatively small areas with fewer than 500,000 inhabitants, with only one in eight living in the 28 mega-cities (10 million inhabitants or more). Many of the fastest growing cities in the world are relatively small urban settlements [10]. To meet the projected demand, it is estimated that approximately one new city of a million people each will have to be built every 5 days until 2050 [12]. Furthermore, it is expected that the majority of people living in these cities will be in the developing world where the most rapid rates of urbanisation are expected to occur [9]. This process is already unfolding in some countries such as India where some urban centres today were only small villages one or two decades ago [12].

2.2.3 Impacts of urban growth and wealth on mobility and transport infrastructure

The economic consequences of urbanisation over the next 15 years will also lead to a shift in the concentration of wealth around the world. Today, only 600 urban centres generate about 60% of the global GDP [1]. Although it is still expected that 600 cities will continue to account for the same share of global GDP in 2025, the membership of this group of cities will be different. Over the next 10 years, the centre of gravity urbanisation is expected to move south and more decisively,

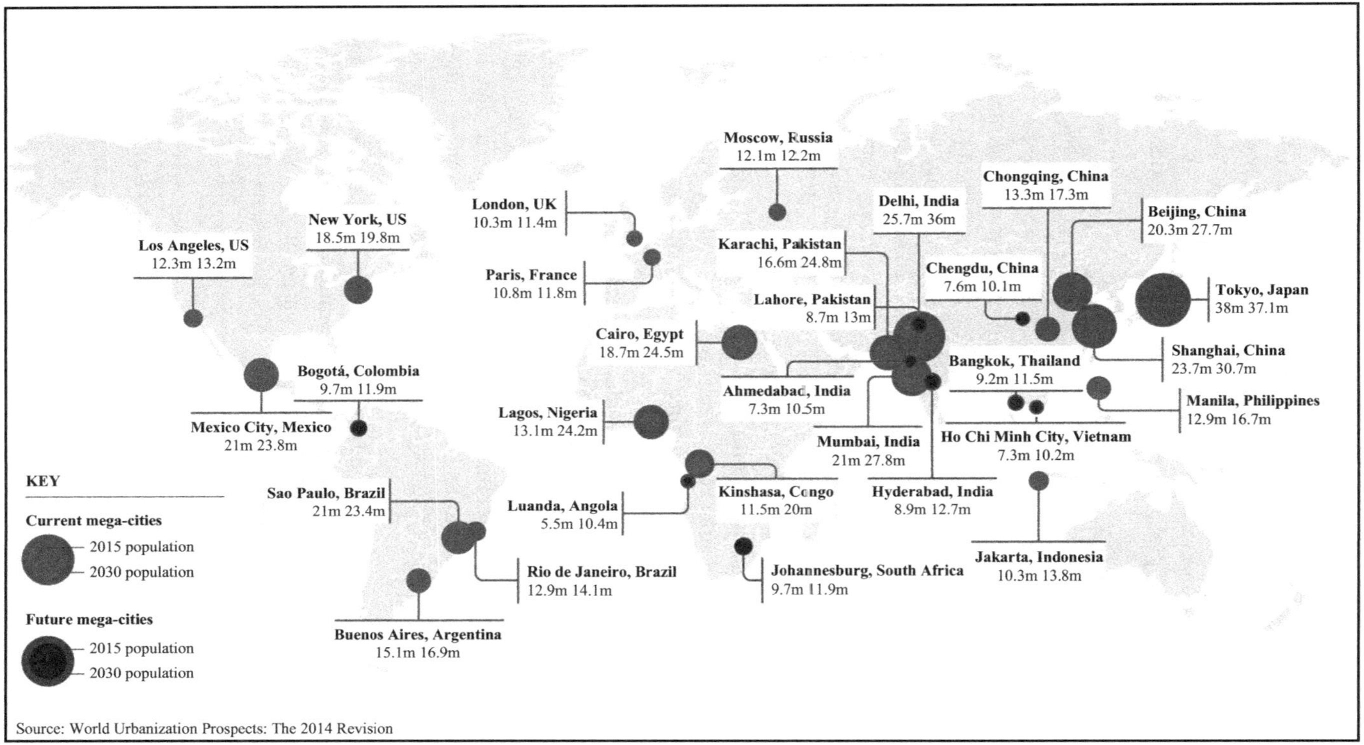

Figure 2.2 Selected current and future mega-cities 2015–30 [11] (Royalty free: Source https://www.allianz.com/en/press/news/studies/151130_the-megacity-of-the-future-is-smart/)

east [1]. For example, it is expected that 136 cities will join the top 600 wealthiest cities by 2025. Around 100 of these cities will be within China and 13 within India. Those top 600 cities are estimated to be producing 60% of the globe's total GDP by 2025, with the top 100 cities alone contributing 35% of total GDP [13]. The increase in wealth as a result of this expansion will occur mainly in developing nations and will enable more than one billion people to have high enough incomes to become significant consumers of goods and services by 2025. These consumers are expected to stimulate the global economy by contributing an extra $20 trillion a year in spending [13].

With this rapid growth in urban population and wealth, access to jobs, services and opportunities is also expected to grow. For example, between 2000 and 2010, the world's urban population increased by roughly 650 million people. The International Energy Agency (IEA) estimates that urban passenger travel increased by nearly 3 trillion annual passenger kilometres during that same period [14]. The IEA expects that with the current trends, global urban passenger mobility will double by 2050 and increase as much as 10-fold between 2010 and 2050 in rapidly urbanising, fast-growing regions in Southeast Asia and the Middle East. This would have substantial implications for the global annual urban transport energy consumption which, by 2050, would increase by more than 80% over 2010 levels, despite improved vehicle technology and fuel-economy enhancements.

The rapid growth in urban population and wealth is also expected to increase the global car fleet which is estimated to reach around 1.7 billion vehicles by 2030 [13]. This will mainly occur in the developing world as there is evidence to suggest that the total vehicle-kilometres travelled (VKT) per capita has plateaued and sometimes decreased in some of the developed countries (Chapter 4). In the emerging economies, the areas where substantial increases are expected include China and India which are the world's two most populated countries. These two countries have not only experienced rapid population growth but have also enjoyed strong economic development resulting in significant increases in the number of motor vehicles on their roads. Motor vehicle growth in both countries is heavily concentrated in urban areas, where 54% of China's population and 32% of India's currently live.

The flexible and comfortable personal mobility which private motor vehicles are seen to provide in these countries has led to a growth of individual, car-based transport which have resulted in negative impacts on an increasing scale – from congestion and traffic fatalities to greenhouse gas emissions that have consequences far beyond the cities themselves and contribute to global climate change.

2.2.4 Futrure directions to successful sustainable urbanisation

These urbanisation trends can lead to some significant benefits if they are well managed [10]. A successful urban planning agenda should include urban centres of all sizes. These urban centres will offer important opportunities for economic development and for providing access to basic services. Infrastructure provision in these cities (and similar dense urban population centres) would be cheaper and less environmentally damaging than providing similar level of services to a dispersed rural population. Access to the latest data on global urbanisation trends and city growth would be vital for setting policy priorities to promote inclusive, equitable and sustainable development for tomorrow's future cities.

2.3 Road crashes and injuries

Nearly 1.2 million people die in road traffic crashes worldwide annually with millions more sustaining serious injuries and living with long-term adverse health consequences [5]. The traffic fatalities worldwide translate into more than 3,000 deaths per day. This is the equivalent of 15 wide-body aircrafts, each with a capacity of 200 passengers, falling out of the sky every single day and killing everyone on board. This wouldn't be accepted in air travel, and it is disturbing that it continues on our roads today. In addition to deaths on the roads, up to 50 million people incur nonfatal injuries each year as a result of road traffic crashes [5].

Globally, road traffic crashes are the leading cause of death among young people, and the main cause of death among those aged 15–29 years (Figure 2.3). Road traffic injuries are also currently estimated to be the ninth leading cause of death across all age groups globally and are predicted to become the seventh leading cause of death by 2030 if the current trends continue. This rise is driven by the escalating death toll on roads in low- and middle-income countries, particularly in emerging economies in which urbanisation and motorisation accompany rapid economic growth [15].

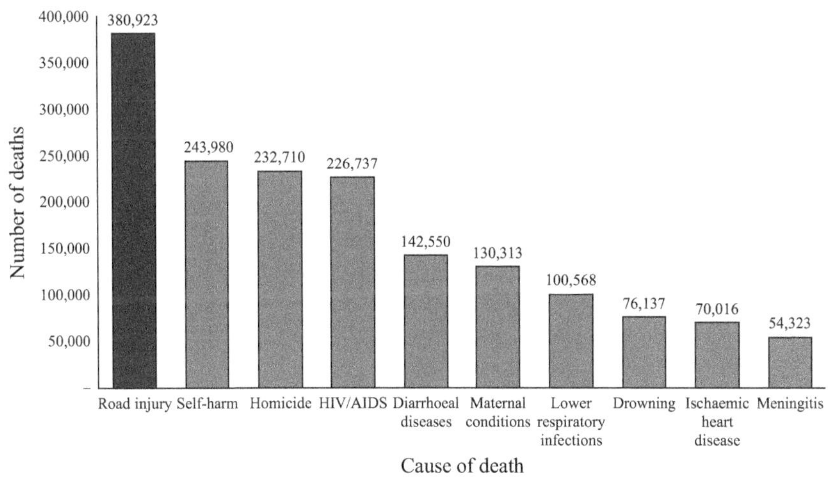

Figure 2.3 Top 10 causes of death among people aged 15–29 years (2015). Data source: [15]

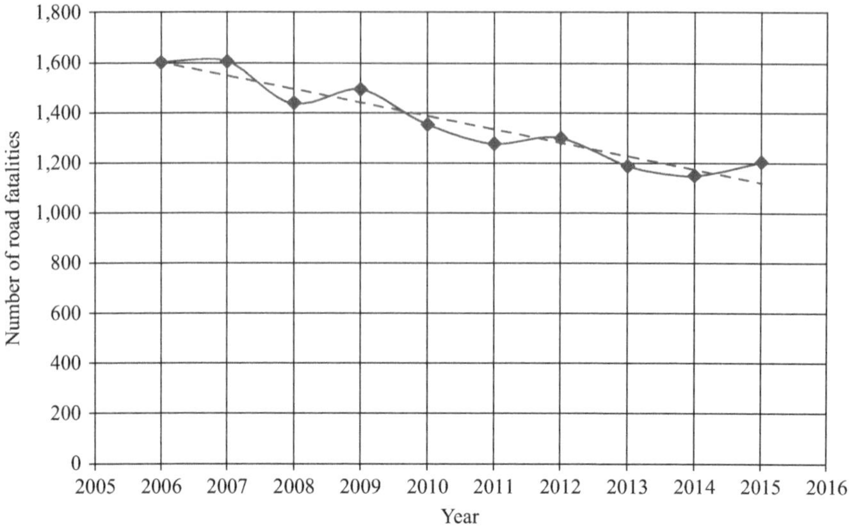

Figure 2.4 Annual road fatalities in Australia, 2006–15. Data source: [16]

In many of these developing countries, the road infrastructure is not designed to international safety standards, and the levels of enforcement have not kept pace with the increasing vehicle use. In contrast, many developed countries have managed to break the link between rising motorisation and road traffic deaths, with some managing to reduce such deaths by making infrastructure safer, improving the safety of vehicles and implementing other interventions which are effective at reducing road traffic injuries. Figure 2.4, for example, shows the annual road crash fatalities for the period 2006–15 in Australia [16]. Available data suggest that the annual road deaths have declined during this period at an average trend rate of 3.7% per annum. Figure 2.5 also demonstrates that the annual road fatalities per 100,000 people over the same time period also declined but at a higher average trend rate of 5.3% per annum [16].

As vehicle ownership grows, particularly in developing nations, cities will face the combined problems of traffic congestion and rising vehicle emissions, resulting in higher rates of respiratory illness. There are also additional indirect health consequences that are associated with this (Chapters 8 and 9). Rising car ownership in these countries will also result in reduced physical activities such as walking and cycling, with associated health consequences [5].

In addition to the pain, suffering and unnecessary loss of life, road traffic crashes have associated economic costs which strike hard at a national level, imposing a significant burden on health, insurance and legal systems. This is particularly true in developing countries which struggle with rapid urbanisation and other development needs, and where the quality of the road infrastructure does not meet safety standards.

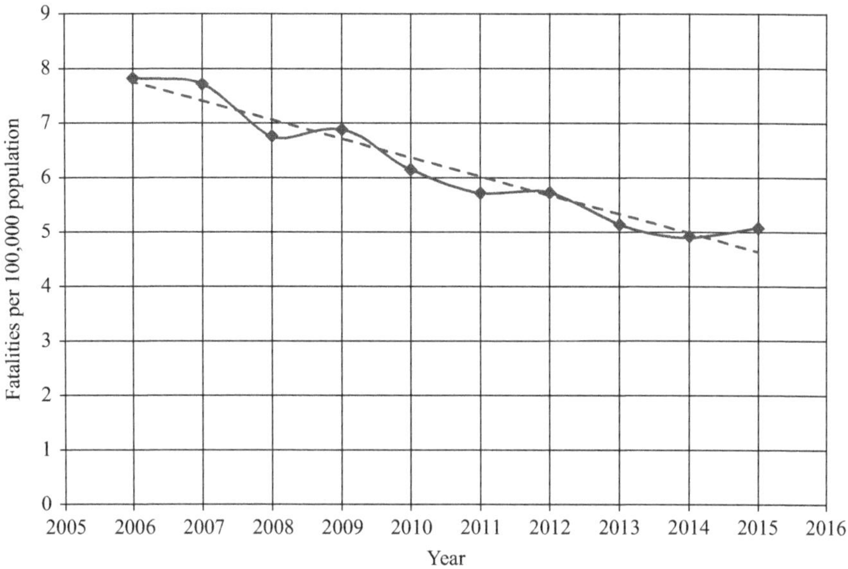

Figure 2.5 Annual road fatalities per 100,000 population in Australia, 2006–15. Data source: [16]

The consequence of such urbanisation leads to poor quality transport infrastructure that has a high safety risk. In these countries, the investment in road safety is also not commensurate with the scale of the problem [5]. The available data suggest that road traffic deaths and injuries in low- and middle-income countries are estimated to cause economic losses of up to 5% of these countries' GDP [5]. Globally, it is estimated that road traffic deaths and injuries cost countries between 1% and 3% of their GDP – more than $500 billion each year globally [5].

Road traffic deaths and injuries also impose a heavy burden on families and households. In low- and middle-income countries, road traffic crashes particularly affect the working-age groups, or those set to contribute the workforce [5]. This causes many families to slide deeper into poverty by the loss of a working family member, or by the expenses of medical care, or the added burden of caring for a family member who is disabled from a road traffic injury.

Staggeringly, some 70–90% of motor vehicle crashes are caused at least in part by human error [17]. A large proportion of these crashes could be avoided by using semi-automated and automated vehicles and there are currently very rapid developments aimed at removing humans – the key source of distraction and collisions – from the driving equation by providing increasingly sophisticated technologies in vehicles (Chapter 11). The biggest challenges across the globe remains for improving safety in low- to middle-income countries that have rapidly urbanised, whereas transport infrastructure has lagged behind.

2.4 Traffic congestion

The past few years have seen remarkable improvements in the collection and analysis of international traffic congestion data. These data make historical comparisons between different cities around the world more consistent and transparent. The main source of these data is provided from the detailed annual reports by the navigation and mapping company TomTom. The company's Traffic Index report [18] provides analysis of congestion on the road networks of 295 cities around the world. The Traffic Index provides detailed information on the impact of congestion on travel times, and the findings are used by governments and road authorities to measure the performance of their networks and identify locations where improvements can be made. The Traffic Indices are calculated annually, allowing cities to track the yearly progression of travel conditions by observing the periodic change in congestion for an average day in the morning and afternoon peak periods. Over the past few years, the company has compiled a database with more than 14 trillion historical travel time measurements with detailed actual traffic measurements for more than 50 countries.

The Traffic Index results are based on congestion level percentages which represent the measured amount of extra travel time experienced by drivers across the entire year, compared to measured travel times during uncongested conditions. The report calculates the overall congestion level (all day) in addition to the morning and evening peak hour congestion levels for each city where data are available. For example, an overall congestion level of 36% means that an average trip takes 36% longer than it would under uncongested conditions. The report takes into account local roads, arterials and highways. The sample size for each city is expressed in terms of total vehicle distance travelled for the period. All data are based on actual GPS measurements from historical traffic databases. Because the indices are calculated individually for cities, they take into account local conditions that could influence congestion, such as road hierarchy, population density and travel distances. In addition, since the TomTom Traffic Index is calculated as a ratio of local conditions per city, the individualised city congestion information is rendered comparable to other cities, thereby providing the opportunity to compare congestion levels between cities.

The annual Traffic Index results are based on analysis of field data from the previous year. The latest available results are based on data from 2015 (Tables 2.1 and 2.2). The findings show that congestion levels have increased globally by 13% compared to congestion levels in 2008. Significantly, the data show substantial differences between continents. Although North America's traffic congestion has increased by 17%, Europe has only increased by 2%. The smaller number for Europe could be heavily influenced by southern European countries such as Italy (−7%) and Spain (−13%) where there has been a marked drop in traffic congestion in the past 8 years.

Table 2.1 Congestion levels in the top 10 congested cities

City	Country	Overall congestion level[#]	Congestion level morning peak	Congestion level evening peak
Mexico City	Mexico*	59	97	94
Bangkok	Thailand*	57	85	114
Istanbul	Turkey*	50	62	94
Rio de Janeiro	Brazil*	47	66	79
Moscow	Russia*	44	71	91
Bucharest	Romania*	43	83	87
Recife	Brazil*	43	72	75
Salvador	Brazil*	43	67	74
Chengdu	China*	41	73	81
Los Angeles	United States	41	60	81

*Classified as developing countries [19].
[#]For example, an overall congestion level of 59% means that an average trip takes 59% longer than it would under uncongested conditions.

2.4.1 The ten most congested cities (2015)

The 2015 results (Table 2.1) show that while congestion is a global problem, nine of the ten most congested cities were in developing countries [19]. As described before, this may be attributed to the challenges that these cities face in addressing rapid urbanisation and the limited funding available for upgrading roads and public transport networks to deal with increased mobility in their urban centres.

The data also provide some insights into regional rankings (Table 2.2). The results in Table 2.2 summarise the findings for regions with four or more urban areas with over 1 million population. The worst performance was in mainland China, with an average Traffic Index of 34. Eastern Europe and Brazil were close behind with traffic indices of 33 and 31, respectively. The lowest Traffic Indexes were in the United States (20) and Canada (26) where generally higher road capacities are available with lower population densities and greater dispersion of economic activity [20].

2.4.2 Cost of congestion

Traffic congestion imposes a substantial cost to society when delays and environmental emissions are taken into consideration. Traffic congestion also dampens prosperity by slowing people and commercial vehicle movements resulting in lost productivity and wasted time.

The economic costs of congestion include the opportunity cost associated with the lost time spent in congestion and the financial costs associated with sitting in traffic, such as fuel consumption [21]. In the United States, the cost of congestion

Table 2.2 Regional rankings based on average traffic index

Region/country	Average Traffic Index			Number of urban centres		
	Overall	Urban centres with population over 1 million	Urban centres with population over 5 million	Total	Urban centres with population over 1 million	Urban centres with population over 5 million
China (Mainland)	34	34	35	22	21	14
Brazil	31	31	38	9	9	2
Europe (Eastern)	31	33	42	20	7	2
Turkey	28	29	50	9	7	1
Australia	25	28		10	5	
Europe (Western)	25	28	32	122	39	4
Canada	23	26	28	12	6	1
South Africa	23	25	27	6	4	1
United States	18	20	27	71	46	9
Other*	32	37	42	14	9	5
Total				**295**	**153**	**39**

*Regions with fewer than four urban centres with population over 1 million.

in American cities is reported to have reached $160 billion in 2014 [22]. The European Union estimates that congestion costs the community 1% of GDP [7], whereas in Asia it is estimated to cost the Asian economies around 2–5% of GDP [23]. Traffic congestion in some of the world's mega-cities like Egypt's Cairo is also estimated to cost the country 3.6% in GDP, with the annual cost of congestion estimated at $7,972 million USD in 2010 [21]. In Australia, the cost of congestion is estimated to have climbed from $12.8 billion in 2010 to around $16.5 billion in 2015 [24].

It is not clear whether these costs will continue to climb in the future given the growing momentum on provision of public transport and active transport, and the more recent evidence of peak-car phenomena which has seen the total kilometres of travel per capita decline in recent years (Chapter 4).

Congestion on a global scale has grown by 13% since 2008, with Australia being no exception to the trend [18]. Australia's population is projected to increase by 6.4 million people by 2031. The major four cities Perth, Melbourne, Sydney and Brisbane are expected to absorb 5.9 million people from this projected population growth [25]. This will exacerbate current congestion costs which have been estimated around $16.5 billion in 2015, having climbed from $12.8 billion since 2010.

2.5 Emissions

Transport activity currently accounts for about half of the global oil consumption and is also a major source of atmospheric pollution [26]. In 2010, 25% of greenhouse gas (GHG) emissions were released by the energy sector, 24% (net emissions) from

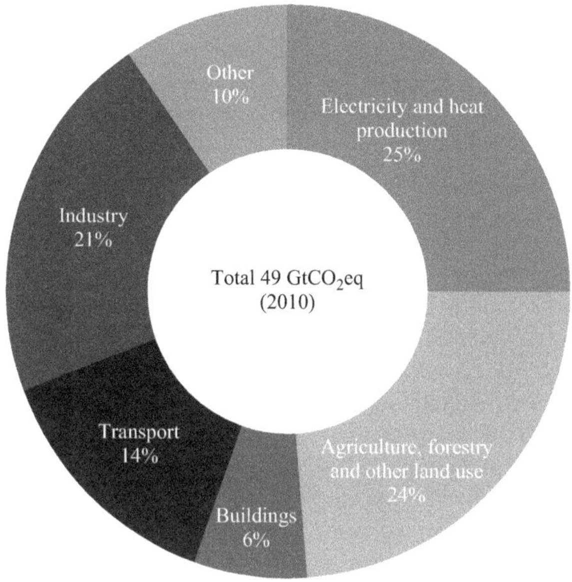

Figure 2.6 Total global anthropegenic greenhouse gas (GHG) emissions by economic sector (gigatonne of CO_2-equivalent per yerar), in 2010. Data source: [30]

agriculture, forestry and other land use, 21% by industry, 14% by transport and 6.4% by the building sector (Figure 2.6).

Emissions from transport have been shown to be growing more rapidly than those from other anthropogenic activities [27]. In the time period 1990–2007, the EU-15 CO_2-equivalent emissions from land transport increased by 24% [27]. In the year 2000, emissions from land transport comprised 74% of the global CO_2 emissions from all transport activities [28,29]. This growth is expected to continue in the future, due to increasing world population, economic activities and related mobility [27]. In the year 2000, there were roughly 625 million passenger light-duty vehicles (PLDVs) around the world [14]. By 2010, that number had reached nearly 850 million PLDVs. Modelling by the International Council on Clean Transportation [26] predicts a doubling of the world's motor vehicle population over the next 20 years. Modelling by the International Transport Forum, using carbon dioxide equivalent measures, projects that transport emissions would grow by 9 $GtCO_2eq$ by 2030 and increase by a further 110% above 2010 emissions levels by the year 2050 [8]. These forecasts underscore the importance of current and future policies that target reductions in oil consumption and GHG emissions from the transport sector.

The effects of growing travel demand and increasing shifts to private motorisation are particularly evident in urban areas in developing countries. Motorised vehicle traffic has significant adverse effects on environmental quality and health. The IEA expects global travel (in terms of passenger and freight-tonne kilometre) to double by 2050 and corresponding transport energy use and emissions to increase 70% between 2010 and 2050, despite expected vehicle technology improvements. Global motorised vehicle stock is expected to double, and subsequent roadway occupancy levels are projected to increase as much as sixfold in some countries [14].

Globally, modelling forecasts a threefold increase in both fuel demand and CO_2 emissions for the period between the years 2000 and 2030 [28]. The modelling also showed that emissions of CO_2 from land transport and shipping accounted for 13% of the total anthropogenic CO_2 warming (year 2005).

In addition to long-lived greenhouse gases, ground-based vehicles also emit aerosol particles as well as a wide range of short-lived gases, including also aerosol precursor species [31]. Atmospheric aerosol particles have significant impacts on climate, through their interaction with solar radiation. In populated areas, they also affect air quality and human health [31–33].

Transport also accounts for half of the global oil consumption and nearly 20% of world energy use, of which approximately 40% is used in urban transport alone [14]. The IEA expects that increased mobility will impose new challenges and anticipates urban transport energy consumption to double by 2050, despite ongoing vehicle technology and fuel-economy improvements. Attention to urgent energy-efficiency policies will be required to mitigate associated noise, air pollution, congestion, climate and economic impacts, all of which can cost countries billions of dollars per year.

The situation across the geographies of the world differs widely. Examples of the situation in Australia, India and China are provided below.

2.5.1 Australia

In Australia, transport activity is one of the major sources of emissions related to the combustion of fossil fuels [34]. Transport contributed 83.2 Mt CO_2 or 15.3% of Australia's net emissions in 2010 (Figure 5.1, Chapter 5). Road transport was the main source of transport emissions, accounting for 71.5 Mt CO_2 or 86% of national transport emissions. Passenger car usage in urban areas was the largest transport source, contributing 8.5% of Australia's net emissions in 2010 and thus accounting for around 39.7 Mt CO_2. Transport emissions are also one of the strongest sources of emissions growth in Australia. Emissions from this sector were 32% higher in 2010 than in 1990 and have increased by about 1.6% annually on average. The Bureau of Infrastructure, Transport and Regional Economics [35] estimates that emissions from the domestic transport sector in 2020 are projected to be around 70.3% above 1990 levels (at 105.2 million tonnes of CO_2 equivalent).

2.5.2 India and China

Urban transport emissions forecasts for China and India also pose some challenges. The current level of CO_2 emissions from the transport sector in China is estimated at about 7% of its total CO_2 emissions. Like Australia, around 86% of transport emissions come from road transport. Between 1994 and 2007, emissions from transport increased by 160% across China's cities. This is expected to continue to increase due to the growing number of vehicles [36].

Despite its low ownership rate of 39 light-duty vehicles per 1,000 persons (compared to the United States' 680 vehicles per 1,000 persons) in 2010, China is already the world's largest CO_2 emitter and has recently become the top global importer of crude oil. With the transport sector over 90% dependent on oil due to the dominance of internal combustion engines, China's demand for oil will only increase as its transport sector grows, which it inevitably will. Increased CO_2 emissions will follow, unless a comprehensive range of policies and measures is implemented to alter the course of development.

In India, increasing household incomes, a growing urban population and surging vehicle sales are also leading to a more motorised society. India's urban population increased from 62 million in 1951 to 285 million in 2001, and as cities grow and new urban agglomerations develop, the demand for transport and total distance travelled by Indians will grow as well, with cars and motorised two-wheelers increasingly becoming the preferred mode choice. The number of passenger vehicles on India's roads grew from 16 million in 1990 to almost 40 million in 2000 and reached 131 million in 2014.

Going forward, the focus will increasingly shift to mitigation measures for road transport. Between 1970 and 2010, the road sector was the largest contributor to the growth in GHG emissions and is still the largest polluting sector within the transport sector. The share of road transport in the overall transport GHG emissions has increased from 59.85% to 72.06% during the same period, making it historically the largest polluting sub-sector in transport. By comparison, all of the other transport sectors have grown from approximately 1–2 $GtCO_2eq$ for the same time period, with each sector growing only slightly in the amount they emit. Therefore, the road sector has to be the focus of any future plans to curb pollution if any meaningful reduction in GHG emissions from the transport sector is to occur [30].

2.6 Ageing assets and the infrastructure investment gap

Today, the world's ever-expanding infrastructure consists of 64 million kilometres of roads and 4 million kilometres of railways [37].

Inadequate or poorly performing infrastructure impedes economic growth and presents major challenges for governments around the world. Without well-maintained and resilient transport infrastructure – roads, rail and public transport – cities cannot meet their full growth potential and economic targets. At the same time, investment in new infrastructure and urgent maintenance of existing assets comes at a time when many governments operate in constrained budgetary environments and have competing demands on their scarce resources.

This issue affects developing countries and also many nations in the developed world. From the United States through Europe to the emerging world, the backlog of projects includes upgrade of existing assets and proposals for new projects to drive economic growth. One of the pressing issues in both developed and emerging markets is the need to invest in transport infrastructure that provides connectivity and ease of access to jobs and opportunities. Although technology has a big role to play in asset optimisation and making better use of existing assets, many cities still suffer from a lack of integrated and connected networks of transport systems.

Although governments and industry bodies agree that an infrastructure deficit exists in many cities, there is less agreement about the size of the global infrastructure gap. The World Economic Forum estimates a global need for $3.7 trillion in infrastructure investment each year, whereas only $2.7 trillion is annually invested, mostly by governments, around the world [38]. The McKinsey Global Institute [39] estimates the infrastructure gap at around $57 trillion over the next 14 years (up to 2030). This includes transport (roads, rail, ports and airports), power, water and telecommunications with transport collectively accounting for around 23.8 trillion (Figure 2.7). This comes at a time when governments have only planned to invest around $37 trillion in all infrastructure assets over the same time period [39]. The same study, however, suggests ways to reduce the required investment and details a number of approaches to achieve an improvement in infrastructure productivity delivering savings of 40% (e.g. optimising project portfolios, delivering projects more efficiently and getting more out of existing assets as an alternative to building new ones). PwC estimates the gap at around $78 trillion in infrastructure needs and forecasts a need for capital project and infrastructure spending more than $9 trillion per year by 2025, up from $4 trillion per year in 2012 [40]. It is widely estimated that the deficit sits around $20 trillion to 2030 [40].

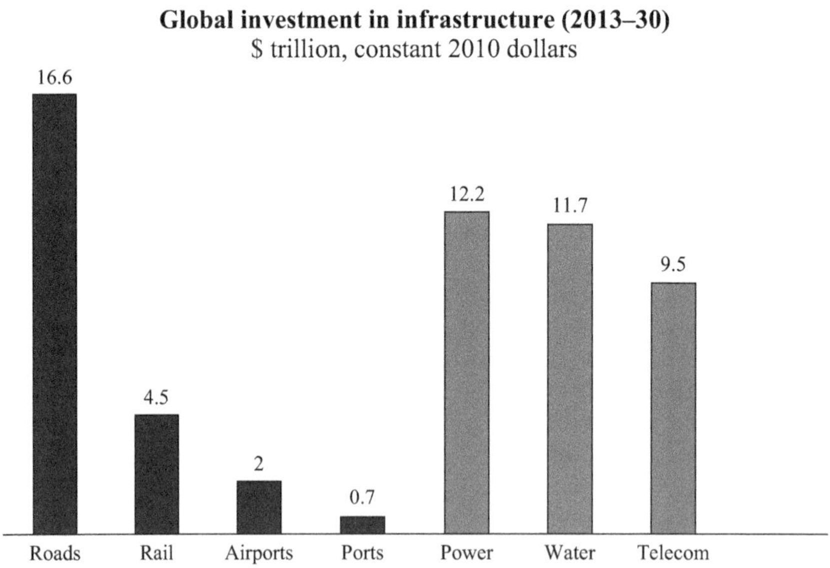

Figure 2.7 Demand for global infrastructure (2013–30). Data source: [39]

The barriers to investment vary between the developed and emerging markets. In developed markets, the barriers are primarily public discomfort with privatised or partly privatised models, and governments taking measures to reduce debt amid fiscal constraints. This is likely to become more challenging in the future as government revenues from transport activities becomes under more pressure. For example, a number of peak bodies in Australia, including the Productivity Commission, have argued that there has been a steady decline in the real-net revenue from the fuel excise in recent years. One of the reasons for this decline in revenue is based on the increased credits and grants that are given to heavy vehicles. The other major factor that is contributing to this shortfall is the increasing fuel efficiency of new vehicles. With more fuel-efficient and electric vehicles coming into the market in future years, it appears that the fuel excise will become increasingly less effective as a mechanism to raise revenue.

In emerging markets, the barriers are more about skills and economic capability. Countries that lack developed capital markets find it challenging to provide long-term finance and currency-exchange protections that investors require.

The global need for infrastructure is significant, particularly in emerging markets where the levels of service are not adequate, and connectivity is largely missing. Going forward, more emphasis needs to be given to maintenance of existing assets and also more deployment of technological solutions to enhance the performance of existing infrastructure, while reducing reliance on building new assets. Future investments should also prioritise projects which promote low carbon mobility including walking, cycling and public transport.

2.7 Resilient infrastructure

The resilience of a city's critical infrastructure, such as transport networks, affects its liveability and economic growth. For the purposes of this chapter, resilience is defined as the ability of a transport system (exposed to hazards) to resist, absorb, adjust to and recover from the effects of a hazard in a timely and efficient manner, including initiatives to preserve and restore essential structures and functions [41].

There are two types of disruptions that affect transport infrastructure in cities. Chronic stresses are long-term problems such as population growth, increasing pressures on infrastructure assets and the gradual effects of climate change. Acute stresses or 'shocks' are immediate short-term incidents such as natural disasters or human-induced stresses such as road crashes [42]. Both types of disruptions impact the reliability of a city's transport system and its ability to meet travellers' needs.

A city's critical infrastructure is highly vulnerable to, and a major casualty of, short-term incidents. Repairing or replacing infrastructure assets after an incident or disaster is often difficult and costly, which can exacerbate the suffering of affected communities. For example, between 2002–03 and 2010–11, more than $450 million was spent each year by Australian governments to restore critical infrastructure after extreme weather events [41]. This equates to about 1.6% of total public infrastructure spending. In addition, it is estimated that $17 billion (in net present value terms) will be needed to directly replace critical infrastructure between 2015 and 2050 due to the impact of natural disasters [41].

In many countries around the world, there is growing national awareness of these issues. For example, the Productivity Commission in Australia undertook a public inquiry into the effectiveness of natural disaster spending [41]. The inquiry found that governments in Australia have overinvested in post-disaster reconstruction and underinvested in mitigation measures and interventions that would limit the impact of natural disasters in the first place. As such, natural disaster costs have become a growing, unfunded liability for governments.

More specifically for transport operations, travel-time reliability is a good measure of the resilience of a city's transport system. Travel-time reliability measures the consistency of trip times within expected time ranges and service levels. Consistent trip times can be achieved by minimising total travel-time variability. Travel time variation is caused either by short-term or long-term stresses within the transport network.

There are three categories of events that have influence on travel time variability [21,43]. These include: (1) traffic influencing events (e.g. traffic incidents and crashes, work zone activity and weather and environmental conditions); (2) traffic demand (e.g. fluctuations in temporal travel demand and special events); and (3) physical road features (e.g. at-grade railway crossings and geometric bottlenecks).

Some of the intervention measures that can improve the resilience of transport systems include investment in smart infrastructure technologies that enable operators to respond quickly to short-term stresses, and also long-term infrastructure investments to increase the system's capacity and connectivity to prevent disruptions and make the system more robust.

There is a strong case that can be made for greater consideration of, and investment in, resilience. Mitigating disruptions and disaster risks is a priority for both existing and future assets. Some guidance and principles for infrastructure planners and decision makers to embed resilience in their projects are available [41]. These include (1) identifying disaster risks; (2) applying robust methodologies for cost–benefit analysis; (3) coordinating, centralising and making available critical data and information; (4) strengthening approval processes; and (5) embedding on-going monitoring of resilience.

2.8 Limitations of traditional approach – 'predict and provide'

The traditional approach to providing transport infrastructure – through expansion of capacity to meet travel demand – has met with limited success over the past few decades. In this approach, forecasts of economic growth are translated into projections for increased demand for transport, and therefore the need for investment in increasing the capacity of the transport network. The rationale behind this approach ('predict and provide') is that the lack of investment in infrastructure would result in more congestion which would impede economic growth [44,45].

One of the main limitations of this approach, where demand is predicted and then the infrastructure is provided, is the generation of induced demand [46]. The theory of induced demand is now well understood and accepted. Road improvements which reduce travel times attract more trips from other routes and also encourage new more frequent travel which would have not occurred if the road has not been improved. This additional vehicle traffic consists in part of induced travel, which refers to increased total VKT compared with what would otherwise occur [47]. If road capacity is increased, peak-period trips also increase until congestion delays discourage additional traffic growth [47]. Research shows that the additional traffic generated would often fill a significant proportion of the capacity added [47].

2.9 The opportunities

There is increasing recognition and acceptance that addressing transport issues through building additional road capacity is not sustainable, and that it does not solve traffic congestion or improve mobility in cities.

Sustainable transport policies and intervention measures provide opportunities to meet the needs and demands of citizens and businesses in urban environments. Setting a city on a course towards sustainable transport requires a roadmap and a holistic vision which incorporates different strategies to meet the demand for travel, including public transport and active transport policies. In recent years, technology has also been playing a big part in enhancing the performance of existing assets and thereby reducing the need for building additional infrastructure.

A sustainable transport system is one that meets the mobility and accessibility needs of people while supporting the community's long-term social, environmental and economic goals and aspirations. The Centre for Sustainable Transportation at the University of Winnipeg in Canada offers a comprehensive definition: a sustainable transportation system is one that accomplishes the following [48]:

1. Allows the basic access needs of individuals and societies to be met safely and in a manner consistent with human and ecosystem health, and with equity within and between generations.
2. Is affordable, operates efficiently, offers choice of transport mode and supports a vibrant economy.
3. limits emissions and waste within the planet's ability to absorb them, minimises consumption of non-renewable resources, limits consumption of renewable resources to the sustainable yield level, reuses and recycles its components, and minimizes the use of land and the production of noise.

Sustainable transport and low carbon mobility are closely related and linked. Mobility is the total amount of travel that is undertaken on all modes of transport and relates to the physical movement from an origin to a destination [49]. When mobility is undertaken using motorised forms of transport, it results in harmful emission of carbon and other pollutants. Transport, on the other hand, is broader than mobility and includes all modes of transport in addition to the supply of transport. For physical movement to take place, transport infrastructure needs to be supplied. Without the physical infrastructure, travel cannot happen. Infrastructure supply does not generate movement but allows it to take place. In this context, mobility can be seen as being situated between the demand for transport and the infrastructure that allows this demand to be realised [49].

2.9.1 Low carbon mobility

Low carbon mobility is defined as 'mobility that results in substantially lower levels of carbon' [49]. There are four key strategies to achieve this, which will be the focus of Chapter 3 in this book. The first one is the 'Avoid' strategy which implies the need to change the social norms and travel behaviour to ones that require less mobility (e.g. telecommuting and living nearer to shops and services). The second is to 'Shift' travel from energy-intensive modes to different forms of transport (e.g. from car to train or cycling or walking). The third is to 'Share' transport and mobility resources (e.g. ride-sharing or car-sharing). The fourth is the 'Improve' strategy which calls for improving the fuel efficiency and emission of vehicles (e.g. the use of hybrids and electric vehicles) and also maximising the efficiency of the physical infrastructure that is required for the movement of people and goods.

The topic of low carbon mobility is not new. Researchers and policy think tanks have tried to advance this agenda before but progress has been slow. The renewed interest and opportunity today stems from the new interpretations that have begun to focus on both the supply and demand sides of travel, and a better understanding for traveller behaviour. This has also been facilitated through the convergence of a number of forces such as shared mobility, digital innovations and disruptive mobility models which are introducing new options for travellers through ride-sharing and car-sharing through easy to use and reliable technology platforms. The proliferation of Intelligent Transport Systems throughout transport infrastructure also means that disruptions and incidents are detected quickly. This information is then provided to travellers with more ease and speed than ever before, providing them with options to change mode of travel or time of departure.

2.9.2 Planning for sustainable transport solutions

Traditional transport planning approaches focused on the provision of vehicular traffic without equal attention to other modes of transport. Today, transport professionals increasingly recognise that the infrastructure decisions that a city makes today will have a profound impact on the shaping of that city and its people's travel behaviour in the future. In many cities around the world, greater emphasis is being given to integrated transport planning and sustainable modes of mobility such as public transport, cycling and walking. Such policies play a key role in reducing greenhouse gas emissions and pollution. They also have a wide range of benefits to the community and have been shown to enhance the fabric of urban environments and make cities more liveable. Well-planned mobility solutions for tomorrow's cities also improve accessibility to jobs and opportunities, which are preconditions for sound economic development and musts for mitigating emissions and pollution in urban centres.

2.9.3 Conventional approaches versus sustainable mobility solutions

Cities that have been successful in implementing sustainable transport solutions have adopted simple but radical approaches to meeting the travel needs of their citizens. Rather than focusing on the infrastructure required to facilitate the movement of private vehicles, the emphasis was shifted towards the movement of people and goods, regardless of the mode of transport. And instead of focusing on operational strategies that promote longer travel and through movements of traffic, the focus was shifted towards providing access and accessibility to all groups of society [48].

Table 2.3 compares the basic premises between sustainable mobility solutions and the traditional approaches which focused on the dominance of motorised transport.

2.10 Conclusion

This chapter highlighted how the dominance of motorised transport has resulted in high levels of mobility and urban sprawl which translated into high volumes and distances needed for travel to access economic opportunities and services, leading to high carbon mobility. Traditionally, transport investment has been based on the lack of capacity in the system, and a misconception that congestion can be solved by building additional infrastructure capacity. Infrastructure investments were based on travel-time savings resulting in interventions that led to longer journeys using more energy and more dependence on motorised transport.

Addressing the challenges facing our cities in the twenty-first century requires new approaches. Efforts to reduce carbon emission from transport fall under four categories: avoid, shift, share and improve. These strategies, if well planned and communicated to the public, have strong potential to produce the desired outcomes. There is now sufficient evidence from cities around the world that low carbon mobility measures under these strategies have helped to reduce motor vehicle traffic and promote more efficient and environmentally friendly travel. The policy measures implemented in these cities have also resulted in increases in transport efficiency, improved passenger mobility, safer roads, reduced congestion, improved health and better air quality.

These strategies are addressed in more detail in the next chapter.

Table 2.3 Shifting urban transport towards low carbon mobility

Conventional approaches (transport planning and engineering)	**Sustainable and emerging low carbon mobility approaches**
Focus on supply and building additional infrastructure and capacity	Focus on demand management, maximising efficiency, reliability and resilience of transport systems
Physical dimensions	Social dimensions (mobility benefits are equally and fairly distributed; fair access to transport infrastructure and services for all income groups)
Focus on mobility or physical movement from an origin to a destination	Accessibility*: focus on the mobility required for access to employment, opportunity, goods and services
Large in scale	Local scale – precinct level
Street as road for vehicles	Street as space to be shared between all modes
Vehicle-oriented	People-oriented and customer-focused. Balanced development of all transport modes and shifting towards cleaner and more sustainable modes such as public transport and active transport
Motorised transport	All modes of transport in a hierarchy with priorities for walking and cycling
Transport modelling approaches	Scenario development and modelling
Traffic forecasting	Visioning on cities
Focus on reacting to congestion and disruptions	Focus on positive business and operational outcomes
Travel as a derived demand	Travel as a valued activity as well as a derived demand
Minimisation of travel times	Reliability of travel times
Key performance indicators: traffic throughput and speeds	Key performance indicators: accessibility, sustainability, social equity, environmental quality, health and well-being and quality of life
Planning by experts	Planning through transparent and comprehensive stakeholder consultations
Segregation of people and traffic	Integration of people and traffic
Economic evaluation driven by transport efficiency gains	Multi-criteria analysis to take into account environmental and social concerns
Funds raised through petrol taxes, vehicle registration and licensing fees	Congestion and road pricing, and user-pay models
Private car ownership	New business models that challenge car ownership, promote public and active transport and a shift to car-sharing and ride-sharing solutions enabled by technology platforms

Table 2.3 (*Continued*)

Conventional approaches (transport planning and engineering)	Sustainable and emerging low carbon mobility approaches
Spending on physical infrastructure	Spending on information technology solutions, data fusion, predictive analytics, integration, decision support systems and adaptive tools
Emphasis on 'knowing and seeing', and measuring past performance against key performance indicators	Emphasis on 'predicting and anticipating' in order to improve resilience and avoid disruptions

Source: Author – adapted from [45,49].
*The terms *access* and *accessibility* offer distinct meanings. Accessibility is 'the potential for reaching locations where services, goods and opportunities are available'. Access is 'the realisation of this potential' [50].

Acknowledgements

The author would like to acknowledge the contributions of a number of research students who made useful interventions while the chapter was in preparation. These include Farid Javanshour, Jack Hill and David Leonard from Swinburne University of Technology.

References

[1] Dobbs, R., Remes, J., Manyika, J., Roxburgh, C., Smit, S., Schaer, F. (2012). *Urban world: mapping the economic power of cities*, Seoul, South Korea: McKinsey & Company.

[2] Wilson, M. (2012). *By 2050, 70% of the world's population will be urban. Is that a good thing?, Fast Company*. Available: http://www.fastcodesign.com/1669244/by-2050-70-of-the-worlds-population-will-be-urban-is-that-a-good-thing. Accessed 17 May 2016.

[3] Winston, C., Mannering, F. (2014). Implementing technology to improve public highway performance: a leapfrog technology from the private sector is going to be necessary, *Econ. Transp.*, 3 (2), 158–165.

[4] Neumann, C. (2015). *Big data versus big congestion: using information to improve transport*, McKinsey & Company. Available: http://www.mckinsey.com/industries/infrastructure/our-insights/big-data-versus-big-congestion-using-information-to-improve-transport. Accessed 17 May 2016.

[5] World Health Organisation (2016). Road traffic injuries, United Nations, Fact Sheets, fs358.

[6] FIA Foundation (2016). *Road safety fund. FIA Foundation*. Available: http://www.fiafoundation.org/our-work/road-safety-fund. Accessed 13 May 2016.

[7] European Commission (2011). *Roadmap to a single European transport area, European Commission: Mobility & Transport*. Available: http://ec.europa.eu/transport/strategies/facts-and-figures/transport-matters/index_en.htm. Accessed 13 May 2016.

[8] International Transport Forum (2010). *Transport Outlook 2010, 'The 2010 International Transport Forum: Transport and Innovation: Unleashing the Potential'*, Leipzig, Germany: Organisation for Economic Co-operation and Development, 26–28 May 2010, pp. 1–28.

[9] Armstrong, B., Davidson, G., de Vos Malan, J., Gleeson, B., Godfrey, B. (2015). *Delivering sustainable urban mobility final report*, Melbourne, Australia: Australian Council of Learned Academics.

[10] United Nations (2014). *World urbanization prospects: the 2014 revision*, Department of Economic and Social Affairs, Population Division CD-ROM Edition.

[11] Allianz (2015). *The mega city of the future is smart*. Available: https://www.allianz.com/en/press/news/studies/151130_the-megacity-of-the-future-is-smart/, accessed 15 September 2016.

[12] International Geosphere-Biosphere Program (IGBP) (2012). *The rise and rise of urban expansion*. Available: http://www.igbp.net/news/features/features/theriseandriseofurbanexpansion.5.705e080613685f74edb800014.html. Accessed 22 June 2015.

[13] Dobbs, R., Remes, J., Manyika, J., Roxburgh, C., Smit, S., Schaer, F. (2012). *Urban world: cities and the rise of the consuming class*, McKinsey & Company, The McKinsey Global Institute (MGI).

[14] International Energy Agency (IEA), 2013. *A tale of renewed cities*. Paris, France: OECD.

[15] World Health Organisation. *Projections of mortality and causes of death, 2015 and 2030*. Available: http://www.who.int/healthinfo/global_burden_disease/projections/en/.

[16] Bureau of Infrastructure, Transport and Regional Economics (2016). *Road Trauma Australia Annual Summaries*. Available: https://bitre.gov.au/publications/ongoing/road_deaths_australia_annual_summaries.aspx.

[17] National Highway Traffic Safety Administration (NHTSA) (2015). *Critical reasons for crashes investigated in the National Motor Vehicle Crash Causation Survey*. Available: https://crashstats.nhtsa.dot.gov/Api/Public/ViewPublication/812115.

[18] TomTom (2016). *TomTom Traffic Index: measuring congestion worldwide*. Available: https://www.tomtom.com/en_au/trafficindex/list.

[19] Andersen, J., Bruwer, M. (2016). *TomTom 2015 Traffic Index: independent analysis report*. University of Stellenbosch. Available: http://www.tomtom.com/.

[20] Wendell Cox (2016). *TomTom 2015 Traffic Index: independent analysis report*. Available: http://www.tomtom.com/.

[21] Downs, A. (2004). *Still stuck in traffic coping with peak hour traffic congestion*. Washington, DC: Brookings Institute.

[22] Schrank, D., Eisele, B, Lomax, T., Bak, J., 2015. *2015 urban mobility scorecard*. Arlington, TX: Texas A&M Institute and INRIX.

[23] Asian Development Bank (2016). *Transport: facts and figures*, viewed 21 June 2016. Manila, Philippines: Asian Development Bank.

[24] BITRE (2015). *Traffic and congestion cost trends for Australian capital cities*. Department of Infrastructure and Regional Development. Available: https://bitre.gov.au/publications/2015/files/is_074.pdf.

[25] Infrastructure Australia (2015). *Australian infrastructure audit, our infrastructure challenges report – 1*. Canberra, Australia: Commonwealth of Australia.

[26] International Council on Clean Transportation (2012). *Global transportation energy and climate roadmap*. Available: http://www.theicct.org/global-transportation-energy-and-climate-roadmap.

[27] Righi, M., Hendricks, J., Sausen, R. (2015). *The global impact of transport sectors on atmospheric aerosol in 2030 – Part 1: Land transport and shipping*. Available: http://www.atmos-chem-phys.net/16/4481/2016/.

[28] Uherek, E., Halenka, T., Borken-Kleefeld, J., *et al.* (2010). Transport impacts on atmosphere and climate: land transport, *Atmos. Environ.*, 44, 4772–4816, doi:10.1016/j.atmosenv.2010.01.002.

[29] Eyring, V., Isaksen, I. S. A., Berntsen, T., *et al.* (2010). Transport impacts on atmosphere and climate: shipping. *Atmos Environ.*, 44, 4735–4771, doi:10.1016/j.atmosenv.2009.04.059.

[30] Intergovernmental Panel on Climate Change (2014). *Climate change 2014 synthesis reoport*. Available: https://www.ipcc.ch/report/ar5/syr/.

[31] Forster, P., Ramaswamy, V., Artaxo, P., *et al.* (2007). Changes in atmospheric constituents and in radiative forcing, in: *Climate Change 2007: The Physical Science Basis, Contribution of Working Group I to the Fourth Assessment Report of the Intergovernmental Panel on Climate Change*, edited by: Solomon, S., Qin, D., Manning, M., *et al.*, Cambridge, UK and New York, NY: Cambridge University Press, 129–234.

[32] Pope, C. A., Dockery, D.W. (2006). Health effects of fine particulate air pollution: lines that connect, *J. Air Waste Manage.*, 56, 709–742, doi:10.1080/10473289.2006.10464485.

[33] Chow, J. C., Watson, J. G., Mauderly, J. L., *et al.* (2006). Health effects of fine particulate air pollution: lines that connect, *J. Air Waste Manage.*, 56, 1368–1380, doi:10.1080/10473289.2006.10464545.

[34] National Green House Gas Investory (2011). *Australian national green house gas accounts*. Available: http://bit.ly/1x0Qeec, accessed 20 November 2014.

[35] Bureau of Infrastructure, Transport and Regional Economics (BITRE) (2009). *Greenhouse gas emissions from Australian transport: projections to 2020 working paper 73*. Australia: Department of Infrastructure, Transport,

Regional Development and Local Government Canberra. Available: https://bitre.gov.au/publications/2009/files/wp_073.pdf.

[36] International Transport Forum (2015). *Low-carbon mobility for mega cities: what different policies mean for urban transport emissions in China and India*. Available: http://www.itf-oecd.org/low-carbon-mobility-mega-cities, accessed 5 December 2016.

[37] Khanna, P. (2016). *How megacities are changing the map of the world. TED Talk*. Available: https://www.ted.com/talks/parag_khanna_how_megacities_are_changing_the_map_of_the_world.

[38] World Economic Forum (2016). *An 8-point plan for closing the infrastructure gap*. Available: https://www.weforum.org/agenda/2015/09/an-8-point-plan-for-closing-the-infrastructure-gap/.

[39] Dobbs, R., Pohl, H., Lin, D., *et al.* (2013). *Infrastructure productivity: how to save $1 trillion a year*. McKinsey & Company, Seoul, South Korea: The McKinsey Global Institute (MGI).

[40] PwC (2014). *Capital project and infrastructure spending Outlook to 2025. Research by Oxford Economics*. Available: http://www.pwc.com/gx/en/industries/capital-projects-infrastructure/publications/cpi-spending-outlook.html.

[41] Deloitte Access Economics (2016). *Building resilient infrastructure*. Available: http://www2.deloitte.com/au/en/pages/economics/articles/building-australias-natural-disaster-resilience.html.

[42] City of Melbourne (2016). *Resilient Melbourne: viable, sustainable, liveable, prosperous*. Available: http://resilientmelbourne.com.au/.

[43] Bhoite, S., Braulio, S., Birtill, K., *et al.* (2013). *City resilience index. Understanding and measuring city resilience*. New York: ARUP and Rockefeller Foundation.

[44] Giovani, M. and Banister, D. (2013). *Moving towards low carbon mobility*. Cheltenham, UK: Edward Elgar Publishing.

[45] Banister, D. (2007). The sustainable mobility paradigm, *Transp. Policy*, 15, 73–80.

[46] Goodwin, P.B. (1996). Empirical evidence on induced traffic: a review and synthesis, *Transportation*, 23, 35–54.

[47] Litman, T. (2016). *Generated traffic and induced travel implications for transport planning*. Victoria Transport Policy Institute. Available: www.vtpi.org.

[48] The Centre for Sustainable Transportation (2016). *Vision of sustainable transportation*. Available: http://cst.uwinnipeg.ca/about.html.

[49] Federal Ministry for Economic Cooperation and Development (2014). *Urban mobility plans: national approaches and local practice*. Available: http://www.eltis.org/resources/tools/urban-mobility-plans-national-approaches-and-local-practice.

[50] Sclar, E., Lonnroth, M., Wolmar, C. (2016). *Improving urban access: new approaches to funding transport investment*. Oxford, UK: Routledge.

Glossary

AFOLU	Agriculture, forestry and other land use
CDE	Carbon dioxide equivalent
GDP	Gross domestic product
$GtCO_2eq$	Gigatonnes carbon dioxide equivalent
IEA	International Energy Agency
MGI	McKinsey Global Institute
PLDV	Passenger light-duty vehicles
VKT	Vehicle-kilometres travelled

Further Reading

[1] Jaber, A., Glocker, D. (2015). *Shifting towards low carbon mobility systems. International Transport Forum*. Available: http://www.itf-oecd.org/shifting-towards-low-carbon-mobility-systems.

[2] Banister, D. (2008). The sustainable mobility paradigm, *Transp. Policy*, 15, 73–80.

[3] Newman, P., Kenworthy, J. (1989). Cities and automobile dependence.

[4] Newman, P., Kenworthy, J. (1999). *Sustainability and Cities: Overcoming Automobile Dependence*. Washington, DC: Island Press.

[5] Newman, P., Kenworthy, J. (2011). Peak car use: understanding the demise of automobile dependence, *World Transp. Policy Pract.*, 17 (2), 32–42.

[6] Nakamura, H., Hayashi, Y., May, A.D. (Eds.) (2004). *Urban Transport and the Environment: An International Perspective*. Oxford: Elsevier.

[7] Nakamura K. and Hayashi Y. (2013). Strategies and instruments for low-carbon urban transport: an international review on trends and effects, *Transp. Policy*, 29, 264–274.

Chapter 3

Policy principles for low carbon mobility

Hussein Dia[1]

Abstract

Achieving low carbon mobility gains requires major policy, behavioural and technological changes. The relevant policy directions available to policy makers can be grouped into four broad categories: those that allow travel to be 'avoided'; those that 'shift' travel to more efficient modes; those that promote 'sharing' of vehicles and rides; and those that 'improve' the efficiency of vehicles and infrastructure. This chapter examines these strategic pathways and presents policy instruments that can help decarbonise transport while also addressing urban mobility challenges such as congestion and poor air quality. The chapter relies on best evidence from the literature and presents a broad range of measures that could be used to influence behaviour.

Keywords

Low carbon mobility; sustainable transport; transport policy

3.1 Introduction

The efficiency, productivity and environmental sustainability of urban transport systems continue to present big challenges for city leaders and decision makers around the world. These challenges, however, are not insurmountable. Despite the enormous difficulties, from growing urban populations to congestion and through to fiscal policy constraints, many cities have successfully achieved energy-efficiency improvements in urban transport and have demonstrated how substantial gains can be achieved through well-planned policies and stakeholder consultations [1]. Although the urban form and transport characteristics of each city are different, and policy responses vary, there are a number of policy pathways available to help decision makers in charting a path for sustainable mobility in their cities. To realise the

[1]Centre for Sustainable Infrastructure, Swinburne University of Technology, Melbourne, Australia

desired outcomes of sustainable mobility, policy makers must consider a holistic and long-term planning to deal with urban transport challenges. Governments should also move beyond isolated projects and consider strategies to build and retrofit the world's cities which will accommodate nearly 6.3 billion people by 2050 [1].

The development of low carbon urban transport systems requires a conceptual leap [2] and a reframing of the objectives of transport policies. At the centre of this shift is a recognition that the purpose of transport is to provide consumers with access to destinations, services, activities, goods and opportunities. Therefore, 'access' becomes the core objective that should be pursued by policy instruments. As a result, urban design should consider how to bring people and places together, by developing transport and mobility solutions that focus on access and accessibility, rather than reliance on increasing the capacity of transport infrastructure. Urban form and integrated land-use and transport planning therefore become a major focus of any policies that aim to achieve sustainable and low carbon urban mobility.

Today, transport planners increasingly recognise that the infrastructure decisions that a city makes will have a profound impact on shaping its future and people's travel behaviours.

Traditional transport planning approaches in most cities have been mainly driven by the economic dimensions of the challenges described in Chapter 2. The traffic gridlocks experienced on roads and motorways have been the basis for the development of most urban transport strategies and policies. The solutions prescribed in most of these strategies have been to build more infrastructures for vehicles, with a limited number of cities paying equal attention to managing the demand for travel or improving other modes of urban mobility (e.g. public transport) in a sustainable manner. But this has changed in the recent past (although progress is slow) and continues to unfold in many cities around the world, whereas greater emphasis is being given to sustainable modes of mobility such as public transport, cycling, walking and in more recent years also to disruptive mobility solutions that rely on car-sharing and ride-sharing enabled by technology platforms.

The policy instruments that promote this transition play a key role in reducing greenhouse gas emissions and pollution. They also have a wide range of benefits to the community and have been shown to enhance the fabric of urban environments and make cities more liveable. Well-planned mobility solutions for tomorrow's cities also improve accessibility to jobs and opportunities which are preconditions for sound economic development and for mitigating emissions and pollution in urban centres.

The current chapter presents these policy instruments and describes some case studies which demonstrate how the policies have been effective and successful in unlocking low carbon mobility benefits in cities around the world.

3.2 Urban transport system energy efficiency

An efficient transport system is what makes urban centres competitive [3]. It produces access to jobs, services, healthcare, education and other activities. It also favours the movement of people over the movement of a particular form or mode of transport.

The policy pathways presented in this chapter aim to help policy makers improve their cities' 'transport system energy efficiency'. The International Energy Agency (IEA) [1] defines transport system energy efficiency as 'the maximisation of travel activity with minimal energy consumption through combinations of land-use planning, transport modal share, energy intensity and fuel type'.

The transport energy efficiency can be improved in a number of ways. Cities with high dependence on motorised transport can promote shifts to cycling, walking and public transport. They can also adopt higher vehicle fuel-economy standards and establish regulations to govern land-use transport interactions. In practice, however, realising energy-efficiency improvements in transport is much more complex. Transport activity and mobility are highly dependent on travel demand, which is very challenging to improve in the short term and needs longer term policies that focus on behavioural changes. The use of more efficient modes of transport also depends on the choices available to consumers and the ease of reaching those modes (accessibility). They also depend on people's travel choices and preferences as influenced by culture, values and socio-economics factors (Figure 3.1).

The energy-efficiency policies can therefore be generally defined as a combination of (a) shaping the development of land-use and the physical transport network; (b) providing travellers with easy access to travel mode options and improving their travel choices; and (c) application of travel demand management strategies which in the long term can influence travel behaviour and mode choice preferences.

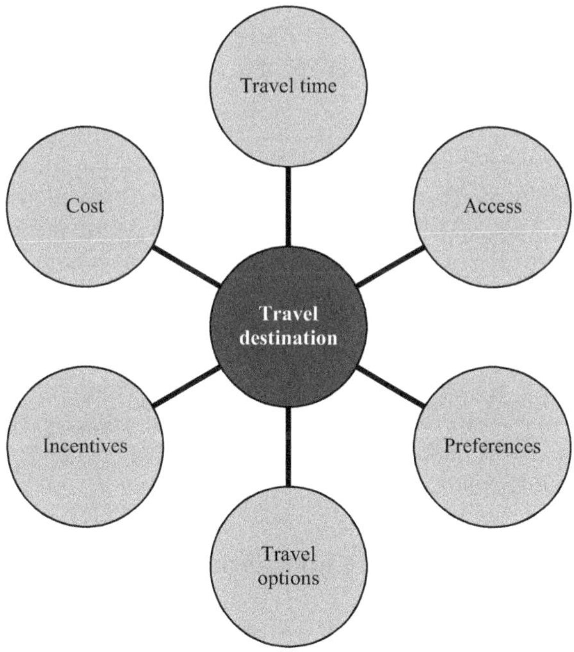

Figure 3.1 Transport activity decision matrix. Source: Adapted from [1]

Under this framework, energy efficiency in transport can be achieved through influencing transport activity and travel decisions from an energy and context-based perspective. This in turn can lead to a broader definition of urban transport system energy efficiency which can be articulated as the 'maximisation of the urban geography-transport network interface so as to provide optimal access and choice and maximise the efficiency of travel activity through combinations of modal share, energy intensity and fuel type' [1,4]. Under this definition, energy efficiency in transport systems can be achieved through prioritising land-use transport integration such that it reduces the need for energy-intensive travel options [1,4].

3.3 Policy instruments

Rigorous yet flexible policy instruments play a key role in shaping low carbon mobility strategies. In the absence of such policies, provision of transport and mobility will continue to favour motorised transport modes which are not energy efficient [1,4]. This would result in poor prioritisation of investments, which would encourage private vehicle usage while giving less weight to more efficient modes of travel.

The IEA has done substantial work on this and recommends several policy instruments to address low carbon mobility. These policies include demand management (e.g. congestion and road pricing), regulatory policies (e.g. parking restrictions) and supply-side strategies (e.g. introducing on-demand public transport). The IEA also suggests that incentives for private motorised travel are eliminated and replaced with taxation regimes to reflect the full range of external costs of fuels and vehicles [4].

Weak regulatory frameworks can also result in inadequate transport provisions. The most recent example of this is demonstrated by the wide range of policy responses to the ride-sharing services such as Uber and Lyft. Policy and decision makers in many cities around the world are still struggling to regulate the new shared business models and markets that are emerging, which are increasingly playing a crucial role in people's mobility in cities. Yet, these services are still considered illegal in many cities and blunt policy instruments are used to discourage rather than support disruptive modes of transport that have potential to curb private vehicle use. By improving or introducing regulatory policies that support innovations in transport provision, policy makers can increase transport system efficiency, improve the quality of transport services while influencing the shift to more efficient travel modes.

3.4 The 'avoid, shift, share and improve' framework

A package of measures, collectively known as the 'avoid, shift and improve' approach, have been proposed over the past 12 years as necessary policy instruments to achieve sustainable transport improvements [1,4]. This framework was proposed with the aim of (1) avoiding motorised travel when possible; (2) shifting travel to more efficient modes; and (3) increasing the energy efficiency of vehicles, fuel technologies and maximising the utilisation of existing infrastructure [5].

In this chapter, the framework is extended to include the recent developments in car-sharing and ride-sharing services (Figure 3.2). Recent evidence suggests that these collaborative mobility-on-demand services have started to influence car ownership models and are reducing the total number of vehicles required to meet people's demand for travel (Chapter 10). In the case of ride-sharing, there is also increasing evidence that 'car-pooling' types of ride-sharing services are increasingly being introduced in cities around the world and resulting in substantial benefits in terms of reductions in the total number of vehicle-kilometres of travel and reducing emissions and pollutions [6].

Together, these policies can help to achieve significant reductions in emissions, while also addressing urban transport issues such as congestion and access to services and employment. The key characteristics of these policies are outlined next.

3.4.1 Avoid

Avoid policies aim to slow travel growth through integrated land-use transport planning and travel demand management [1,4]. These policies include (1) virtual mobility programs (e.g. tele-working); (2) initiatives to reduce trip length (high

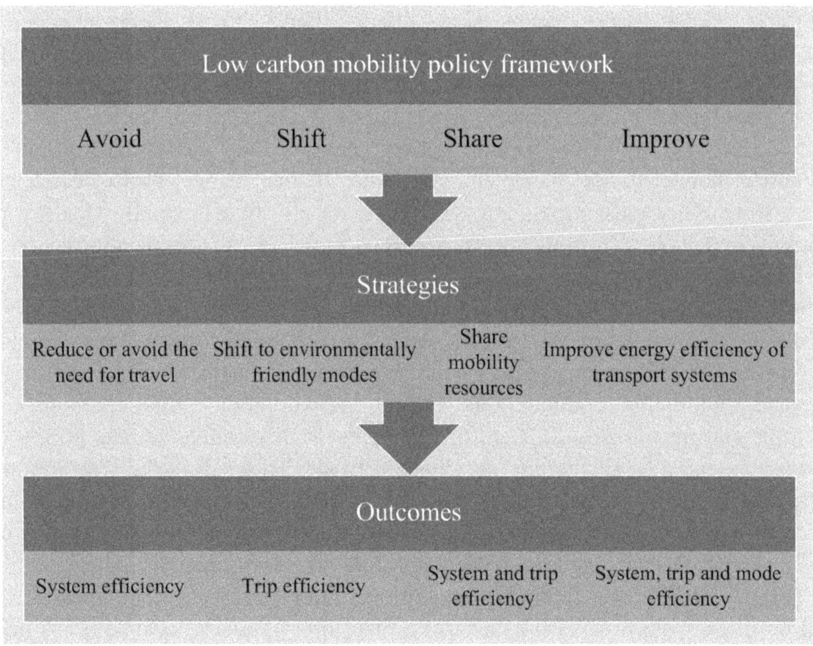

Figure 3.2 The avoid, shift, share, improve framework. Source: Adapted from [1]

density and mixed land-use developments); and (3) initiatives to reduce the need or desire for travel (congestion pricing, promotion of car-sharing and ride-sharing schemes). Other examples include on-line purchasing and similar business models which help consumers to avoid shopping trips altogether. Different forms of this policy principle are analysed in Chapters 4–6.

3.4.2 Shift

These policies encourage travellers to shift their travel from private motorised vehicles to more efficient modes such as public transport, walking and cycling. Policies under this category include integrated public transport and land-use planning, improved bus routes and services, pricing strategies to discourage private vehicle use (road use pricing, congestion charging, vehicle quotas or bidding systems for license plates such as the Certificate of Entitlement (COE) in Singapore), and road space allocation (dedicated lanes for cycling or bus lanes). Increasing the reliability and affordability of public transport, for example, can also promote their use over private vehicles which improves access to destinations. The same applies for incentives which make electric vehicles (EVs) more affordable. This policy principle is analysed in more detail in Chapters 4–7.

3.4.3 Share

Share policies enable and promote a shift from car ownership models and private vehicle use towards car-sharing, ride-sharing, bike-sharing (and even vehicle-parking space sharing in resource restrained urban centres). Consumers today, especially younger generations, are no longer interested in the vehicle as a 'status symbol' nor owning an expensive asset that stays parked for more than 90% of the time. Consumers are also increasingly demanding a higher level of service, more efficient and less polluting mobility options that suit their modern-day lifestyles. Disruptive mobility trends (Chapter 13), enabled by increasingly sophisticated app-based technology platforms, have the potential to fundamentally change the relationship between the consumer and automobile. The rise of the collaborative or sharing economy, popularised over the past few years by companies such as Airbnb, Zipcar and Uber, looks set for rapid growth over the next decade [7]. Access to mobility rather than to car ownership will enable customers to be more selective in choosing from the door-to-door mobility services offered by 'mobility operators' for intercity, suburban as well as 'last kilometre' travel solutions. The transport sharing economy is therefore an opportunity to move people, goods and services collectively by sharing the transport mode. This will in the long run help improve congestion, reduce emissions and air pollution and enable a new form of crowd-sourced on-demand mobility. With mobility offered as a service and on-demand (regardless of fixed bus routes or public transport schedules), consumers in the future will have the flexibility to choose the best solution for a specific trip purpose during any time of the day using their smart phones. Chapter 10 is dedicated in full to this issue and how shared mobility is impacting urban transport in cities around the world.

3.4.4 Improve

The 'improve' policies include initiatives which promote efficient fuels and more efficient combustion engine vehicles. They also include policy responses such as the application of transport and urban information technologies, introduction of EVs and adoption of intelligent transport systems (ITS), eco-driving, low carbon electricity generation and smart grids for EV charging stations. This policy principle has been well embraced in recent years with a shift in the thinking on how to provide the infrastructure required to support our mobility needs. Instead of building additional road capacity, there is more reliance nowadays on using technologies to optimise the performance of existing infrastructure and sweating of assets, in addition to improving vehicle performance and energy efficiency. Examples of policy interventions are provided in Chapters 6, 7 and 10–13.

3.5 Ten key principles from the 'avoid, shift, share, improve' framework

The following principles summarise the practical approaches for the application of the 'avoid, shift, share, improve' framework [8]:

1. **Planning dense cities**. In 1989, an influential study by Newman and Kenworthy [9] compared the density of 32 cities across the world. They found that denser cities have lower car use than sprawling cities. This is largely accepted today. Planning dense, walkable cities are now the new paradigm in city planning. Central to this concept is the integration of urban and transport development (Chapters 4–6). This approach supports the creation of small urban fabrics, mixed-use city quarters and car-free housing.
2. **Develop transit-oriented cities**. Transit-oriented developments are a successful strategy that addresses a range of issues including traffic congestion, increased urban sprawl, air pollution and increased energy use [10,11]. It is a strategy aimed at the grouping of high density and mixed use commercial and residential developments near or around public transport hubs or corridors (rail, light rail, bus). It is characterised by designs that encourage the use of alternate modes of transport such as walking, cycling or public transport, with lower parking supply, shopping malls and pedestrian amenities being prominent design features [12]. These are analysed in Chapters 4–7.
3. **Optimising the road network and its use**. This strategy enhances urban connectivity and reduces the distances and times of travel between origins and destinations. It also uses transport technologies and ITS solutions to optimise performance through network management systems, incident management and traveller information systems to keep travellers informed of traffic conditions especially in the case of delays due to incidents and other unexpected events (Chapters 5, 6 and 10–13).
4. **Implementing public transport improvements**. This strategy looks at enhancing existing public transport to ensure high quality of services for

public transport users. It can include examples such as introduction of bus rapid transit (BRT) and other high performance public transport services, integration of interchanges to make it easier for travellers to switch between modes, and integration of car-sharing, taxis and ride-sharing services into the mix of public transport options especially at interchanges. This topic is the focus of Chapter 7 but also receives coverage and analysis throughout Chapters 4–13.

5. **Encouraging walking and cycling**. This strategy includes creation of complete urban cycling networks, and high-quality design standards for sidewalks and cycle paths, ensuring safe facilities for cyclists and pedestrians. It also includes enhancement of traffic signal control strategies and intersection designs to improve mobility and safety for pedestrians and cyclists at road junctions. This topic is covered in more detail in Chapters 4–9.
6. **Controlling vehicle use**. This includes strategies such as congestion charging and user-pays road pricing based on charge per kilometre or time of day, and incentives to commute by bike or public transport and teleworking. Some cities around the world are also experimenting with total car bans in city centres. The case studies in this chapter provide more details, and some further analysis is also covered in Chapter 13.
7. **Parking management**. This includes strategies to establish parking fees, limit parking duration, better enforcement of parking rules, parking information systems including mobile-based apps that show parking availability or shared parking spaces, and providing incentives to property developers to replace in-development car parks with fewer parking spaces dedicated to car-sharing (Chapters 4–6, 11 and 13).
8. **Promoting clean vehicles**. This includes introducing incentives and rewards for EVs and clean-fuel vehicles, and promoting low-emission zones in city centres. It also includes rolling out EV charging infrastructure and improved methods for taxing transport fuels (Chapters 4 and 6).
9. **Stakeholder consultations and communication of results**. This includes stronger engagement with the public and decision makers, marketing campaigns for better public transport, car-sharing and ride-sharing. This is covered in a later section of this chapter.
10. **Pathways to comprehensive deployment**. This includes creating bodies that are responsible for overseeing low carbon mobility initiatives, integrating transport into climate change action plans, developing low carbon urban mobility plans and monitoring implementation and performance. This is also covered later in this chapter.

3.6 The benefits

A number of cities around the world have already taken successful measures to reduce motorised transport and promote more efficient forms of travel. This resulted in increases in transport efficiency, improved passenger mobility and access to opportunity, safer roads, reduced congestion, and improved health and air quality. According to IEA estimates, improving the energy efficiency of transport systems could save the world economy $70 trillion by 2050 [1]. The IEA estimates that the combined 'avoid' and 'shift' policies have the potential to reduce the global transport sector expenditure (to year 2050) by nearly USD 30 trillion. When further combined with 'improve' policies, an 'avoid, shift and improve' approach could lower expenditure by nearly USD 70 trillion by 2050 [1]. The IEA study is based on examples from more than 30 cities around the world and shows how improvements in transport efficiency and energy use can be achieved through better land-use transport integration planning, travel demand management and infrastructure technologies. Chapter 13 provides an analysis of the value of some of the technology-based disruptive mobility interventions.

3.7 City contexts and urban needs

Different cities will require different policy responses. The interventions will depend on the type of urban centre, existing base conditions, immediate transport challenges and the population's drivers of travel demand. The IEA identifies four city contexts which describe general travel characteristics and challenges facing cities across the globe [1].

3.7.1 Developing cities

These cities are characterised by high demand for transport services, and rapid growth in private vehicles. They often have inadequate transport infrastructure, especially for walking and cycling, and weak public transport services. The urban sprawl in these cities encourages growth in motorised private vehicles leading to increasing roadway congestion, rising road-induced crashes and injuries, and harmful emissions. These cities are also characterised by inequities in access to transport, employment and social services.

3.7.2 Sprawling cities

These cities generally have low densities with poorly defined urban cores. Commercial and business hubs spread throughout the metropolitan areas. The share of travellers using public transport, cycling and walking is generally low, whereas private motorised transport dominates as the primary mode of travel. Long distances between destinations make it difficult to provide efficient public transport services. Peak-hour congestion is high, and road infrastructure requires heavy investments and maintenance due to ageing assets. Emissions and road-induced crashes are major concerns [13].

3.7.3 Congested cities

These cities experience high volumes of traffic especially during peak hours. They have medium-to-high densities and strong urban cores. They generally have good public transport systems and high public transport modal shares. The cities are characterised by daily gridlock due to heavy traffic and dependence on private vehicles. Emissions and road crashes and injuries are also major concerns [5,14].

3.7.4 Multi-modal cities

These cities are characterised by high public transport, cycling and walking mode shares. They also have strong urban cores and high densities. Their transport networks are interconnected and well developed which encourages more efficient travel. Their mixed land-use developments and high-quality public transport services provide travellers with good access to different mode choices. Their non-motorised infrastructure is also well developed and includes cycling lanes and adequate pedestrian spaces. These cities feature prominent public transport terminals, including train stations and bus interchanges. Some of these cities would also have implemented demand management strategies which discourage driving (e.g. road pricing, congestion charging and parking restrictions) [15,16].

3.8 Matching policies to city needs

The 'avoid, shift, share, improve' policies are applicable in numerous contexts but the degree or intensity of response may vary to suit the specific local needs and challenges [1].

3.8.1 Developing cities

In these cities, some of the potential policies include regulations that limit sprawling developments and land-use strategies that prioritise dense urban cores, such as transit-oriented development. Transport infrastructure upgrades, such as spaces for pedestrians and public transport networks, can also help to improve access to destinations. To maximise benefits, the infrastructure and land-use policies should be co-ordinated with travel demand management to ensure that improvements are competitive with private vehicles. Other actions include better regulation of public transport operations and increasing service quality and frequency of public transport.

3.8.2 Sprawling cities

The long-term objective of policy interventions in sprawling cities is to transition them to denser urban centres that support more efficient transport. Medium and long-term goals include travel demand management and land-use policies that favour denser developments. Shorter term policies include improving existing systems and prioritising strategies that shift travellers away from private motorised travel. These policies include parking reform, road pricing and traffic management tools that help to maximise efficiency of existing assets such as adaptive traffic signal control.

3.8.3 Congested cities

A range of policies apply to these cities. These include travel demand management to influence a shift to more energy efficient travel modes, in addition to policies that discourage vehicle ownership (e.g. Singapore's vehicle quota and COE). ITS solutions including real-time travel information can also help to improve mobility, whereas car-sharing and ride-sharing incentives can encourage additional shifts to more efficient travel modes. Shorter term policies could address the gaps in network connectivity which also helps to encourage shifts away from private vehicles. These interventions would be more effective when they are paired with travel demand management measures.

3.8.4 Multi-modal cities

These cities can benefit from policies that improve traffic flows and access to more travel options. This includes dedicated facilities for energy-efficient modes (e.g. bus and cycling lanes) and investments in vehicle technology improvements. Multi-modal cities can also benefit from travel demand management policies. The wide-spread use of ITS and on-demand shared mobility can also improve urban mobility including real-time information on road conditions and public transport arrival and departure times.

3.9 Can policy measures be combined to increase impact?

Improvements which focus on transport efficiency can have increased impacts when they are planned using a co-benefits approach [17] which addresses the relationship between transport and other urban issues, such as health and well-being (Chapter 9). By developing such interactions between transport and other urban policies, decision makers can influence changes in the transport sector, which can help to secure broader support for transport initiatives (Table 3.1).

3.10 Stakeholder consultations

The success of urban transport policies relies to a large extent on the support they receive from the public and the wider stakeholders impacted by the policies. Soliciting the input and agreement of the stakeholders ensures the long-term success and continued support for these policies into the future. The stakeholders involved include city and regional elected officials, civil servants and national governments (if they are involved in the planning, financing and implementation of projects that extend beyond municipal jurisdictions). The stakeholders also include non-governmental bodies, professional organisations, consumer groups and private companies. Local government officials, particularly mayors, are very important in urban transport improvements. Cities often have direct ownership of city roads and have the capacity to set and enforce policies related to urban transport systems. This control enables mayors to introduce cycling lanes, congestion charging and road pricing, BRT (Busways) and EV charging infrastructure. Often mayors also

Table 3.1 Example benefits of sustainable urban transport policies

Policy options	Benefits				
	Reduction in GHG emissions	**Improved air quality and health**	**Reduced congestion**	**Increased accessibility**	**Improved road safety**
Bus rapid transit (BRT) or Busways	Medium	Medium	High	High	Medium
Light rail or metro rapid transit	Medium	Medium	High	Medium/high	Medium
Rail	Medium	Low	Medium/high	Medium	Low
Low-carbon vehicles	Medium/high	High	Low or negative	Low or negative	None
Non-motorised transport	Low	Medium	Medium/high	Medium/high	Medium
Land-use planning	Medium	Medium/high	High	High	Medium

Source: [12] .

exert control over the major public transport services through regulation and policy setting [18].

3.11 Case studies

Low carbon mobility has many social and economic impacts that can transform cities around the world. In this section, a number of case studies are presented to demonstrate the use of some of the policy instruments described in this chapter. These case studies demonstrate how city investments in low carbon transport strategies result in more affordable, efficient and safe travel connecting people to more employment, social and business opportunities. They also demonstrate how cities can accelerate the provision of social and economic benefits for city residents by choosing low carbon transport policies and projects.

3.11.1 Congestion charging in London, Stockholm, Gothenburg, Singapore and Milan

Congestion charging has long been proposed by planners, economists, policy makers and peak bodies as a likely solution to improve mobility particularly in congested urban centres [19]. In some countries like Australia, road pricing and congestion charging are also the focus of an ignited debate on tax reform and the need to rethink how city dwellers pay for the transport infrastructure that supports their mobility needs. The current transport funding models, which are largely based on collecting revenue from fuel excise, vehicle registrations and license fees, is inequitable and is unlikely to generate enough funds to meet future infrastructure expenditure requirements. These models will be under more stress in the future when revenue from fuel excise declines as consumers move towards EVs and more fuel-efficient cars.

Congestion pricing is also the most likely solution to curb transport emissions in congested urban areas [19]. Through this approach, drivers would pay directly, through a user charge, for travel into congested areas (a road user driving in a regional area and not contributing to congestion in the city would not pay). The charge can be dynamic and would change to reflect the current congestion levels (higher charges would apply during peak hours). This would be part of an overall tax reform that would remove or reduce other transport taxes (e.g. vehicle registration which is essentially a 'property' tax as it is not a charge related to the amount of travel).

The scheme would also provide incentives for car-pooling, car-sharing and other strategies to encourage people not to drive into congested areas. Measures would also be put in place to provide more public transport and improve services. The revenue generated from the user charges would go towards maintaining and operating roads and transport services, and improving the travel options available for travellers.

Overseas, a recent report by the Royal Academy of Engineers in the United Kingdom [20] identified congestion pricing as the single best option to tackle road congestion. In Australia, road pricing has been paraded as a key approach to fix transport problems over the past few years. The Henry tax review in 2009 [21], the Productivity Commission review of public infrastructure in 2014 [22] and the Harper Review in 2014 [23] all urged governments to consider road pricing as a matter of priority. These reviews also recommended that governments conduct pilot studies to demonstrate the benefits to road users.

A number of cities overseas have already implemented congestion pricing. These include London, Stockholm and Gothenburg in Sweden, Singapore and Milan.

The benefits reported for these cities are compelling. In the 12 years since London started the scheme, traffic congestion was reduced by around 10% in the central city area [24]. This included a 34% drop in private cars entering the area, and 28% increase in cyclists. The scheme also resulted in a 16% reduction in emissions within the charging zone, amounting to 30,000 t annually. A recent study has also found that the programme resulted in reducing traffic collisions by 40% between 2000 and 2010. In addition to saving money and lives, congestion pricing is also reported to raise more than $300 million every year which would go towards improving the city's transport services.

In Stockholm, the congestion pricing scheme removed 20% of peak hour vehicles and stimulated around 9% increase in public transport ridership [25]. In the densely populated city centre, emissions decreased by around 10%–14%. The system is also reported to have cut down particulate matter by 9% and mono-nitrogen oxides by 7%. It has also been generating daily revenues of $500,000–$2.7 million for the city. These funds are then reinvested for expanding bike lanes and new bus services.

Sweden's lesser-known congestion charging programme in Gothenburg is also reported as a success [26]. The scheme was effective in reducing traffic by 12% during the peak hours, with many travellers switching to public transport.

Singapore's experience with congestion pricing dates back to 1975. This was in the form of an area license system (ALS) which was based on a flat rate charge. The ALS reduced traffic by 45% and vehicle crashes by 25%. The ALS was replaced by an electronic road pricing system in 1998 resulting in a further 15% decrease in traffic entering the central business district area [27].

These overseas success stories provide a unique insight into those cities' remarkable journeys in implementing congestion charging. Although the schemes initially proved to be decisive and controversial, more people have gradually become accepting of the approach particularly as the benefits became more visible.

3.11.2 Bus rapid transit in Guangzhou, Ahmedabad and Lagos

BRT or Busway projects can help reduce travel, make journeys more efficient and cut greenhouse gas emissions.

The city of Guangzhou launched its BRT in 2010. The system carries 850,000 passengers every day and has succeeded in shifting 10%–15% of trips away from private vehicles. In addition to generating annual savings of US $14 million and reducing 86,000 tons of CO_2 every year, passengers are reported to personally save $103 million in out-of-pocket travel costs and 30 million hours of travel [28].

In Ahmedabad, the BRT has carried over 140,000 daily passengers since its establishment in 2009 and is expected to reduce the city's CO_2 emissions by 30% before 2040 [28]. A number of cost-effective projects such as designated entry and exit points to the system have led to 79% positive safety ratings for the system.

Nigeria's Lagos BRT system is also reported as a success story. It provides an efficient and affordable service to the local population through which it has helped reduce the income spent by poor households on public transport from 17% to 11%. The BRT fares are structured such that they are 30% lower than those for traditional public transport [28].

3.11.3 Bike share system: Paris

Paris' bike share program has about 20,600 bikes at 1,451 stations spread throughout the city. The municipal government is supporting the programme through road-safety campaigns, speed restriction zones and about 125 miles of new bike lanes to improve connectivity. This has helped increase ridership by 70% since the introduction of the programme, cutting emissions down by 32,330 t of CO_2 per year [28].

3.11.4 Car-Free urban centres: Plans for Oslo, Milan, Dublin, Paris, Madrid and Brussels

Plans to ban private vehicles in parts of central business districts are being considered by some European cities [29]. These plans are very controversial and are seen by many people as an intrusion on their freedom of choice. Nevertheless, some cities are using them as a last resort to curb vehicular traffic from central business areas. Since the start of 2014, at least six cities have announced ambitions to convert parts of their central districts into pedestrian-only spaces.

3.11.4.1 Oslo

The Norwegian capital plans to ban private cars in the central district by 2019. The city plans to enhance its bike infrastructure and improve its public transport services [29].

3.11.4.2 Milan

Milan has an existing congestion charging scheme in the city's centre. In July 2016, the city announced plans to restrict cars from its historical centre. The plan proposes for Piazza della Scala to join two other downtown pedestrian zones and gradually stretch outward to other parts of the city [29].

3.11.4.3 Dublin

In 2016, the Irish capital proposed a €150 million plan to ban private vehicles from parts of the city centre as early as 2017. The city hopes to reach a point where 25% of people would commute into the central business district using private cars. The city is targeting increasing the mode share of public transport to (55%), cycling (15%) and walking (10%) [29].

3.11.4.4 Paris

The French capital trialled a one-day ban on odd-numbered private vehicles in the spring of 2014. The trial resulted in a substantial decrease in traffic congestion and pollution. This was followed by another temporary car-free day in 2016. The city is reported to be considering more trials and studies in future years [29].

3.11.4.5 Madrid

One of the world's most ambitious plans to ban cars from the city centre can be found in Madrid. In 2015, the Spanish capital began issuing tickets to drivers who either do not live in the central areas or do not have a guaranteed parking space in an official parking space. The city is hoping to make more areas of the city strictly for pedestrian use [29].

3.12 Policy pathways

Policy pathways are intended to provide a practical guide to help national and local policy makers improve their response to urban transport system energy efficiency. Achieving energy-efficiency improvements in urban mobility is not always easy, but many cities around the world have used the policy instruments described in this book and achieved substantial gains in energy efficiency as was discussed in Chapter 2.

The policy pathways include detailed steps for supporting the development, financing, implementation and evaluation of the described policies. These have been identified by the IEA based on the findings from a study [1] they conducted which documented success stories in more than 30 cities around the world. These pathways include four stages (Figure 3.3): Plan, implement, monitor and evaluate, with ten critical steps which have been developed on the basis of lessons learnt from successful policy implementations from cities around the world.

The four stages in the policy pathway and the detailed ten steps identified by the IEA [1] are summarised below.

3.12.1 Plan

The planning stage includes administrative, technical and financial preparation to set the stage for implementation of the proposed policies. Stakeholder consultations and engagement with the public and decision makers are an important characteristic of this stage. This helps to ensure that the policy addresses the wider views of the community including other aspects such as transport accessibility, economic, social and environmental needs which also helps to ensure acceptance of the policies once they are introduced.

This stage includes steps 1–4 of the ten critical steps [1]:

1. **Identify the transport needs and the aims and objectives of the policy**. This includes identifying the issues and needs of the community where the policy is being proposed; defines the vision, aims and objectives, and clarifies the policy responses most adequate to meet these needs.

   When identifying the user needs, data and information are collected/collated about the mobility needs of households; relationships that may exist between socio-economic status and access to services and opportunities; limitations of existing transport provisions; future needs; existing and desired modal shares;

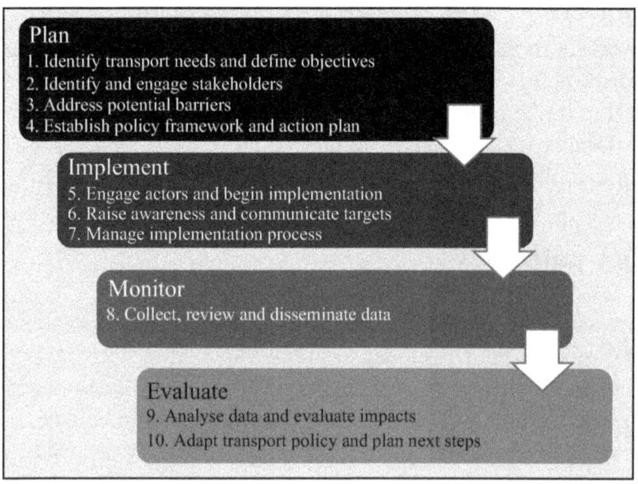

Figure 3.3 Policy pathways to low carbon mobility. Source: Author, adapted from [1]

the status of the transport infrastructure; availability of public transport services; security and reliability of services; land-use plans for the area; the nature of the urban economic structure; governance; and population and demographics trends.

Once the user needs and issues are identified, policy makers need to articulate a vision and define the aims and objectives to address these issues. At this stage, both short-term and long-term objectives can be identified under the single vision. The projected improvements in the levels of service which the public can expect to see as a result of the policies are also articulated and expressed to the community.

When identifying the policy responses, more than one specific policy can be implemented on the basis of experience and lessons learnt from other implementations. This includes consideration of co-benefits and complementary measures by implementing a multiple-benefits approach that capitalises on common threads between multiple urban issues [1]. For example, the pairing of road-safety programmes and travel demand policies could lead to improvements in the safe movement of people and goods, while also addressing congestion and energy efficiency.

2. **Identify and engage stakeholders**. Stakeholders provide support and feedback that can have valuable impact on the project. The number of engaged stakeholders varies but would generally include road and transport agencies, local city councils, public transport operators and professional organisations representing their members. This step identifies the stakeholders who are likely to contribute to the decision making; who have authority and power to support interventions; who are affected by the policies; and what motivates their decisions. Genuine engagement and incorporating feedback from stakeholders is key to the success of any new policy interventions.
3. **Address potential barriers**. Regardless of the policy instruments, barriers will exist to effective policy implementation. Identifying the potential barriers from the start, for example financial constraints, legal restrictions or inadequate regulatory frameworks, can help to manage their impacts if they are anticipated and planned for in advance. The steps involved include identifying the potential barriers; preparing to respond to the barriers; and engaging with the stakeholders to respond within specific time frames. For example, initiating legislative approvals and outreach to disfranchised groups can be pre-emptively taken before barriers hinder progress [1]. It is also important in this step to identify the technical, institutional and financial resources required for successful implementation.
4. **Establish policy framework and action plan**. This includes identifying the tasks that need to be completed, roles and responsibilities, time frames, and expected outcomes [1]. This helps to create a step-by-step framework to implementing and achieving the policy goals. The action plan should have direct input from the implementing bodies and stakeholders to ensure that the steps and timeframes are feasible. This step also includes preparation of a robust analysis of economic impacts; contingency plans and a framework for

how progress will be measured and the key performance indicators that will be used to judge the efficacy of the policy. This could include measurement of impacts on energy use, car ownership and public transport ridership, parking spaces and requirements, population density, emissions, exposure to noise and accidents, and government revenues.

3.12.2 Implement

This stage includes steps 5–7 of the ten critical steps:

5. **Engage actors and begin implementation**. This includes defining the contracting and project management procedures that need to be put in place to allow for measuring performance and setting clear lines of communications, roles and responsibilities. This helps to ensure an implementation process that is smooth and successful. Once the planning process is complete, the policy is then launched. The policy launches are important events to inform the public of changes and can also be used to raise awareness of policy objectives.
6. **Raise awareness and communicate targets**. This is a vital step to ensuring public acceptance of policy changes and initiatives. By communicating the goals and objectives of the policy measures, policy makers can raise awareness and encourage people to support the policy initiatives. Campaigns can also help to dispel any misconceptions and promote the real benefits of the proposed policies through effective communication. This step is also an opportunity for policy makers to communicate to the public the true cost of their mobility choices thereby encouraging behavioural shifts to more energy-efficient modes of transport.
7. **Manage implementation process**. This is a project management task that ensures the parties responsible for project execution are adhering to their roles on the project, and that progress is being made. Poorly managed projects can undermine confidence in a government's capacity to implement change [1], and such projects can potentially derail the policy goals. This step also includes verification of progress and ensuring compliance and timely delivery of project deliverables. Finally, this step is also an opportunity to provide technical training and build capacity to provide all relevant stakeholders with the support they need to execute the policy changes and disseminate information to the public.

3.12.3 Monitor

This stage includes step 8 of the ten critical steps:

8. **Collect, review and disseminate data**. To establish the impacts of the policy initiatives, data are collected before and after the policy initiation to measure the impacts. Effective and meaningful data are necessary to evaluate not only the benefits, but also any negative impacts not initially foreseen during the planning stage. The data collection process and goals should also reflect institutional needs and resource limitations, including budgetary concerns and

financing. Data collection and analysis can be costly and therefore considerable planning needs to be undertaken to ensure the correct types and amounts of data are being collected to measure performance. Once the data are collected, it needs to be reviewed, pre-processed and analysed according to established key performance indicators by which the project is going to be judged a success. All analyses should be verified, and the results should be reproducible. There is an increasing expectation nowadays that the data collected should be open to the public and other interested parties.

3.12.4 Evaluate

This final stage includes steps 9 and 10. This stage is aimed at establishing the effectiveness of the programme of works in meeting the objectives of the policy. This in turn helps to determine how the policy can be improved for future implementations; what were the success/failure factors; and how can they be mitigated:

9. **Analyse data and evaluate impacts**. Once the data have been collected and pre-processed, it can then be analysed and used to assess the effectiveness of the policy. There are a number of key performance measures against which the policy can be evaluated. These include [1] relevance (i.e. whether the project has met the policy objectives), effectiveness (the degree to which the objectives were met), impact (improvements that have been achieved), efficiency (cost-effectiveness), and sustainability (benefits can be maintained in the long term). The evaluation also identifies the success or failure factors; the strengths of the project which should be replicated; and the weaknesses that should be avoided. Another important element of this step is communication of results and information dissemination, which are vital to raising awareness of the benefits and increasing public support.
10. **Revise policy and plan future directions**. This step includes regular reviews of the policy to ensure that it continues to meet the desired objectives even as the system changes and evolves. This helps to provide insights into the interplay between the demand for travel and the policy initiatives. It also helps one to understand the public's willingness to change with respect to these initiatives. Planning the next steps also helps to continue development of existing programmes.

3.13 Conclusions

This chapter provided policy guidance for developing low carbon mobility solutions. It identified the emerging trends and policy instruments that can be used to steer our cities on a sustainable course of urban mobility. These solutions are grouped into four broad categories using the 'avoid, shift, share, improve' dimensions. The chapter also examined the strategic pathways to low carbon mobility and presented the policy principles that can help achieve efficiency improvements. These include planning dense cities, developing transit-oriented and pedestrian-oriented cities, optimising the road network and its use, public transport improvements, and planning for walking and cycling.

Central to these strategies is a recognition of the crucial role of integrated land-use transport planning (Chapter 5), which involves the optimal distribution of facilities and services to allow residents in urban areas to access services and economic opportunity while minimising the use of private vehicles. This helps to ensure effective integration of transport and urban development policies.

The case studies presented in this chapter demonstrate the benefits that can be achieved, not only in reducing emissions and pollution, but also in reducing congestion and improving access to jobs and opportunities. The chapter argued that accessibility should be the central focus of any policies for achieving an urban form that is environmentally sustainable and socially equitable.

Finally, the chapter included steps for practitioners – based on lessons learnt from successful cities around the world – to support them in the development, financing, implementation and evaluation of low carbon mobility policies and practices.

References

[1] International Energy Agency (2013). "A Tale of Renewed Cities. A Policy Guide on How to Transform Cities by Improving Energy Efficiency in Urban Transport Systems. OECD/IEA". Available from http://bit.ly/1rPpWPq.

[2] UN-Habitat (2012). "Planning and Design for Sustainable Urban Mobility: Global Report on Human Settlements 2013". Available from http://unhabitat.org/planning-and-design-for-sustainable-urban-mobility-global-report-on-human-settlements-2013/. Accessed 17 September 2016.

[3] GIZ (2013). "Urban Mobility Plans: National Approaches and Local Practice". Available from http://www.eltis.org/resources/tools/urban-mobility-plans-national-approaches-and-local-practice. Accessed 17 September 2016.

[4] Banister, D. (2007). "The Sustainable Mobility Paradigm", *Transport Policy*, vol. 15, 2008, pp. 73–80.

[5] IEA (2012), *Energy Technology Perspectives: Tracking Clean Energy Progress*, OECD/IEA, Paris.

[6] Dia, H. (2016). Mapping the value of collaborative mobility. *Proceedings of the ITS World Congress*. Melbourne, Australia.

[7] Kaas, H.-W. *et al.* (2016). "Automotive Revolution – Perspective Towards 2030". McKinsey & Company. Available from http://www.mckinsey.com. Accessed 17 September 2016.

[8] Bongardt, D. (2016). "10 Principles SUT". Available from https://prezi.com/7ufnp8crzc1l/10-principles-sut/. Accessed 18 September 2016.

[9] Newman, P. and Kenworthy, J. (1989). *Cities and Automobile Dependence: An International Sourcebook*. Gower Publishing, Brookfield, VT, USA.

[10] Cervero, R., Ferrell, C. and Murphy, S. (2002). *Transit-oriented development and joint development in the United States: a literature review*. Research Results 52, Transit Cooperative Research Program, Washington, DC.

[11] Nahlik, M.J. and Chester, M.V. (2014). "Transit-Oriented Smart Growth Can Reduce Life-Cycle Environmental Impacts and Household Costs in Los Angeles", *Transport Policy*, vol. 35, pp. 21–30.

[12] Chatman, D.G. (2013). "Does TOD Need the T?", *Journal of the American Planning Association*, vol. 79, no. 1, Winter 2013, pp. 17–31.

[13] Benfield, K. (2012). "Can Nashville Shake Its Sprawling Past?", Atlantic Cities. Available from www.theatlanticcities.com/commute/2012/02/can-nashville-shake-its-sprawling-past/1241/.

[14] UTIP (International Association of Public Transport) (2012). "PTx2 Showcase: Doubling the Market Share of Public Transport Worldwide by 2025". Available from www.ptx2uitp.org/.

[15] Wien (Stadtentwicklung Wien) (2006). "Transport Master Plan Vienna 2003, City of Vienna, Municipal Department 18 (MA 18) – Urban Development and Planning, Wien". Available from www.impacts.org/cities/vienna/smart-moves.pdf.

[16] Winkler, A. and Hausler, D., eds. (2009). "Smart Moves: Strategics and Measures of Vienna's Transport Policy, City of Vienna, Municipal Department 18 (MA 18) – Urban Development and Planning". Available from www.impacts.org/cities/vienna/smart-moves.pdf.

[17] Ang, G. and Marchal, V. (2013). "Mobilising Private Investment in Sustainable Transport: The Case of Land-Based Passenger Transport Infrastructure", OECD Environment Working Papers, No. 56, OECD Publishing. Available from http://doi.org/10.1787/5k46hjm8jpmv-en. Accessed 17 September 2016.

[18] ARUP (2011). "Climate Action in Megacities: C40 Cities Baseline and Opportunities". Available from www.greenbiz.com/sites/default/files/Climate%20Action%20in%20Megacities.pdf.

[19] Crawford, D. (2015). "Road Pricing Best for Congestion, Say Engineers". Available from http://www.transport-network.co.uk/Road-pricing-best-for-congestion-say-engineers/12372#.VmmErjwVmaQ.linkedin. Accessed 18 September 2016.

[20] Dia, H. (2015). "How to Fix Traffic Congestion in Australia". Available from http://www.gizmodo.com.au/2015/12/how-to-fix-traffic-congestion-in-australia/. Accessed 18 September 2016.

[21] Department of Treasury and Finance (2010). "Australia's Future Tax System". Available from http://taxreview.treasury.gov.au/.

[22] Productivity Commission (2014). "Public Infrastructure Inquiry". Available from http://www.pc.gov.au/inquiries/completed/infrastructure.

[23] Commonwealth of Australia (2014). "Competition Policy Review". Available from http://competitionpolicyreview.gov.au/files/2014/09/Competition-policy-review-draft-report.pdf.

[24] Davies, A. (2015). "London's Congestion Pricing Plan Is Saving Lives". Available from https://www.wired.com/2015/03/londons-congestion-pricing-plan-saving-lives/.

[25] Jaffe, E. (2012). "Traffic Jams, Solved". Available from http://www.citylab.com/commute/2012/12/traffic-jams-solved/4160/.

[26] Börjesson, M. and Kristoffersson, I. (May 2015). "The Gothenburg Congestion Charge. Effects, Design and Politics", *Transportation Research Part A: Policy and Practice*, vol. 75, pp. 134–146, ISSN 0965-8564. Available from http://dx.doi.org/10.1016/j.tra.2015.03.011.

[27] DAC & Cities (2014). "Singapore: The World' First Digital Congestion Charging System". Available from http://www.dac.dk/en/dac-cities/sustainable-cities/all-cases/transport/singapore-the-worlds-first-digital-congestion-charging-system/.

[28] Yadav, N. and Lefevre, B. (2016). "Beyond Emissions: 5 Cities Achieve Social and Economic Development by Reducing GHG Emissions from Transport". Available from http://thecityfix.com/blog/beyond-emissions-5-cities-development-reducing-ghg-emissions-transport-neha-yadav-benoit-lefevre/.

[29] Jaffe, E. (2015). "6 Big European Cities with Plans to Go Car-Free". Available from http://www.citylab.com/cityfixer/2015/10/6-european-cities-with-plans-to-go-car-free/411439/.

Glossary

BRT	bus rapid transit
COE	Certificate of Entitlement (Singapore)
EV	electric vehicles
GHG	greenhouse gas
IEA	International Energy Agency
ITS	intelligent transport systems
TOD	transit oriented developments

Further Reading

[1] Changing Course: A New Paradigm for Sustainable Urban Transport (ADB)

[2] Climate Action in Megacities: a comprehensive analysis of what mayors in the C40 megacities are doing to tackle climate change (ARUP)

[3] Eco-Efficient and Sustainable Urban Infrastructure Development Toolkit (UNESCAP)

[4] Guidelines and Toolkits for Urban Transport Development (ADB)

[5] Principles for Transport in Urban Life (ITDP)

[6] Sustainable Transport: a sourcebook for policy makers in developing cities (GIZ)

[7] Sustainable Urban Transport Project: guides for policy makers (GIZ)

[8] Urban Passenger Transport: framework for an optimal modal mix (ADB)
[9] Urban Transport and Energy Efficiency: a sourcebook for policy makers in developing cities (GIZ)
[10] Urban Transport Policy and Planning Documents (GIZ)
[11] Arthur D. Little (2014). The future of urban mobility 2.0. Available from http://www.sutp.org/en/news-reader/the-future-of-urban-mobility-20.html.

Chapter 4

Low carbon mobility and reducing automobile dependence

Peter Newman[1]

Abstract

Low carbon mobility in our cities will need to be a combination of structural change that reduces automobile dependence as well as technological change. To determine the potential mechanisms for accelerating this transition, the emerging trends in reduced automobile dependence are outlined in the world's cities together with emerging technological trends. The policies outlined can enable 100% oil-free cities by 2050 and 100% renewably fuelled cities by 2100.

Keywords

Transport; oil; automobile dependence; peak car; urban fabrics

4.1 Introduction

The last 70 years of the twentieth century was the era of cities based on the Internal Combustion Engine (ICE) and oil; this has created the phenomenon we have labelled automobile dependence [1,2]. The twenty-first century is unfolding rapidly as the period of dramatic decline in oil as both technology and city structure are changing away from the need for oil. The domination of ICE technology however will take some time to phase out. The limitations of ICE technology have exercised the minds of many technologists and regulators who have struggled to make cleaner and greener cities that have less smog. But now the changes are being driven by an economy that not only needs to move away from all fossil fuels but also needs to find ways of getting more people out of their cars.

We have been collecting data on cities, transport, energy and land use for nearly 40 years [3,4]. In recent data, we began to see an interesting trend downwards in car-use growth with the 1990s showing a plateau in car use per capita

[1]Curtin University Sustainability Policy (CUSP) Institute, Curtin University, Perth, Australia

growth after decades of rapid growth. Then it became really interesting as from the early part of this century in the United States, Europe and Australia, we saw a reduction in car use per person for the first time in 100 years (give or take a few disasters) [5–7]. In Australian data, every city including the congestion-free cities of Canberra, Darwin and Hobart all went into reverse on car use per person.

The first signs of this can also be seen in emerging cities like those in Eastern Europe, Latin America, China and India [7]. As a result, we are seeing oil consumption slowing and reversing its historic growth trends in the world, and in particular oil is decoupling from economic growth [7,8]. These trends seem to be suggesting structural changes that are influencing the nature of our cities.

My reading of the trends in technology, global climate change governance, city planning, urban economics and urban cultural change suggest that we have at this point in history a convergence towards a new kind of city building that is based around renewables and electric transport and a significant reduction in automobile dependence [7]. It promises to create much lower carbon mobility cities than could have been imagined before. These cities can now be oil-free and their regions and the links between cities will follow. This chapter will outline further detail about these trends and how I see this unfolding.

4.2 The historical opportunity

At each point in industrial history, the different waves of innovation have shaped our cities. In Figure 4.1, the waves of innovation, as set out by Hargroves and Smith [9], can be seen to rise and then fall with a major economic downturn punctuating each of the industrial phases. Different transport and associated fuels are associated with

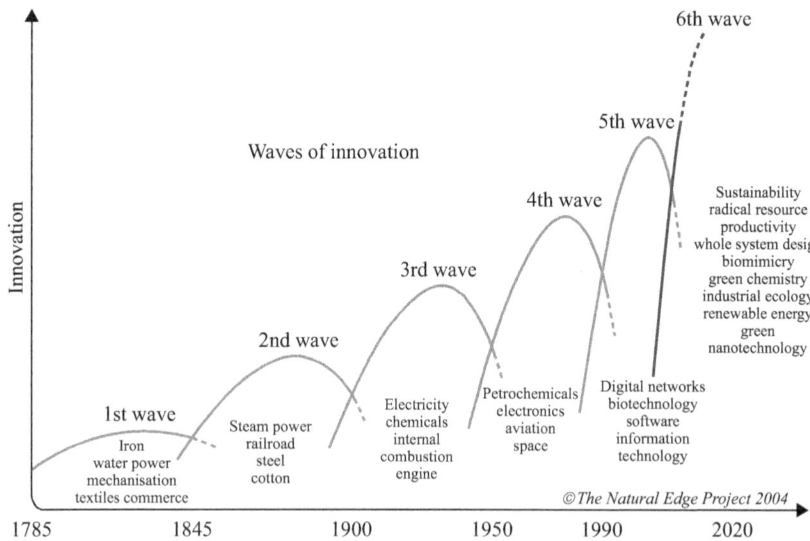

Figure 4.1 Waves of innovation (Source: Hargroves and Smith [9])

each of the waves, and these in turn change the nature of cities and hence facilitate these economic eras. It is possible to see emerging the sixth wave of innovation being related to the green economy with transport and fuel that enable the next era of oil-free cities and regions.

Cities began as walking cities with dense, mixed land use based around narrow winding roads and buildings hard up against their property lines. This Walking City urban fabric lasted from early times including the early stages of the industrial revolution and continue through to today whenever an informal, slum settlement is built as they take on this form. The first wave of industrial innovation began in the old Walking Cities that were linked by water transport and began using water power to make industrial products.

By the 1840s, the limits to this began to be obvious in space and materials; thus, the next phase of innovation, in the second wave, saw the arrival of the steam engine and railway. Cities began to spread out along the rail tracks and to build much bigger production based on steam power and steel. Then the third wave came in the 1890s when the incredible pollution of living on top of steam boilers, and their coal-based air pollution collapsed into a global depression from which came the amazing innovation of electricity. This enabled electric trams and electric trains to spread the city along corridors and the production systems to be separate from their power source – as well as the delights of lighting a city. They increased coal-fired power enormously with growing consequences. They created the Transit City with its own linear, medium density urban fabric based around trams and trains.

By the 1930s, these cities were reaching their limits, and a new era was created around the automobile, cheap oil and highways that enabled cities to spread in every direction and much further out. Thus the Automobile City urban fabric was formed with its characteristics of low density, separated in land-use functions with big setbacks and large, high capacity roads. Now these cities are reaching their limits just as previous cities reached theirs. As outlined below, these cities can no longer afford the costs of their spatially inefficient urban sprawl, their traffic speeds and travel time budgets are becoming dysfunctional and hence cities are moving away from cars and automobile dependence with oil becoming a commodity that is phasing out. New kinds of cities are emerging.

4.3 Twenty-first century cities

The growth and decline of automobile dependence has been traced in our Global Cities Database from the 1960s [4]. The role of urban form (especially density) and the level of transit in shaping this car dependence have been shown to be the main statistical explanation for how twentieth century cities have grown [10]. Both of these factors have plateaued and are now reversing.

4.3.1 The reversal in density decline

Cities are structured by transport. Marchetti [11] and Zahavi and Talvitie [12] were among the first to show that there is a universal travel-time budget of a little over one hour on average per person per day, which is around 30 min on average for the journey to work. This Travel Time Budget has been found to apply in every city studied; we found it holds true in all 100 or so cities in our Global Cities Database [3], as well as in data on UK cities for 600 years [13].

The Travel Time Budget helps us to understand how the three urban fabrics outlined above have been structured, and how they can be transformed for agendas such as low carbon mobility [2,7,14]. Cities grow to being about 'one-hour wide' based on the speed by which people can move in them. If cities go beyond this, they start to become dysfunctional and therefore begin to change infrastructure and land use to adapt again to this fundamental principle [15].

Thus the three main city types or urban fabrics are based on this principle: Walking Cities grew outwards about 3–4 km due to walking speeds and hence were very dense with narrow streets; Transit Cities could grow outwards for around 20 km due to the speed of transit enabling long medium density corridors of urban development with dense walkable centres at stations; and Automobile Cities grew outwards with the popularity of the automobile extending cities out 50 km or so. This last era ushered in the phenomenon of low density automobile dependence.

Most cities today have a mixture of all three urban fabrics as shown in Figure 4.2.

Those cities built mostly in the automobile era are heavily oil dependent as outlined in Figure 4.3 that shows there is an exponential link between car use/fuel use and urban density.

However, there is a new world emerging in terms of this fundamental structuring of cities: re-urbanization. Cities have been reducing in density for the last 100 years as the benefits from faster movement have rebuilt urban development opportunities around mechanised transport, first rail-based then road-based.

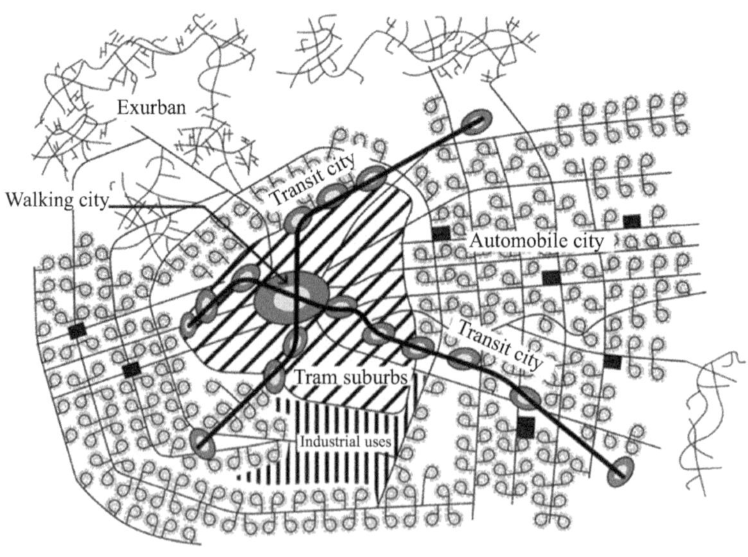

Figure 4.2 Growth of three city types based on transport modes (Source: Author)

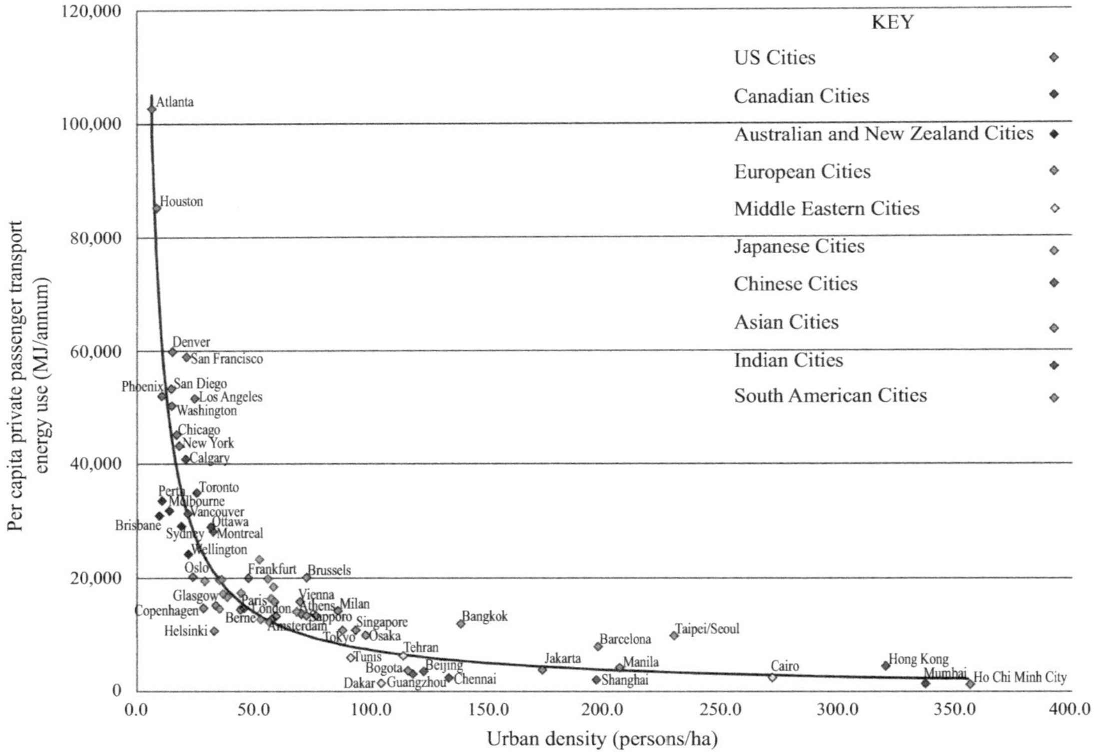

Figure 4.3 Transport fuel use and urban density in global cities (Source: Newman and Kenworthy [7])

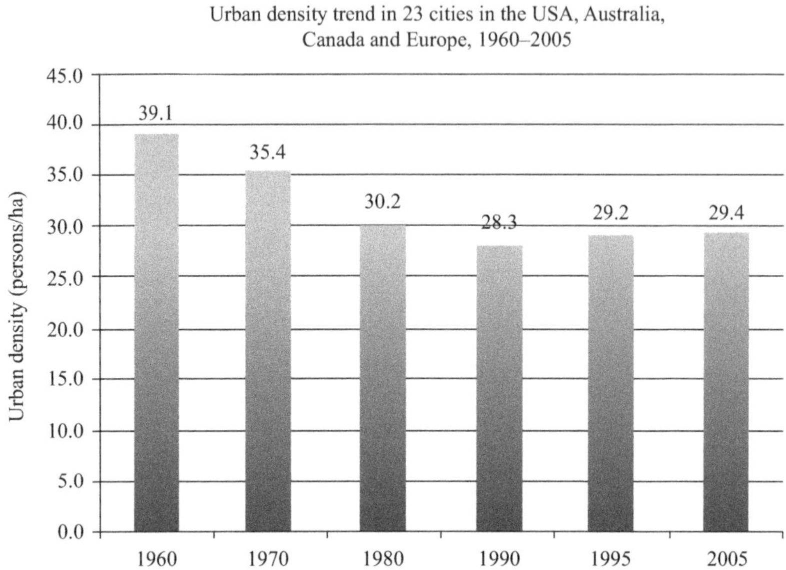

Figure 4.4 Trends in urban density from Global Cities Database, 1960–2005

Figure 4.4 shows the data from our Global Cities Database and how cities have stopped declining in density, they have plateaued and now are going up.

If a city begins to increase its density, then in structural terms it will reduce its car use and fuel use. In fact, it will reduce it exponentially. Re-urbanization is a structural move supporting low carbon mobility and can explain the peak car and transit growth phenomenon. But why is this happening?

In Figure 4.5, the growth in the travel-time budget in Australian cities due to urban sprawl and traffic congestion is very evident. All these cities are now structurally dysfunctional as they have gone beyond the 30 min average travel time for the journey to work.

It is not hard to see why cities would respond to such increases in travel time by

1. Enabling re-urbanization so that people can live nearer their work and so save time. This suggests that work is centralizing so residential opportunities, and transport opportunities need to recognize this.
2. Enabling rail projects that can go around, over or under the traffic, and generate some real travel time savings.

4.3.2 The reversal in transit decline

Urban rail went into a hiatus during the expansion of the car-based city and the freeway era. However, starting in the 1990s and accelerating this century, there has been a dramatic increase in urban rail [17]. The rapid growth of urban rail can be seen in the traditional Transit Cities of Europe and in the recent Automobile Cities of North America and Australia, as well as the emerging cities in the Middle East,

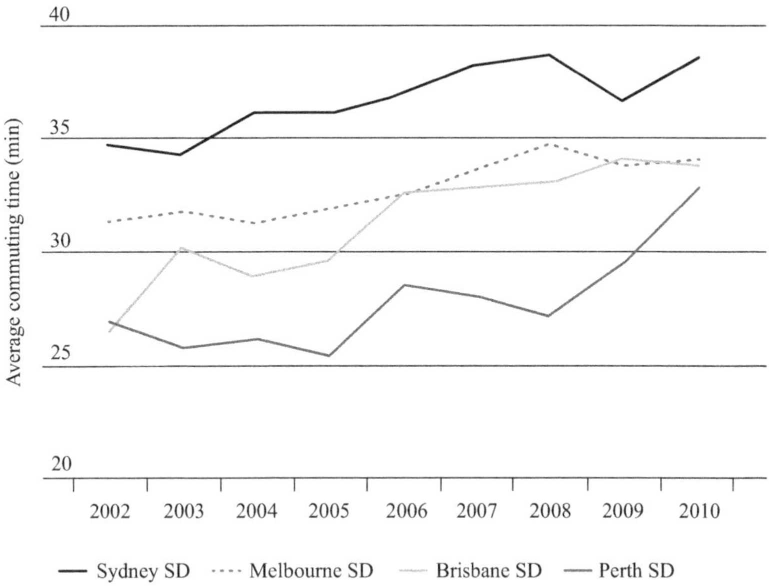

Figure 4.5 Trends in journey to work times in Australian cities (Source: BITRE [16])

Asia and Latin America. In our Global Cities Database, the growth rates in transit passenger-kilometres between 1995 and 2005 were European cities 22%, Asian cities 20%, Australian cities 11%, Canadian cities 12% and American cities 16%. Such growth rates have not been seen for many decades, indeed they have mostly been falling.

The growth has been based on massive new investment in transit. In Europe, 65 cities added light rail to their transit systems with now a total of 160 cities recognizing that light rail can bring back to life Transit City urban fabric. Expanded heavy rail and dramatic growth in high-speed rail across Europe has also been a feature of the rapid growth in transit. In the dense emerging cities of the Middle East, Asia and Latin America even more investment in new transit is happening. Most Middle East cities, even in Saudi Arabia, are now building fast Metro systems across their cities. There are now 81 Chinese cities building Metros and 52 Indian cities after President Modi announced every city over 2 million would have a Metro.

In the Automobile Cities there has also been a big commitment to investment in transit with most Australian, Canadian and US cities committing to new urban rail lines in their cities. In the United States light rail grew 190% between 1993 and 2011 based on expansion or new building in 18 cities with dozens more underway. Heavy rail grew 68% during this period though bus patronage dropped 3% suggesting that transit speed has become critical in these automobile dependent cities that are now stuck in traffic.

Data from our Global Cities Database are summarised in Tables 4.1 to show how automobile speeds have slowed in relation to transit and especially where there is a rail option.

The ratio of overall public transport system speed compared to general road traffic has increased from 0.55 to 0.70 between 1960 and 2005; the ratio of rail system speed to general road traffic has gone from rail being slower than cars in 1960 (0.88) to a situation in 2005 where rail was on average faster (1.13). And this trend has been a steady increase.

Within the different global regions, it can also be seen that even today in American and Canadian cities, public transport overall is barely half as fast as general traffic speed, whereas their rail systems are about 90% to 95% as fast, meaning that in many cases they are competitive with the car, especially into dense urban centres where clogged highways are losing out to fast rail systems time and again. Australian cities do a little better with public transport overall now being only about 25% slower than cars, whereas the rail systems are now on average about 8% faster and have generally been improving their competitive position since the 1960s and 1970s. When major corridors into city centres are considered, the rail systems are at a clear advantage.

European cities have generally had quite competitive rail systems in terms of speed, but they have hit a high in this 45-year perspective by reaching an average of 28% faster than cars. Their overall public transport systems, as a result of their fast rail systems, are on average 90% as fast as the car. It is interesting to note, however, that their rail speed relative to traffic dropped from 1.07 to 0.80 during the 1960s when they opened up to the car and built a lot of roads. A similar thing can be seen in Australian cities, but after that, rail speeds relative to the car rose quite consistently.

Table 4.1 Ratio of overall average transit system and rail speed to general road traffic speed in cities, 1960 to 2005 (Source: Newman and Kenworthy [7])

Comparative speeds in global cities	**1960**	**1970**	**1980**	**1990**	**1995**	**2005**
Ratio of overall public transport system speed to road speed						
American cities	0.46	0.48	0.55	0.50	0.55	0.54
Canadian cities	0.54	0.54	0.52	0.58	0.56	0.55
Australian cities	0.56	0.56	0.63	0.64	0.75	0.75
European cities	0.72	0.70	0.82	0.91	0.81	0.90
Asian cities	–	0.77	0.84	0.79	0.86	0.86
Global average for all cities	**0.55**	**0.58**	**0.66**	**0.66**	**0.71**	**0.70**
Ratio of metro/suburban rail speed to road speed						
American cities	–	0.93	0.99	0.89	0.96	0.95
Canadian cities	–	–	0.73	0.92	0.85	0.89
Australian cities	0.72	0.68	0.89	0.81	1.06	1.08
European cities	1.07	0.80	1.22	1.25	1.15	1.28
Asian cities	–	1.40	1.53	1.60	1.54	1.52
Global average for all cities	**0.88**	**1.05**	**1.07**	**1.11**	**1.12**	**1.13**

Asian cities in the sample have very fast rail systems compared to their crowded road systems, whereas in 2005 rail speeds were some 52% higher. This has fluctuated somewhat over the decades, but rail speeds on average were never less than 40% better than cars, even in the 1970s. Such data give hope to the many emerging cities struggling with the car, which rail systems can help them to develop a very different kind of overall urban transport system that is not so overwhelmed by the automobile. In the emerging cities, traffic speeds were very slow (around the low 20 km/h), and bus speeds are always slower than this. Bangkok had a traffic speed of 14 km/h and a bus speed of 9 km/h in 1990. Thus, as these emerging cities build rail (often above or below the traffic), the data will be reflected in dramatic transit speed improvements.

An example of how emerging cities are beginning the transition away from car use growth is given by Beijing where Figure 4.6 shows that the rapid growth of cars hit the wall around 2005 and is now in decline as the Beijing Metro and associated bus system have grown. The Beijing Metro now has 9 million passengers a day.

Fast rail between cities is having a rapid growth phase – especially in China and Europe. Japan built their system in response to the early oil crises and now other countries like the United States and Australia are recognizing that they must also though this is painfully slow in a time where major infrastructure investment is difficult.

The reversal of density and transit trends would suggest that a significant structural change is underway in our cities. It seems like an era is ending particularly as travel times go into dysfunctional areas beyond 30 min for the journey to work. It is not hard to see why cities would respond to such increases in travel time by (a) enabling rail projects that can go around, over or under the traffic and generate some real travel time savings, and/or (b) enabling re-urbanization so that people can live nearer their work and so save time. This suggests that work is centralizing so what are the economic factors driving these changes?

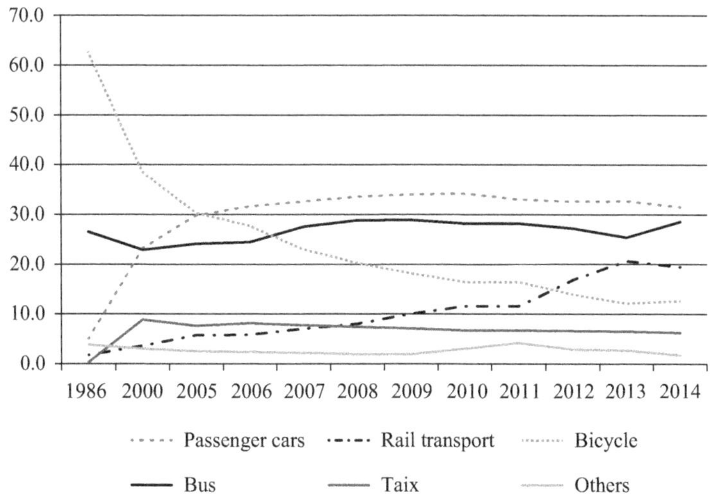

Figure 4.6 Beijing transport trends (Source: Gao and Newman [18])

4.4 What are the economic drivers of reducing automobile dependence?

If cities are shaped by transport infrastructure to enable reasonable travel times, what are the drivers that are locating work and housing? Two factors seem relevant: the rise of the knowledge economy and the rise of digital technology in changing culture.

4.4.1 Knowledge economy and city form

The biggest change in the economy during the period leading up to and including the period of car-use decline and rail growth has been the digital transformation and the consequent knowledge/service economy. Despite this being global and enabling long distance communication, it has in fact been a concentrating force in terms of city structure and fabric. The data on global cities outlined above show a universal increase in density, and this is even more obvious when it is seen how knowledge economy jobs are focussing in city centres and sub centres [7,19].

The knowledge economy and digital jobs are focused in city centres as these are where the creative, face-to-face synergies occur between people [20]. Old central business district's (CBD's) have been transformed back into functional Walking Cities and those which have done this best have attracted the most capital and young talent to work there [21,22]. The six most walkable cities in the United States have 38% higher gross domestic product [23]. In Boston, 70% of the knowledge economy workers live in walkable locations.

Transit systems and walking are the most spatially efficient forms of transport. If 1 km of a lane of road was considered as a unit of travel then car traffic can fit about 800 people per hour down that lane in a suburban street, a freeway up to 2,500, a busway around 5,000, a light rail between 10,000 and 20,000 and a heavy rail up to 50,000. These striking differences in spatial efficiency are translating into competitive advantage based on the need to bring people together in centres.

Transit City corridors are also regenerating because of these efficiencies enabling the linkages between centres and sub-centres with their need to provide housing and work locations within the 30 min journey to work. The revival of the Transit City is mostly around the need to link universities, health campuses and IT job clusters that have created their own centres for jobs and the need to attract housing and transit to link them together efficiently. In all these places, the waste of urban space to parking and large capacity roads has become a major issue impacting on their competitive advantage. Only sustainable transport solutions can help resolve this.

Other parts of the economy, such as manufacturing, small and large industry, freight transport and storage, have remained car-based and are outside this new knowledge economy. They will remain as part of the consumption economy, but they are not where the growth in jobs or the growth in wealth is happening. Thus, the Automobile City economy and culture has become somewhat distinct from the new regenerated urban economy of knowledge/services with their basis in Walking and Transit City locations.

4.4.2 Digital technology, culture and city form

As with many economic changes, there is also a cultural dimension to this change that perhaps explains the rapidity of the changes observed above as well as the demographic complexion of the change. Young people (especially those involved in knowledge economy jobs) are moving to reduce their car use and switch to alternative transport faster than any other group (Chapters 10, 11 and 13). This has been recognised by a few commentators and related to the use of social media devices [24]. On transit or walking (and even to an extent while biking), young people are already connected by their smart technology phones and tablets. They are hardly usable while driving a car. The Davis *et al.* [52] report shows that the mobile phone is a far more important device than a car for younger people. This is a cultural revolution that partly underlies the rail revolution as well as the re-urbanization of cities.

Baby-boomers gained freedom and connection with a car, Gen Y's are not needing one. They like to save time on a fast train, but they also like to use the time constructively relating to their friends and work through smart technology devices. They are not put off by density as they know it enables them to live closer to the city where all the action is as well as the all the best jobs. They are also more inclined to be part of the shared economy with innovations such as UberPool which claims to have reduced London's Vehicle Miles Travelled by 700,000 in 6 months from its opening in November 2015.

Thus the structural expression of this change is that younger people are moving to live in the Walking City or Transit City as these locations more readily enable them to express the kind of urban experience and culture that they aspire to [24,25] as well as save precious time. Thus they feed the market that enables peak car, the rail revival and city centre renewal to continue. They are the leaders in ending automobile dependence.

4.5 What comes next?

Three factors are considered to be shaping the future: the urban economy, oil and climate change.

4.5.1 Urban economy

There seem to be no limits to how much the knowledge economy will grow as the basis for the future. One group of innovations that are part of this are associated with the green economy and are based around the provision of renewable energy, high efficiency housing and building and low carbon mobility. These are part of the sixth wave of innovations described earlier and few commentators would suggest that they are going to diminish in their significance. Thus, from an economic perspective, there is likely to be a continuing drive to reduce automobile dependence and move towards decarbonising transport.

4.5.2 Oil

Reducing automobile dependence means reducing demand for oil that is now reflected in the price of oil. The limit of oil with its associated need to dig deeper and more dangerous deposits has been debated for many years. Murray and King [26] suggest peak oil happened in 2008 (at least for conventional oil sources) when oil prices tripled and the areas where cities had spread or indeed scattered beyond normal commuting times simply crashed as they had no local services, no local jobs and no public transport. The sub-prime mortgage meltdown affected areas with the greatest car dependence far more significantly. Many of these suburbs are now abandoned, and some cities are even contemplating bulldozing urban areas that are beyond any conceivable repair. Perhaps financial institutions will start taking risks on urban developments that are highly car dependent, but they mostly would have seen that such investment is not very resilient.

4.5.3 Climate change

Climate change governance at every level is now requiring plans to be implemented to reduce greenhouse gases; the global goal is 80% by 2050 and 100% by 2100. This has undermined most long-term planning based on the fourth wave economy in which fossil fuel use expanded exponentially. Since the Paris Accord in December 2015 and the New York ratification by 170 nations in April 2016, the commitment to phase out oil has been given significant momentum.

The next phase in the innovation waves according to Hargroves and Smith in Figure 4.1 is a combination of the digital economy and the sustainability economy. In terms of transport, it seems that the main innovations are electric transit and electric vehicles based on renewable energy. What are the options and what will they mean for our cities and regions?

4.6 Options for the low carbon mobility transition

In combination with technology including electric transit and electric vehicles, our cities will be shaped by practices including travel demand reduction, demand management, and investment in regional biofuels and renewable natural gas.

4.6.1 Electric rail transit investment

Electric transit powered by renewable energy can bring about dramatic reductions in car use and their associated need for oil. When cities provide a combination of transportation and land-use options that are favourable for green modes and offer time savings compared to car travel, then the switch to transit is inevitable. This means wherever transit is reasonably competitive with traffic in major corridors then people will use it. Those cities where transit is relatively fast are those with a reasonable level of support for it as shown above. The reason is simple – they can save time.

Perth began this transition to modern fast rail first in Australia in response to its rail system being closed down in 1979. The city has 172 km of fast suburban rail with the southern Rail carrying eight lanes of traffic equivalent after just 10 years of operation; this success was based on bus integration [27].

With fast rail systems, the best European and Asian cities with the highest ratio of transit-to-traffic speeds have achieved a transit option that is faster than the car down the main city corridor. Rail systems are faster in every city in our 84-city sample by 10–20 km/h over bus systems, as buses rarely average over 20–25 km/h. Busways with a designated lane can be quicker than traffic in car-saturated cities, but in lower density car-dependent cities it is important to use the extra speed of rail to establish an advantage over cars in traffic. This is one of the key reasons why railways are being built in so many US cities that shut down high quality rail in the beginning of the Fourth Wave and are now regretting their unbalanced transport systems.

Rail has a density-inducing effect around stations, which can help to provide the focused centres so critical to overcoming car dependence. New funding and financing alternatives use this land value capture approach to provide significant private investment in new urban rail systems [28,29]. Thus, transformative change of the kind that is needed to rebuild car-dependent cities comes from new electric rail as they provide a faster option than cars and can help build transit-oriented centres.

We need to stop increasing road capacity and provide major increases in transit capacity if we are going to help this transition to a low carbon mobility city and its associated economy. This is not easy given the funding systems that have developed around the car and truck.

How much is it possible to change our cities? It is possible to imagine an exponential decline in car use in our cities that could lead to 50% less passenger km driven in cars? The key mechanism is a quantitative leap in the quality of public transport accompanied by an associated change in land-use patterns.

Figure 4.7 shows the relationship between car passenger km and public transport passenger km from the Global Cities Database. The most important thing about this relationship is that as the use of public transport increases linearly the car passenger km decrease exponentially. This is due to a phenomenon called *Transit Leverage* whereby one passenger km of transit use replaces between three and seven passenger km in a car due to more direct travel (especially in trains), trip chaining (doing various other things like shopping or service visits associated with a commute), giving up one car in a household (a common occurrence that reduces many solo trips) and eventually changes in where people live as they prefer to live or work nearer transit.

Using this relationship, we have shown that cities can project their continued transit growth and expect significant reductions in car use [30]. Indeed it is feasible that each city could set a target of increases in passenger km per capita for public transport in order to achieve 50% reductions in car use as part of their commitment to reaching the global goal of 80% reduction in greenhouse gases by 2050.

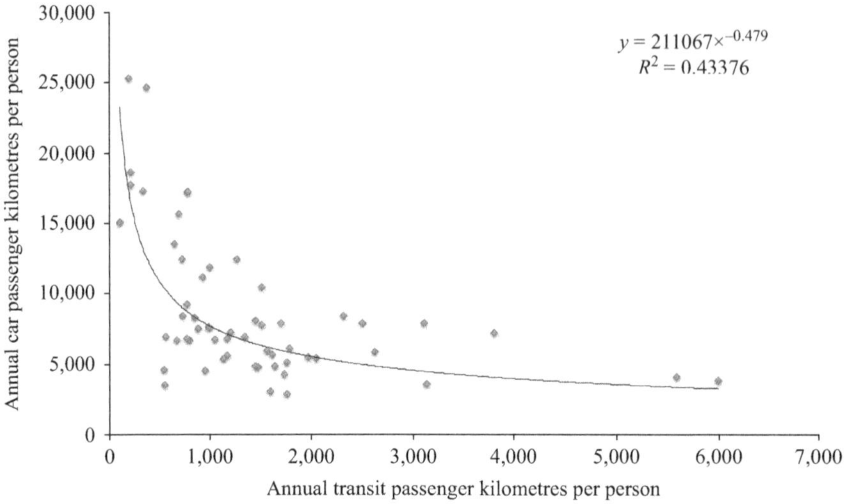

Figure 4.7 The exponential relationship between transit and car use (Source: Global Cities Database)

The biggest challenge in an age of radical resource efficiency requirements will be finding a way to build fast rail systems back into scattered automobile-dependent suburban areas. Rail reserves are not available so many cities are choosing to tunnel or go above ground on elevated structures – both are expensive. The solution may well be provided by Perth and Portland, Oregon that have built fast rail systems down freeways deep into car-dependent suburbs. To build fast electric rail down the middle of roads is easier than anywhere else as the right of way is there, and engineering in terms of gradients and bridges is compatible. If a median is not available then a lane may need to be replaced by rail, but in an era of declining car use, this should be possible. They are not ideal in terms of ability to build transit-oriented developments (TOD), but it can still be done using high-rise buildings as sound walls. Linkages from buses, electric bikes, and park and ride are all easily provided so that local travel to the system is short and convenient.

Speed of the transit system is key and in Perth the new Southern Railway has a maximum speed of 130 km/h (80 m/h) and an average speed of 90 km/h (55 m/h) that is at least 30% faster than traffic. The result is dramatic increasing in patronage far beyond the expectations of planners who see such suburbs as too low in density to deserve a rail system. The Southern Suburbs Rail line opened in December 2007 and is now carrying 75,000 people a day, whereas the bus system carried just 14,000. There is little else that can compete with this kind of option for creating a future in the car-dependent suburbs of many cities.

Light rail (and heavy rail) can be built back into automobile-dependent suburbs by using land value capture [8,31]. More recently, we have suggested a new way called the Entrepreneur Rail Model which is based on urban regeneration [29]. In this approach, urban regeneration is planned first so that the best sites are determined for new urban centres in the suburbs. The ability to fund a light rail that can unlock the value in this land for redevelopment is then assessed. This is best done by seeking bids from consortia who can build, own, operate and finance the light rail through their investment in the urban regeneration. This is how urban rail is built in Japan, Hong Kong and is beginning in some Canadian cities. It is how urban rail was first built in the era of the Transit City. It would seem that the new market for urban rail could unlock the kind of urban development of a new polycentric twenty-first century city.

4.6.2 Reducing travel

There are many ways that technology can help us to reduce the need to travel (Chapter 3). Transport to meet people by long distance or even short distance trips within cities may be considerably reduced once the use of broadband-based telepresence begins to make high-quality imaging feasible on a large scale. There will always be a need to meet face-to-face in creative meetings in cities, but for many routine meetings, the role of computer-based meetings will rapidly take off. Cities that are attractive places to meet (walkable, safe and lively) will thrive even more in these conditions, whilst suburban areas with little more than scattered houses will find their economies being undermined.

The relationship between density and car use is also exponential as shown above. If a city begins to slowly increase its density, then the impact can be more extensive on car use than expected. Density is a multiplier on the use of transit and walking/cycling, as well as reducing the length of travel. Increases in density can result in greater mixing of land uses to meet peoples' needs nearby. This is seen, for example, in the return of small supermarkets to the CBDs of cities as residential populations increase and demand local shopping opportunities within an easy walk. Overall, this reversal of urban sprawl will undermine the growth in car use.

The need to develop city-centres is now shifting to the need to develop sub-centres across the whole car-dependent city in order to help get people out of cars. This process aims to create TOD. This will ultimately lead to the Polycentric City which is a series of sub-centres or small cities in the suburbs all linked by quality transit but each providing the local facilities of an urban centre.

The facilitation of TODs has been recognised by all Australian cities and most American cities in their metropolitan strategies [32]. The major need for TODs is not in the inner areas as these have many from previous eras of transit building, but in the newer car-dependent suburbs. There are real equity issues here as the poor increasingly are trapped on the fringe with high expenditures on transport. A 2008 study by the Center for TOD shows that people in TODs drive 50% less than those in conventional suburbs [33]. In both Australia and the United States of America, homes that are located in TODs are holding their value the best or appreciated the fastest under the pressure of changing real-estate markets. The report suggested that TODs would appreciate fastest in up-markets and hold value better in down markets. This is the rationale for how TODs can be built as Public–Private Partnerships in rail projects as outlined above in the Entrepreneur Rail Model.

Thus TODs are an essential policy for responding to low carbon mobility, especially when they incorporate affordable housing. The economics of this approach have been assessed by the Center for TOD and the NGO Reconnecting America. In a detailed survey across several states, these NGOs assessed that the market for people wanting to live within half a mile of a TOD was 14.6 million households. This is more than double the number who currently live in TODs. The market is based on the fact that those living in TODs now (who were found to be smaller households, the same age and the same income on average as those not in a TOD) save some 20% of their household income by not having to own so many cars – those in TODs owned 0.9 cars per household compared to 1.6 outside. This frees up on average $4,000 to $5,000 per year [34]. In Australia, calculations have shown similar private savings [35]. Most importantly, this extra income is spent locally on urban services, which means the TOD approach is a local economic development mechanism.

TODs must also be pedestrian-orientated developments (PODs), which is pedestrian-oriented development or they lose their key quality as a car-free environment to which businesses and households are attracted. Urban designers need to assure that public space is vibrant, safe and inviting. Jan Gehl's transformations of central areas such as Copenhagen, New York, London and Melbourne are showing the principles of how to improve TOD spaces so they are more walkable, economically viable, socially attractive and environmentally significant [36–41]. Gehl's work in Melbourne and Perth has been evaluated after a decade of implementation and in both cases indicate substantial increases in walkability, numbers of pedestrians who use the city centre for their work, shopping, education and especially their recreation. It will be important for those green developers wanting to claim credibility that scattered urban developments, no matter how green their buildings and renewable infrastructure, will be seen as failures in a low carbon mobility world unless they are building pedestrian-friendly TODs.

At the same time TODs that have been well designed as PODs will also need to be GODs – green-oriented developments. TODs will need to ensure that they have full solar orientation, are renewably powered with Smart Grids, have water sensitive design, use recycled and low-impact materials, and use innovations like green roofs.

4.6.3 Reducing demand for the automobile

The real test of a 100% oil-free city will be how it can simultaneously be reducing its oil through new technologies and the need to travel through urban design, whilst at the same time changing behaviour to reduce the demand to travel by car. The need for a parallel delivery of physical infrastructure and mental infrastructure will be essential for any city contemplating the delivery of an oil-free city.

Every nation and city has its own way of making the adoption of more planetary lifestyles convenient and easy compared with more consumptive lifestyles. When it comes to cars, however, the more that a city is car-dependent, the harder it is to use tax incentives to change people's lifestyles. European cities have much higher gasoline taxes than American and Australian cities, and accordingly they use cars less [42].

In the car-dominated cities of North America and Australia, the major public policy to reduce the global and local impacts of the car has been through regulations on vehicles that have forced them to become cleaner. Vehicles are now becoming much more energy efficient [43]. Following introduction of these, most urban atmospheres have become cleaner, although fuel use has continued to increase as vehicles became bigger and their use has continued to grow. Regulations also are applied to safety and congestion management, but this will continue to worsen if more and more car use is facilitated. Regulations alone do not change behaviour. The economic principle known as the Rebound Effect or the Jevons Paradox – increasing efficiency means increasing consumption – has been found to apply to car use. If people buy cars that use less fuel, they just drive them more – undermining most gains made possible through the new technology [44].

All these necessary policy approaches will be wasted without education on a changed role for the car. Thus cultural change to help people to want to drive less needs to be part of any city's policy arsenal if it is to face up to the challenge of growing a sustainable low carbon mobility city. TravelSmart is one such program that shows this is indeed possible [45,46].

German sociologist Werner Brög developed TravelSmart based on the belief that behaviour change towards less car dependence can work if it is community-based and household-oriented. After some trials in Europe, TravelSmart was adopted in large-scale projects in Perth. It has since spread across most Australian cities and to other European cities, especially in the United Kingdom, and has now been piloted in six American cities.

Good behaviour change programmes do not use media. So TravelSmart targets individual households directly asking them if they would like to find ways to reduce their car use. The residents who show interest receive information and, for the few who need extra support, a visit from a TravelSmart officer or Eco-coach who encourage people to start with local trips, especially the school trip for children, which is now seen as an essential part of the healthy development of young people's sense of place and belonging in any community as well as a way to reduce obesity.

TravelSmart has been found to reduce the kilometres travelled by vehicle by around 12%–14% across whole communities – a result that seems to last for at least 5 years after the programme ends. Where transit is not good and destinations are more spread out, the programme may only reduce car use 8%, but where these are good it can rise to 15% [47].

Behaviour change programs can develop social capital to support people as they try to get out of cars and this can change local cultures. People show friends how much money it saves as well as making them feel they are doing their bit for climate change and oil vulnerability. There is evidence in Brisbane, Australia that at least 50% more people than those involved in the initial household interviews were actually following the programme when the surveys were done; in other words, people were spreading the message to their friends and colleagues [48]. When people start to change their lifestyles and can see the benefits, they become advocates of sustainable transport policies in general. The politics of change is easier to manage when communities have begun to change themselves.

In Perth the extension of the rail system to the far outer suburbs has been more positive and politically achievable than expected, with a massive 90% support for the last stage, the Southern Suburbs Railway. In parallel to this political process, Perth had some 200,000 households undergoing the TravelSmart programme; the suburbs where TravelSmart had been conducted increased their public transport patronage by 83%, but in areas without TravelSmart it was 59% [49].

TravelSmart has now been applied to more than 450,000 residents in Perth, at a cost to the state of under A$36 per resident. Worldwide, TravelSmart's individualised approach to travel demand management has been delivered to approximately five million people. If you take into account the reductions in the public and private costs of car use that it has achieved, the programme saves $30 for every dollar it costs [50].

Behaviour change programmes can also work in the workplace. TravelSmart was found to work well when a TS Club was formed that enabled people to share experiences, bring in local speakers and lobby for facilities like showers for bike riders and transit passes instead of parking spaces. For example, the natural gas company Woodside in Perth involved their employees in planning their new building; a strong representation from the TS Club at Woodside led to good bicycle facilities being provided. The firm now has more employees biking to work than riding and the subsequent saving in car-parking spaces is considerable [51].

TravelSmart has also used the walking school bus concept to enable children to safely work to school. A variation is the bicycle bus in which adults supervise children riding their bikes to school. Walking or riding to and from school provides valuable exercise for children. It enables them to learn early that walking and cycling are very practical, pleasant and healthy forms of local travel. At the same time, the adult supervision that walking school bus schemes provide ensures that children – especially very young ones – are safe, thus overcoming a barrier that causes parents to prevent their children from walking or cycling. The experience of walking or cycling teaches children about their neighbourhood and environment, as well as gaining road safety skills and equipping them for independent mobility as they get older. Finally, these programmes, if they are supervised by parents or volunteers, involve virtually no costs.

The changes in behaviour being seen amongst young people in cities across the world, especially the United States, suggest that programmes to assist this process may have considerable success in further reducing the demand for car use [52]. Much of this cultural shift appears to be because young people are connected substantially through social media devices rather than cars, and to a growing use of Skype and telepresence to replace the need for travel.

4.6.4 Electric vehicles

Even if we manage to reduce car use by 50% as suggested above, by a rather herculean effort, we still have to reduce the oil and carbon in the other 50% of vehicles being used. The question should therefore be asked: what is the next best transport technology for motor vehicles? The growing consensus seems to be: plug-in electric vehicles (PEVs), possibly adding in the autonomous vehicle capability.

PEVs are now viable alternatives due to the new batteries such as Lithium Ion and hybrid engines for extra flexibility. PEVs are increasingly attractive to the market in the transition period and electric recharging infrastructure is building up in the world's cities. PEVs not only reduce oil and greenhouse gas but they are also becoming a critical component in how renewable energy will become an important part of a city's electricity grid.

Solar-storage is moving rapidly into mainstream household energy supplies, and in Perth, for example, solar photovoltaics (PV) is already the largest power station in the grid [53]. Thus, households are not just becoming zero carbon in their electricity consumption they are producing more renewable power than they need during the day with the one household monitored in Perth producing 24% more power than needed in the house based on just a 3-kW rooftop PV system [53]. Electric vehicles can easily be plugged into this solar-storage system and the excess power used to recharge the vehicle. The PEV's battery can also serve as a storage function for the house or to providing renewables to the grid when the system needs help. Thus electric vehicles are becoming an essential part of how a city can become not just free of oil but free of coal [54].

The same approach can be taken using electric buses, electric bikes, scooters and gophers. But the interesting next phase in low carbon mobility is the shift to using Li Ion batteries in light rail trains. Not only does this mean light rail would not need an overhead wire catenary system, but it can be rapidly recharged through inductive recharging that is linked to solar and battery systems at each station. Hence, the TOD–POD–GOD urban regeneration projects mentioned above can be zero carbon with solar PV and batteries that feed directly into the running of the zero carbon light rail as well.

Signs that this transition to electric transport is underway are appearing in demonstration projects such as in Boulder and Austin, in Google's 1.6-MW solar campus in California (with 100 PEVs) and by the fact that oil companies are acquiring electric utilities.

What sort of immediate impact could there be? According to one study [55], the integration of PEV cars with the electric power grid could reduce gasoline consumption by 85 billion gallons per year. That's equal to

- 27% reduction in total US greenhouse gases,
- 52% reduction in oil imports,
- $270 billion not spent on gasoline.

4.6.5 Renewable fuels

Biofuels have promised a lot, but they have been criticised for their impact on food prices when used to convert grain into fuel. However, they still have a potentially significant role in some areas where there is a crop surplus, and eventually when the technology improves to make them from cellulose materials (agricultural and forestry waste) and from blue green algae. It is likely that biofuels will be used as a do-it-yourself fuel on farms [56]. Thus, biofuels may have a role in agricultural regions as a fuel to assist farmers, but as a widespread fuel for cities it is not an option that can be yet taken seriously.

Aircraft are the hardest to imagine in low carbon mobility. They can reduce their fuel intake by a range of efficiency measures though only biofuels look to be suitable at this stage as a replacement for oil (see Transport Chapter in [8,57]). But the cost of biofuels for aviation appears to make them prohibitive at this stage. Gilbert and Perl [58] suggest a few ways that air travel will adapt but mostly they see little of potential other than regional high-speed rail and a return to ship travel. Perhaps the regional transport technology that could make a come-back is airships. These are able to fly at low levels at speeds of 150–200 km/h and carry large loads with one-tenth of the fuel of aircraft technology [59]. They are already being used to carry large mining loads to remote areas and to take groups of 200 or more eco-tourism ventures similar to a cruise ship. Perhaps this is a possible use for biofuels as they will be a much smaller part of the regional cost of travel than present day aviation.

What do you do with freight transport and regional transport outside of cities where electric grids are not so easily used with vehicles? And what about industries that presently use oil? Can they also go oil-free?

4.6.6 Renewable gas for freight

Freight is also a highly contested area for the future. There are potential reductions in freight as the economy dematerializes, such as how mobile phones now contain over 20 functions replacing 20 other devices [57]. Food miles are likely to reduce also as local food cultures grow. But freight trucks and trains and regional transport will still go on.

There are various futures being predicted for freight including the use of electric trucks powered through batteries or overhead catenaries though this is unlikely given the extent of the infrastructure required. The transitional stage for larger vehicles, industry and regional transport would appear to be to switch to greater use of natural gas supplemented by some biofuels. Trucks and trains and fishing boats can use compressed natural gas or LNG (liquefied natural gas) in their diesel engines (with pay-off times of just a few years due to high diesel costs). Cars for regional transport can also be switched over as well (particularly if the manufacturer makes them standard as occurred in Sweden when the government committed to natural gas cars for their vehicle fleet). The attraction is that natural gas is already in place in terms of infrastructure with almost 80% of the population in most developed countries having access to reticulated natural gas.

Global natural gas production has had a boost in recent years due to the technology of fracturing shale, though not without its environmental controversies. Thus estimates on its peak have been extended out into a long distance future. However, the cost will be increasing with these new sources as well as with the use of more and more off-shore gas similar to oil. But ultimately the issue of carbon reduction will undermine the long-term future of fossil natural gas. Natural gas in reality can only be a small part of the transitional arrangements for oil; it cannot be seen as the long-term replacement. It will, however, be an obvious way to ease the pressure on diesel supplies, and this will be a great advantage to low carbon cities (gas buses have already shown their big advantages over diesel buses in air quality) as well as energy security and taking pressure off the need for oil production in dangerous places.

The benefit of the transition to natural gas has always been seen as an enabler of the long-term transition to hydrogen. However, there is another process that has much greater potential than the widespread use of hydrogen: the use of renewable natural gas. Biogas can be created from biomass via gasification, and this could play a role in the future. However, there is an even bigger process that could assist in creating an oil-free future: the catalytic process known as the Satanalia Process facilitates the joining of CO_2 and H_2O into CH_4 and O_2, just as in the photosynthetic process whereby sunlight uses chlorophyll to catalyse CO_2 and H_2O into carbohydrates. Around the world, a range of research labs are vying to develop catalysts that can make this process into a commercially successful system. Thus, if there can be a development of the hydrogenation of CO_2 using renewable energy, then natural gas can become a renewable fuel in itself which can be fed into the present natural gas grids and even be an export item through LNG [60]. There is large potential in this process, considerably more than 'clean coal' as a totally new infrastructure will not be required for its distribution (most coal fields and hence coal-fired power are long distances away from the deep caverns that can act to absorb CO_2). In the interim years as coal-fired power stations are being phased out, they can have renewable natural gas production facilities attached to them. Eventually CO_2 can be extracted from the atmosphere and used as a renewable fuel.

Freight, industry and regional transport are likely to continue to expand into natural gas and transition into the use of renewable natural gas. Thus, natural gas can be given a long-term future and can be part of the low carbon mobility 'oil-free' transition.

4.7 Conclusions

There are not many guidelines to the future of our cities and regions that take account of what could happen to mobility in response to the challenge of climate change. The current paper has examined what could happen on the basis of the following: first, assessing the trends that are already leading to less oil and car dependence; second, imagining this oil-free future based on what we know about technology, urban planning and behaviour change possibilities available or emerging. The alternatives all require substantial commitment to change in both how we live and the infrastructure priorities we have in our cities and regions. The transition will take time but already it is clear that the early adopters in cities around the world are establishing economic competitiveness.

The first signs of change towards these emerging technologies can now be seen: the re-urbanization of our cities; the dramatic growth in electric transit; the rapid move towards electric vehicles and solar energy with a 30%–40% per annum growth in global renewables; the emerging use of natural gas and biofuels; and new technologies like Skype and telepresence. Their application into large-scale urban demonstrations is now underway in places like Kronsberg and Vauban in Germany, Masdar in UAE and the Low Carbon Cities of China.

The potential for creating oil-free cities is here. The technologies and practices outlined above suggest that we can be oil-free by 2050 and renewably based oil-free by 2100.

References

[1] P. Newman and J. Kenworthy. *Cities and Automobile Dependence*; 1989.
[2] P. Newman and J. Kenworthy. *Sustainability and Cities: Overcoming Automobile Dependence*, Island Press: Washington, DC; 1999.
[3] J. Kenworthy and F. Laube. *The Millennium Cities Database for Sustainable Transport*, UITP: Brussels; 2001.
[4] J. Kenworthy, F. Laube, P. Newman, *et al. An International Sourcebook of Automobile Dependence in Cities, 1960–1990*, University Press of Colorado: Boulder; 1999.
[5] P. Jones. The Evolution of Urban Mobility: The Interplay of Academic and Policy Perspectives, *IATSS Research*, 2014, doi: 10.1016/j.iatssr.2014.06.001.
[6] P. Newman and J. Kenworthy. Peak Car Use: Understanding the Demise of Automobile Dependence, *World Transport Policy and Practice*, 2011;17(2): 32–42.
[7] P. Newman and J. Kenworthy. *The End of Automobile Dependence: How Cities are moving Beyond Car-based Dependence*. Island Press: Washington, DC; 2015.
[8] International Energy Agency. *Decoupling of Global Emissions and Economic Growth Confirmed, Press Release*, IEA: Paris; 2016. https://www.iea.org/newsroomandevents/pressreleases/2016/march/decoupling-of-global-emissions-and-economic-growth-confirmed.html.
[9] C. Hargroves and M. Smith (eds). *The Natural Advantage of Nations*, Earthscan Books: London; 2004.
[10] J. McIntosh, R. Trubka, J. Kenworthy and P. Newman. The Role of Urban form and Transit in City Car Dependence: Analysis of 26 Global Cities from 1960 to 2000. *Transportation Research Part D: Transport and Environment*, 2014;33: 95–110.
[11] C. Marchetti. Anthropological Invariants in Travel Behaviour, *Technical Forecasting and Social Change*, 1994;47: 75–78.
[12] Y. Zahavi and A. Talvitie. Regularities in Travel Time and Money Expenditures, *Transportation Research Record*, 1980;750: 13–19.
[13] Standing Advisory Committee on Transport. 'Trunk Roads and the Generation of Traffic' (Report to the Department of Transport, Government of the United Kingdom), London; 1994.

[14] P. Newman and J. Kenworthy. Urban Design and Automobile Dependence: How Much Development Will Make Urban Centres Viable, *Opolis*, 2006;2: 35–52.
[15] B. Van Wee, P. Rietveld and H. Meurs. Is Average Daily Travel Time Expenditure Constant? In Search of Explanations for an Increase in Average Travel Time, *Transport Geography*, 2006;14: 109–122.
[16] BITRE. *Traffic and Congestion Cost Trends in Australian Cities, Information Sheet 74*. Bureau of Industry and Transport Research: Canberra; 2016.
[17] P. Newman, G. Glazebrook and J. Kenworthy. Peak Car and the Rise of Global Rail, *Journal of Transportation Technologies*, 2013;3(4): 272–287.
[18] Y. Gao, P. Newman and J. Kenworthy. Are Beijing and Shanghai Reaching Peak Car? *Journal of Urban and Regional Research*, under review, 2016.
[19] P. Newman and J. Kenworthy. Peak Car Use: Understanding the Demise of Automobile Dependence, *World Transport Policy and Practice*, 2011;17(2): 32–42.
[20] M. Kane. 'Linking the Knowledge Economy, Urban Intensity and Transport in Post Industrial Cities with a Case Study of Perth', PhD Thesis, Curtin University; 2016.
[21] J. Gehl. *Cities for People*, Island Press: Washington, DC; 2012.
[22] A. Matan and P. Newman. *Green Urbanism in Asia*, World Scientific: Singapore; 2012.
[23] C. Leinberger. 'Foot Traffic Ahead: Ranking Walkable Urbanism in Americas Largest Metros' (report prepared for American Federal, State and Local Policy Makers and Private Clients), The George Washington University, Centre for Real Estate and Urban Analysis, School of Business: Washington, DC; 2014.
[24] R. Florida. *The Great Reset: How New Ways of Living and Working Drive Post-Crash Prosperity*, Harper Collins: New York; 2010.
[25] E. Glaeser. *The Triumph of the City, How Our Greatest Invention Makes Us Richer, Smarter, Greener, Healthier and Happier*, Penguin Press: London; 2011.
[26] D. Murray and D. King. Climate Policy: Oil's Tipping Point has Passed, *Nature*, 2012;481: 433–435.
[27] J. McIntosh, P. Newman and G. Glazebrook. Why Fast Trains Work: An Assessment of a Fast Regional Rail System in Perth, Australia, *Journal of Transportation Technologies*, 2013;3: 37–47.
[28] J. McIntosh, R. Trubka and P. Newman. Can Value Capture Work in a Car Dependent City? Willingness to Pay for Transit Access in Perth, *Western Australia Transportation Research – Part A*, 2014;67: 320–339.
[29] P. Newman, S. Davies Slate and E. Jones. The Entrepreneur Rail Model: Funding Urban Rail through Majority Private Investment in Urban Regeneration, *Research in Transportation Economics*, 2017 (in press).
[30] P. Newman, J. Kenworthy and G. Glazebrook. How to Create Exponential Decline in Car Use in Australian Cities, *Australian Planner*, 2008;45(3): 17–19.

[31] R. Cervero. Transit-Oriented Development in America: Strategies, Issues, Policy Directions. In: T. Haas, ed. *New Urbanism and Beyond: Designing Cities for the Future*, Rizzoli: New York; 2008, pp. 124–129.

[32] C. Curtis, J. Renne and L. Bertolini. *Transit Oriented Development, Making It Happen*, Ashgate: London; 2009.

[33] CTODRA. *Hidden in Plain Sight: Capturing the Demand for Housing Near Transit, Center for Transit Oriented Development and Reconnecting: America*; 2004. www.reconnectingamerica.org.

[34] H. Dittmar and G. Ohland (eds). *The New Transit Town, Island Press: Washington, DC. CTODRA (2004) Hidden in Plain Sight: Capturing the Demand for Housing Near Transit, Center for Transit Oriented Development and Reconnecting: America*; 2004. www.reconnectingamerica.org.

[35] R. Trubka, P. Newman and D. Bilsborough. Costs of Urban Sprawl (2) – Greenhouse Gases, *Environment Design Guide*, 2010;84: 1–16.

[36] J. Gehl. *Life between Buildings: Using Public Space, translated by Jo Koch*, Van Nostrand Reinhold: New York; 1987.

[37] J. Gehl. *Cities for People*, Island Press: Washington, DC; 2010.

[38] J. Gehl and L. Gemzøe. *New City Spaces*, The Danish Architectural Press: Copenhagen; 2000.

[39] J. Gehl and L. Gemzøe. *Public Spaces, Public Life*. The Danish Architectural Press: Copenhagen; 2004.

[40] J. Gehl, H. Mortensen, P. Ducourtial, *et al*. *Places for People: Study Report*. City of Melbourne/Gehl Architects: Melbourne/Copenhagen; 2004.

[41] J. Gehl, A. Modin, J. Wittenmark, *et al*. *Perth 2009: Public Spaces and Public Life: Study Report*. City of Perth/Gehl Architects: Perth/Copenhagen; 2009.

[42] G.P. Metschies. Prices and Vehicle Taxation (Germany: Deutsche Geslleschaft fur Technische Zusammenarbeit GmbH, in: R. Porter, (1999)) *Economics at the Wheel: The Costs of Cars and Drivers*, Academic Press: London.

[43] M. Sivak and B. Schoutl. *What Individual Americans Can Do to Meet the Paris Agreement, UMTRI-2016-7*, University of Michigan: Ann Arbor; 2015.

[44] P. Litman. Changing Vehicle Travel Price Sensitivities: The Rebounding Rebound Effect, *Transport Policy*, July 2013; 28:2–10.

[45] R. Salzman. Travel Smart: A Marketing Program Empowers Citizens to be a Part of the Solution in Improving the Environment, *Mass Transit*, 2008.

[46] R. Salzman. Now That's What I Call Intelligent Transport … SmartTravel, *Thinking Highways*, 2008.

[47] C. Ashton-Graham, C. TravelSmart + TOD = Sustainability and Synergy. In: C. Curtis, J. Renne and L. Bertolini, eds. *Transit Oriented Development, Making it Happen*. Ashgate: London; 2009.

[48] I. Ker. *North Brisbane Household TravelSmart: Peer Review and Evaluation, for Brisbane City Council*, Queensland Transport, and Australian Greenhouse Office, Brisbane; 2008.

[49] Department of Transport. *TravelSmart Household Final Evaluation Report Murdoch Station Catchment (City of Melville 2007)*, Socialdata Australia, Department of Transport: Perth; 2008.

[50] C. Ashton-Graham. *TravelSmart and Living Smart Case Study in Garnaut Climate Change Review*; 2008. www.garnautreview.org.au/CA25734E0016A131/WebObj/Casestudy-TravelSmartandLivingSmart-WesternAustralia/%24File/Case%20study%20-%20TravelSmart%20and%20LivingSmart%20-%20Western%20Australia.pdf.
[51] D. Wake. *Reduced Car Commuting through Employer-based Travel Planning in Perth*, Australia, TDM Review, 2007;(1).
[52] B. Davis, T. Dutzik and P. Baxandall. *Transportation and the New Generation: Why Young People Are Driving Less and What It Means for Transportation Policy*, Frontier Group and US PIRG: Santa Barbara; 2012.
[53] J. Green and P. Newman. *Citizen Utilities: The Emerging Power Paradigm*, Energy Policy; 2016 (in press).
[54] A. Went, W. James and P. Newman. *Renewable Transport, CUSP Discussion Paper*; 2008. www.sustainability.curtin.edu.au/renewabletransport.
[55] M. Kintner-Meyer, K. Schneider and R. Pratt. *Impacts Assessment of Plug-in Hybrid Vehicles on Electric Utilities and Regional U.S. Power Grids, Part 1: Technical Analysis*, Pacific Northwest National Laboratory: US; 2007. DoE, DE-AC05-76RL01830.
[56] L. Mastny (ed). *Biofuels for Transportation*, Worldwatch Institute, GTZ, German Ministry of Agriculture: Washington, DC; 2006.
[57] IPCC. *Climate Change 2014: Mitigation of Climate Change*. Contribution of Working Group III to the Fifth Assessment Report of the Intergovernmental Panel on Climate Change, Cambridge University Press: Cambridge; 2014.
[58] R. Gilbert and A. Perl. *Transport Revolutions: Making the Movement of People and Freight Work for the 21st Century*, New Society Publishers: Vancouver; London; 2006.
[59] D. Bradbury. *Airships Float Back to the Future*. BusinessGreen.com (accessed 2 September 2008).
[60] C. Creutz and E. Fujita. *Carbon Dioxide as a Feedstock in Carbon Management: Implications for R&D in Chemical Sciences*, CPSMA, National Research Council: Washington, DC; 2001.

Glossary

AV	Autonomous vehicle
CBD	Central business district
CNG	Compressed natural gas
GDP	Gross domestic product
GOD	Green orientated development
Ha	Hectares
ICE	Internal combustion engine
LNG	Liquefied natural gas

MJ Mega joules
PEV Plug-in electric vehicle
POD Pedestrian orientated development
PPP Public–private partnerships
SD Statistical district
TOD Transit-orientated development

Further Reading

[1] D. Banister. The sustainable mobility paradigm, *Transport Policy*, 2008;15: 73–80.
[2] *Changing Course: A New Paradigm for Sustainable Urban Transport*, ADB: Metro Manila, Philippines.
[3] *Eco-Efficient and Sustainable Urban Infrastructure Development Toolkit*, UNESCAP: Bangkok, Thailand.
[4] M. Givoni and D. Banister (eds). *Moving Towards Low Carbon Mobility*, Edward Elgar Publishing: Cheltenham, UK; 2013.
[5] International Energy Agency. 'A tale of renewed cities: a policy guide on how to transform cities by improving energy efficiency in urban transport systems'. Policy Pathway Report, International Energy Agency, Paris, 2013. Available: www.iea.org.
[6] A. Jaber and D. Glocker. Shifting towards low carbon mobility systems. *International Transport Forum*, 2015. Available: http://www.itf-oecd.org/shifting-towards-low-carbon-mobility-systems.

Chapter 5

Integrated land-use and transport planning for future cities: and the importance of active transport

Michael A.P. Taylor[1]

Abstract

Land-use-transport integration (LUTI) is concerned with the development, management and operation of urban transport systems that provide for sustainable outcomes for our cities. LUTI is primarily concerned with the optimal distribution of facilities and services in an urban area so that these are available to all of its inhabitants in ways that minimise adverse environmental impacts, whilst maximising opportunities for sustainable economic development and social interaction. LUTI is a key consideration in low carbon mobility. A major means for achieving this is through the encouragement and use of the active transport modes walking and cycling and public transport, because access to public transport generally requires substantial use of the pedestrian and cycle modes. The key concept behind LUTI is in fact the basic driver of urban development: accessibility. In the planning context, accessibility is about the effort required for people and enterprises to engage in everyday life. Integrated planning of transport and land use can offer high levels of accessibility based on the use of active transport. Amongst the many issues facing planners are environmental impacts, both local (e.g. noise and air quality) and global (climate change and greenhouse gas emissions), economic productivity, conservation of resources (especially the use of land), traffic congestion, social severance and public health. Following the established principles and practices of LUTI – while these may not always coincide with what has been past practices in our cities – offers a systematic path to long-term resolution of these issues. This chapter will explain why and how this can be accomplished.

Keywords

Land-use planning; transport planning; integration; accessibility; travel behaviour; environmental impact

[1]School of Natural and Built Environments, University of South Australia, Adelaide, Australia

5.1 Introduction

'New world' cities such as those in Australasia are typically composed of low density, dispersed suburbs, which are highly car dependent. This is resource and carbon emission intensive and therefore unsustainable in the long term [1,2]. There is a pressing need for Australian cities to transition to a low carbon alternative with more efficient form and function with respect to carbon emissions. A low carbon city must improve the energy efficiency of its buildings and transport systems [3] and the connections between them. A key aspect of this transition is transportation; the level and intensity of the demand for which depends greatly on the land-use system and the land-use planning instruments that have been applied. Integration of land-use and transport planning policies and practices directed towards urban sustainability is essential.

Transportation systems guide our mobility, that is, our ability to move from one place to another in order to achieve our needs for participation in activities and utilisation of facilities, the locations of which depend on the distribution and intensity of land uses across a region. Transportation provides access to jobs, education and social interactions, for example, all of which are fundamental to human development [4]. Public awareness and perception are two major barriers for the transition to low carbon transport options. Banister [5] emphasised the need to understand the importance of time and how people want to use time in travel compared to other activities. Aditjandra *et al.* [6] suggested residents have to be well informed of the available opportunities for them to pursue sustainable travel choices.

Passenger car usage in Australian cities is thus a significant source of greenhouse gas (GHG) emissions, and one that can be directly related to the shape and form of the built environment. The National Greenhouse Gas Inventory (NGGI) data [7] indicate that 15.3% of Australia's total GHG emissions came from the transport sector. Analysis of the NGGI data indicates that the use of passenger cars in urban areas was responsible for 55.5% of the transport-based emissions, that is 8.5% of Australia's total GHG emissions (39.3 Mt out of 543.2 Mt CO_2-e) was due to the use of passenger cars in our cities. Figure 5.1 illustrates this breakdown.

The shape and form of our built environment, along with lifestyle choices, strongly influence the use of the private car in urban areas. Newton and Newman [8] addressed this issue in their consideration of the carbon benefits that can accrue from more compact urban forms where public and active transport modes are generally of most benefit. They considered the need for urban design innovation in Australia, largely focusing on energy demands in housing and transport, and available alternative energy technologies and fuel types. The 'post-war' suburbs, that is those areas of our cities subdivided and developed in the latter half of the twentieth century which are heavily car dependent, provide the most challenges in transitioning to a lower carbon future. A discussion of the Newton–Newman framework for low carbon technology interventions based on the consideration of appropriate low carbon technologies applied to housing and transport in suburban and inner urban areas is given in Chapter 6 of this book.

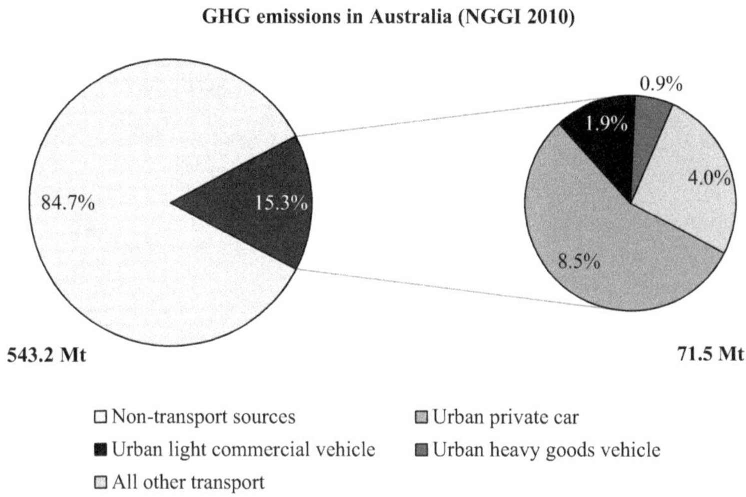

Figure 5.1 *Contribution of passenger car usage in urban areas to Australia's total GHG emissions (derived from data in [7])*

5.2 Sustainable land-use and transport

Amongst the many issues facing planners are environmental impacts, both local (e.g. noise and air quality) and global (climate change and GHG emissions), economic productivity, conservation of resources (especially the use of land), traffic congestion, social severance and public health. The demand for usage of more sustainable forms of travel in urban areas is being accentuated through our society's increasing concerns such as GHG emissions and global warming, social equity in terms of mobility for all demographic groups in the community (especially the young and the elderly), and the need to conserve liquid fossil fuels. The task is to find and fully utilise alternatives to travel by private car that are equitable and improve the environment and our quality of life through more sustainable transport and urban development.

Sustainability in the context of urban transport may be defined in terms of two broad characteristics of sustainability as an objective: (1) that it should include both the welfare of present day society and the society of the very distant future, and (2) that sustainability implies conservation of natural resources. Thus, natural resources should be valued not only as something that may be consumed but also as stocks that benefit us even when not being consumed. Minken [9] defined a sustainable urban transport and land-use system as one which:

- provides access to goods and services in an efficient way for all inhabitants of an urban area
- protects the environment, cultural heritage and ecosystems for the present generation

- does not endanger the opportunities for future generations to reach at least the same welfare level as that of the present generation, including the welfare derived from the natural environment and cultural heritage.

Subsequent research [10] has identified nine broad objectives for sustainable urban transport policies:

1. economic efficiency
2. liveable streets
3. environmental protection
4. equity, social inclusion and accessibility
5. safety and security
6. economic growth
7. finance
8. practicability
9. inter-generational equity.

These objectives and policies and policy mixes designed to accomplish them are described in detail in [10].

The recent initiatives in transit-oriented development (TOD) in Australian cities (e.g. [11,12]) fit under the umbrella of sustainable transport and the 'environment-friendly modes', a collective term for active transport (walking and cycling) and public transport, through the explicit intention to provide enhanced options for the usage of public transport for travel in place of equivalent trips by car. However, there are many interpretations of what transit-oriented development means. In its purest form, it would be the transit village, a compact, mixed-use community centred around the transit station that is designed to invite residents, workers and shoppers to drive their cars less and ride public transport more. The transit village brings the community closer to the transit node, and the station is the gateway of the village. Thus, one could define a TOD expressly as 'a vibrant, relatively dense and pedestrianised mixed-use development precinct, featuring quality public space and immediate access to high-frequency public transit' [11]. An alternate view of TOD is that of a transport interchange serving a region within the metropolis, with associated higher density land-use development, but featuring feeder service connections to the line haul public transit system. The feeder services are likely to include both 'kiss and ride' and 'park and ride', and so the implication is that a considerable part of the facility will be devoted to motor vehicle access and car parking. Although each form of TOD will have its place, they are not the same. Further, it could be expected that a real-world development may have to be a hybrid of the two forms. An important question for urban planners and transport planners is the extent to which TOD can provide the solution to urban sustainability and low carbon mobility (Chapter 4). This chapter suggests that broader initiatives in land-use-transport integration (LUTI) are required.

To understand how LUTI initiatives may be applied to improve urban sustainability and encourage low carbon mobility requires first an appreciation of the travel behaviour of individuals, and the factors which influence that behaviour, and then the urban planning notion of accessibility as a planning goal. LUTI is a key consideration in low carbon mobility.

5.3 Travel behaviour

Travel is a derived demand, in that for almost all purposes it is not undertaken for its own sake. Rather travel is used to access some service or facility, or take part in some activity, at a different spatial location. The decision by an individual to travel is multi-layered, involving number of aspects, some of which are related. The basic decision is whether to undertake a journey, with the alternatives including to make a trip (some time) today, to delay it until another day, to find an alternative means to satisfy the purpose requiring the travel (e.g. through telecommunications or the internet), or to find another individual who can travel in your stead or add your requirement to their own activity schedule.

Once an individual has decided to undertake travel, the subsequent choices include the destination (where can the activity, service or facility be found, and which is the preferred option if there is more than one alternative), the mode of transport to be used to go to the destination, the time of day when the trip is to be made, and the specific route to be taken to reach the destination by the chosen mode.

A number of these choices can be connected, for instance the need to go to a specific destination may then dictate the alternative set of modes that can be used to reach the destination within a specified time period. The choices will also be affected by considerations of the monetary and total costs of the journey (e.g. including travel time, distance, costs of fares and tolls, fuel prices, costs at the destination, such as car parking charges), expected levels of traffic congestion and crowding, and comfort, safety and security issues amongst others.

Travel decisions for regular trips, such as the journey to work, tend to be made at particular times (epoch events) in an individual's life, and then adhered to in the longer term unless some significant change in the travel environment occurs (or is perceived to have occurred, for some changes happen gradually). Epoch events include such things as moving home, changing jobs, changing family situation (leaving home, marriage, parenthood, separation and divorce, etc.). When changes do occur, they may be seen in terms of the following set of options, arranged in (probable) order of ease:

- change in trip timing, so that the start time of a (regular) trip, such as the journey to work, is altered (e.g. a trip starts earlier in the day perhaps in response to longer travel time because of traffic congestion)
- change in route, to the same destination, because of traffic congestion or new transport facilities or services
- change in destination, where alternatives exist, for example, travel to a closer shopping centre seen as less desirable but more accessible in the face of increasing congestion
- change in trip frequency, in which the relative difficulty of accessing a particular destination leads to the decision to make the trip to it more or less often

(the phenomenon of travelling home for lunch, often seen in rural townships but seldom in a metropolis, would be one example. A second might be travelled to a shopping centre for basic purchases such as food)

- change in vehicle occupancy, when two or more individuals travelling to the same or proximate destinations decide to share vehicles
- change in transport mode, due to increased costs and inconvenience of the initial choice of mode, for example, through increasing congestion or changes in travel costs
- change in origin (e.g. place of residence), perhaps because of serious changes in travel conditions or an epoch event in the life of the individual.

Hills [13] provided a detailed discussion of the potential changes in an individual's travel behaviour and the factor that influence these.

There are powerful mathematical tools for modelling and analysis of travel behaviour and travel choices. These tools are perhaps best illustrated by the use of discrete choice models [14] based on the microeconomic theory of utility, in which utility is the intrinsic (perceived) worth of a given option to an individual. The discrete choice models may be used to estimate the probability p_{rn} that individual n will choose option r from a set of R similar choices (e.g. transport mode, or destination, or trip departure time). U_{rn} is the utility of alternative r to individual n, generally defined in the linear form

$$U_{rm} = \beta_{r0} + \sum_j \beta_{jr} X_{jrn} + \varepsilon_{rn} \tag{5.1}$$

where X_{jrn} is the value for n of variable X_{jr} influencing the choice of option r, β_{r0} is a constant for alternative r, β_{jr} is the coefficient for X_{jr} and ε_{rn} is an error term, which includes the perception errors of individual n and accounts for any factors not explicitly included in the set of influencing variables. The variables X_{jr} may relate to the characteristics of the option (e.g. cost of travel by a particular mode) and the socio-economic characteristics of the individual (e.g. age or income). The summation

$$V_{rn} = \beta_{r0} + \sum_j \beta_{jr} X_{jrn} \tag{5.2}$$

in (5.2) is the deterministic component (observable attributes) of the utility function.

Given the utility functions for each alternative r in the choice set R, the probability p_{rn} may then be estimated using a family of choice models [14]. The most common and simplest of these models is the multinomial logit (MNL) model, given by

$$p_{rn} = \frac{\exp(V_{rn})}{\sum_r \exp(V_{rn})} \tag{5.3}$$

The result of (5.3) is the probability that a particular individual will choose one option from a number of similar alternatives. This probability then has to be aggregated to form the proportions that then represent the likely choices of a group of individuals, such as the residents of a given area.

5.4 Accessibility

The key concept behind LUTI lies in what may be taken as the basic driver of urban development – accessibility. In the planning context accessibility is about the effort required for people and enterprises to engage in everyday life. The notion of accessibility in transportation planning can be illustrated through a basic definition: accessibility is 'the ease with which desired destinations may be reached' [15]. In particular circumstances such as those pertaining to low carbon mobility and the availability and use of alternative transport modes, this broad definition may be refined to explicitly include other relevant factors, such as time dependency which may be an issue in, for instance, assessing accessibility levels for public transport users where the level of service of public transport provision varies widely over hours of the day or days of the week. Thus Primerano and Taylor [16,17] defined accessibility as 'the ease for people to participate in activities from specific locations to a destination using a mode of transport at a specific time'. This is an appropriate definition for use in urban studies where different regions within a metropolis have different levels of access to public transport services. 'Wegener's Wagon Wheel' model [18] of the urban development process (and the interaction between the land-use and transport systems) illustrates the primary role of accessibility in the key decisions that underpin urban development and urban activity.

Figure 5.2 shows this model. Urban development – indeed urban evolution – is shown as a continuous process in the figure. The central part of the figure comprises the land-use system and the transport system, connected through accessibility and the activities undertaken in the urban area. The wheel includes a number of decisions made by different players and a number of outcomes stemming from those decisions. The outcomes are indicated in italics in Figure 5.2. The basic outcome is the set of activities. Location decisions of developers and consumers are made on the basis of the relative accessibility of different areas – outcome 'attractiveness' – including their suitability for different activities (services and facilities). The accessibility of a given location is affected by the travel choices of the consumers (when, where and how to travel, and of course a basic decision about private vehicle ownership and use), which lead to the development of congestion and hence the actual (time, distance and cost) separations between localities. These separations affect the accessibility actually experienced by people, hence the ongoing evolution of the system. Choices of location, vehicle ownership and travel depend on (or are influenced by) broader characteristics and influences of the urban area, including its physical environment, history and culture, social and economic environment, and available technology. These factors in the urban milieu serve to influence the choices of players in the urban development system, at macro, meso and micro levels. Although subject to external shocks (such as geo-politics, energy

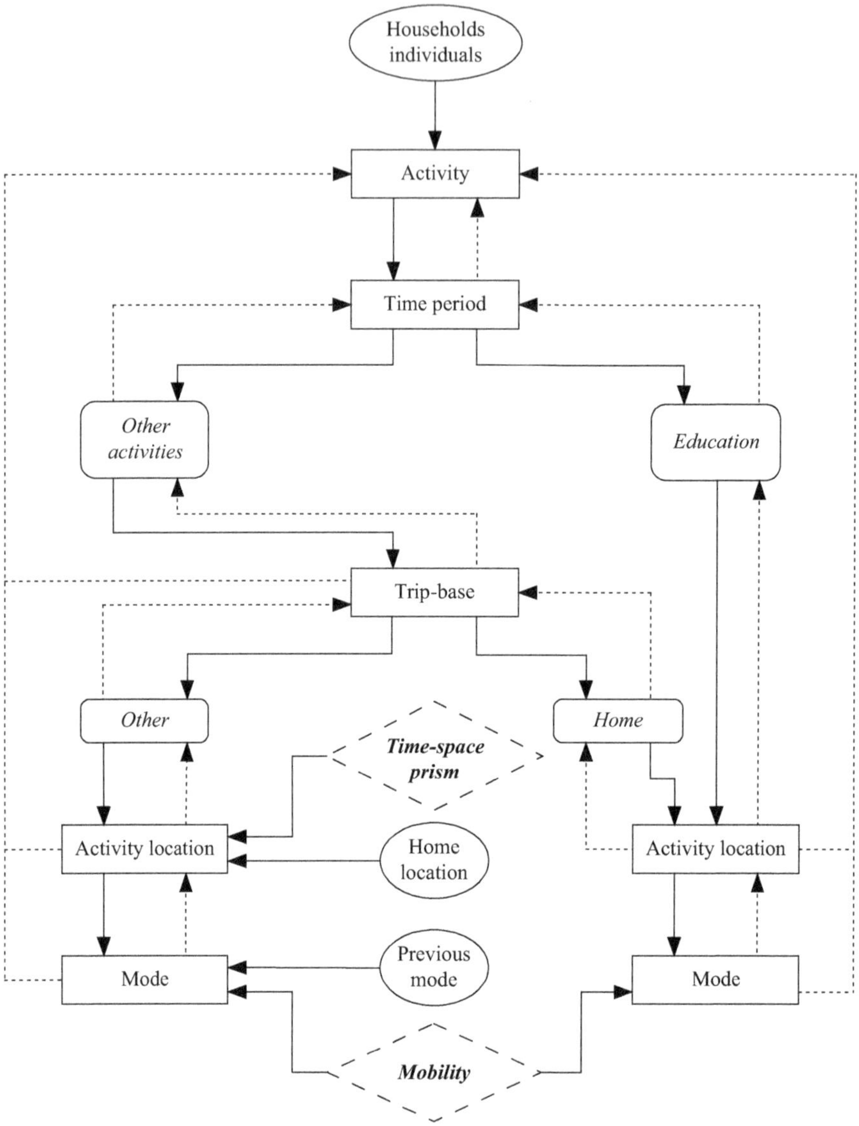

Figure 5.2 Primerano–Taylor accessibility framework for multimodal urban transport systems (the author holds copyright)

prices and overall economic conditions, including competition between regions and cities, and climate change), the system is effectively closed and is evolutionary.

LUTI policies and planning practices may be used to influence location and travel decisions in a city towards achieving goals of sustainability and low carbon mobility, by facilitating use of active transport and public transport and by location of services and facilities close to residences, so that travel distances are reduced for some trip purposes and high capacity public transport is widely available for longer distance travel.

Although there is no physical unit representing accessibility, planners have considered and used a number of alternative indicators that provide some measure of the relative differences in accessibility between different locations or for different individuals and population groups.

5.4.1 Accessibility indices

An integral accessibility metric gives the overall level of accessibility from a specified location to all defined available destinations. Assume that A_i is the integral accessibility metric of an object i for the case when i has a defined spatial location in a study area. The object may be a defined region, a town or city, a sub-area or zone within that town or city or defined region, a group of individuals, a household, or even an individual. The choice of object type will depend on the particular application. At the strategic level, object i is likely to be a town or city, at the urban level it may be a zone, precinct, land parcel, household or individual, and at the regional (rural) level it may be a zone or settlement.

Following the concept developed by Morris *et al.* [19], an integral accessibility index is a measure of the overall accessibility of object i to all relevant activity locations in a study area. Further, the summation $TA = \sum_i A_i$ provides a measure of the overall accessibility level in the study area. In terms of the definition or choice of a suitable index of accessibility, Morris *et al.* [19] specified a set of requirements for an accessibility indicator suitable for use by transport planners:

- the indicator should include an element of spatial separation which is responsive to changes in performance of the transport system
- it should have sound behavioural foundations
- it should be technically feasible and operationally simple, and
- it should be easy to interpret, and preferably be intelligible to lay people.

There are several specific accessibility indices that meet these criteria. The well-known Hansen integral accessibility index [20] is one such metric that has proven suitable for use at the level of national, state and regional transport networks, for example, to consider the accessibility of one city relative to all other cities in the study region (e.g. see [21–24]).

However, the Hansen index is less useful for intra-urban studies. The index is a discrete measure, that is, it assumes that populations can be represented by point locations (i.e. at nodes) in the study area network. This is not reasonable inside a continuous urban area with multiple transport modes for which different population groups have different levels of availability. Other indices are required for these types of applications. An accessibility framework for use in these circumstances, using the Primerano–Taylor definition [16,17] of accessibility presented above, based on the microeconomic theories of utility and consumer surplus (CS), can be applied in this case.

5.4.2 Accessibility measurement in urban areas

Accessibility can be measured in a number of ways [16], including

- *topological accessibility*, concerned with connectivity in networks
- *space–time framework*, considered the constraints imposed on an individual's movement around a region in terms of available time and the speeds of travel possible given the transport infrastructure, modes and services available to that individual in the region (i.e. could an individual reach a given location in the time available using a particular form of transport)
- *opportunity* (or impedance)-*based accessibility*, based on the separation between the individual and the activities (s)he wishes to participate in, and the density or intensity of those activities at their different locations.

Opportunity-based accessibility is suitable for application at the intra-urban level and provides a powerful tool for use in multimodal transport systems. It may be addressed using a number of methods, including behavioural utility (as applied in discrete choice models, for example [14,25]) and the CS value from microeconomic theory. CS is the benefit, in monetary terms, which an individual receives from a consumption choice situation. It can also be referred to as a measure of the willingness-to-pay for a commodity as it is the difference between what a person is willing to pay for a commodity and what they actually pay. The extra (or lesser) value an individual receives above (or below) what they paid is CS. When a change occurs (e.g. a price movement) the margin between what the person is willing to pay and what (s)he actually pays changes. The difference between what (s)he was willing to pay and actual payment under the two scenarios represents the change in CS.

Primerano and Taylor [16,17] used a CS measure of accessibility based on the 'inclusive value' (IV) parameter derived from discrete choice models of multi-dimensional choices to examine both the impacts of transport infrastructure improvements [16] and land-use changes [17], using their accessibility framework. These studies both considered changes in accessibility across metropolitan Adelaide, first as a result of a major road infrastructure project (the new Adelaide-Crafers Highway) and second of a substantial increase in employment opportunities in the north-east region of Adelaide.

The IV and CS measures of accessibility can be applied at a variety of levels of spatial aggregation in a study region, for example meshblock, Statistical Areas Levels (SAs) 1–3 in Australia [26], or aggregations of these smaller zones to wider regional levels. Conceptually – and with the appropriate data – the measures can also be applied to individuals or to the households to which they belong.

The theoretical aspects of the IV and CS measures of accessibility are described in Ben-Akiva and Lerman [25] and Train [27] and their application in the accessibility framework approach is described in Primerano and Taylor [16,17]. An accessibility metric derived from marginal choice probabilities in logit models of multidimensional choice, which is also called the IV or *logsum* [14,25], has the form

$$V_n = \log \sum_{r \in R_n} \exp(V_{rn}) \tag{5.4}$$

where V_n is the deterministic component (observable attributes) of the maximum utility for an individual n, and V_{rn} is the deterministic component of each secondary choice r in the set of choice R_n, as given in (5.2). This measure represents, in a single value, the overall benefit that an individual obtains from a set of alternatives. When using the MNL model (see (5.3)), it is possible to use the IV and a coefficient representing cost to estimate CS as follows

$$E(\mathrm{CS}) = \frac{1}{\alpha}\log\left(\sum_{n=1}^{N}\exp(V_n)\right) + C \tag{5.5}$$

where the logsum part is equivalent to (5.4), α represents the negative of the coefficient of time or cost from the deterministic part of the utility function, and C is an unknown constant that represents the difference between the actual value of CS and the estimated value [27]. The estimated change in CS is then given as

$$\Delta E\left(\mathrm{CS}^{\mathrm{AB}}\right) = \frac{1}{\alpha}\left[\log\left(\sum_{n=1}^{N^A}\exp\left(V_n^A\right)\right) - \log\left(\sum_{n=1}^{N^B}\exp\left(V_n^B\right)\right)\right] \tag{5.6}$$

where the superscripts B and A in (5.6) represent 'before' and 'after' scenarios and the two logsum terms represent the IVs derived from behavioural models under the two scenarios. Equation (5.6) thus gives an estimated change in CS with a measurement unit (e.g. hours or dollars) depending on the choice of α [27].

The accessibility framework shown in Figure 5.3 is described in full elsewhere [16,17]. The framework was developed to combine the strengths of existing accessibility measures for use in transport and urban planning. The aim was to have

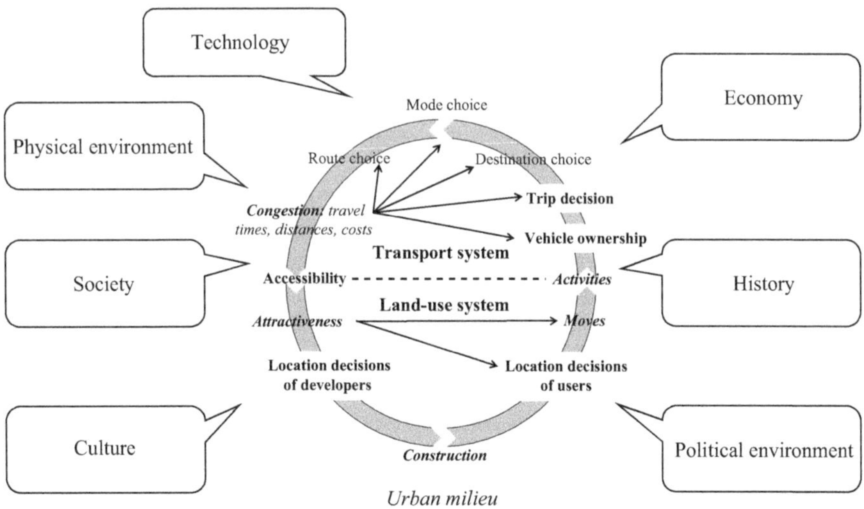

Figure 5.3 'Wegener's Wagon Wheel' conceptual model of urban development and the role of accessibility (adapted from [18])

a framework in which policies related to transport and urban form could be tested and implemented to improve accessibility for all socio-economic groups. The method used to develop the framework was activity-based rather than just location based. This means that the accessibility framework determines the accessibility of an individual to an activity rather than the accessibility between locations. Considering accessibility in this way implies that accessibility is dependent on three components, namely the

- traveller (individual or group)
- transport system (modes, routes, roads and service and congestion levels)
- land use (characteristics of land uses at origins and destinations).

The flow chart presented in Figure 5.3 shows the framework used to capture the choices individuals make that influence their accessibility to activities. The choices are represented in the rectangular boxes, the properties of the traveller are represented by the oval shapes, the alternatives of a choice set are represented by the rounded edge rectangular boxes, and the procedures used to restrict the choice sets of individuals based on their characteristics or their situations anytime during the survey period are represented by the diamond-shaped boxes. Five types of travel choices were modelled in the Adelaide implementation of the accessibility framework. These were: activity choice; time period choice; trip-base choice; location choice and mode choice. All models were MNL with exception of the mode choice models, which were nested logit. The framework and its use is described in full in [16,17].

5.5 Need for integrated policy and practice

LUTI is concerned with the development, management and operation of urban transport systems that provide for sustainable outcomes for our cities. It is primarily concerned with the optimal distribution of facilities and services in an urban area so that these are available to all of its inhabitants in ways that minimise adverse environmental impacts whilst maximising opportunities for sustainable economic development and social interaction.

Given the likely trade-off between housing type and location, with resulting impacts on demand for transport which are largely met at present by the private car in suburban locations, strategies to reduce urban carbon emissions need to be directed at providing substantial alternatives to that mode, based around greater use of the active modes and public transport. To make this direction feasible requires in-depth consideration of built form, relationships between transport modes and land-use development, land-use mixes, the location and intensity for services and facilities and the supporting infrastructure, and precinct planning and design including the relationships between neighbouring precincts and between precincts and major activity centres such as the central business district (CBD). The consideration of these planning issues combination may be collectively described in terms of LUTI, with policies for and implementation of LUTI offering the pathway to urban sustainability.

Sustainable cities require land-use planning decisions that support the use of public transport and the active transport modes of walking and cycling (Chapter 3). Active transport is very suitable for short distance trips and is flexible and timely. Most trips involving public transport usage include active transport access to the transit services, even if this is only walking to a bus stop. It is also vulnerable and erratic. It has also been largely neglected in conventional transport planning, and indeed in urban design – urban spaces and corridors for pedestrian movement need to be (and be seen as) safe, attractive and efficient places to be and to move through. Encouraging active transport – for example by locating destinations closer to origins and so reducing trip lengths – may not only be the key to reducing environmental impacts but also lead to improved public health, through increased physical activity and fitness, and improved productivity, through reduced congestion on the roads and reduced stress amongst people.

Encouraging public transport usage through land-use planning involves the planning of new developments ('greenfield' sites) and the management (or redevelopment, at 'brownfield' sites) of existing land so that (1) conditions for the efficient operation of public transport services can be improved and (2) accessibility to a wider set of destinations for more people can be achieved. These measures then serve to increase the demand for public transport, especially by encouraging mode change from the private car. The general way to approach these changes is by organising the location and mix of land-use types and activities to suit public transport use. Public transport nodes can serve as catalysts for more accessible land use by creating activity centres best suited to pedestrians and cyclists. People in such precincts will tend to travel less by car than those in lower density, more homogeneous areas. Higher densities of public transport trips provide the circumstances for transit service operators to provide higher levels of service (such as higher frequency services) and thus lead to a positive feedback between public transport demand and service supply. There are two specific, interrelated ways in which land-use planning can thus encourage the use of public transport by

- locating trip origins and destinations near public transport routes, which for new developments also means including the provision of efficient public transport services from the first day of use, and
- ensuring trip densities are sufficiently high to justify establishment of efficient services.

The broad impacts of potential changes in travel behaviour on overall GHG emissions may be seen using a set of sensitivity tests using the NGGI methods and data as outlined in the introduction, and supplemented by data from the Sydney Household Travel Survey database on vehicle occupancy, trip generation and trip length distributions [28]. Three *sensitivity tests* were performed:

- a switch of transport mode, in which 10% of person-kilometres of travel by private car transfers to public transport – a net reduction of 0.78 Mt p.a. in CO_2-e emissions

- a 10% decrease in average private vehicle trip length – a reduction of 2.74 Mt p.a. CO_2-e emissions
- the removal of all trips less than 2.0-km length by private car (substituted by use of walking and cycling) – a reduction of 0.35 Mt p.a. CO_2-e emissions.

These sensitivity tests are indicative only and may not fully represent real world expectations. Nevertheless they highlight the importance of trip length as a factor in GHG emission totals – and hence the opportunity through providing closer destinations in reducing GHG emissions.

5.6 Integration – policy and practices

An important means for achieving integration is through the encouragement and use of the active transport modes walking and cycling and public transport, because access to public transport generally requires substantial use of the pedestrian and cycle modes. This requires the planning and design of locations for services and facilities and public transport services, routes, terminals, stations and stops to enable people to make ready use of them.

Urban form can have a significant effect on the carbon intensity of travel, with some urban forms showing a greater capacity than others to reduce the rate of carbon emissions per capita. Newton *et al.* [29] underscored the importance of integrated land-use and transport planning for growing cities, providing examples in which transport planning and development planning have happened independently resulting in reliance on private vehicles or excessive pressure on public transport services. A major transformation in the way transport planning is carried out is required with a new approach towards environmental and liveability aspects and a focus on achieving carbon efficiency in transport and urban precincts [30].

Beyond reducing carbon emissions, changing the way we undertake urban and transport planning offers the potential to generate other important long-term benefits, such as open space preservation, improved air quality and public health, and reduced infrastructure investments, leading to improved quality-of-life in urban areas [4].

The travel decisions people make have been found to be significantly associated with built environment factors such as density, location, mix of land uses and precinct design [31]. Changes are starting to occur in the structure of cities, as the cities are reaching a limit to car-use growth, and there are shifts away from traditional urban culture and economic paradigms [32]. As such there exists the potential for urban design and planning to play a major part in establishing a low carbon transport future.

Many theories exist regarding the ideal layout for low carbon cities; however, there has been found to be a huge gap between theory and practice [31]. The current understanding is that urban form and layout set the 'envelope of possibilities' for travel, and residential layout and supporting facilities have a complex relationship with resident attitudes, preferences and perceptions [6,31]. The 'compact city' approach attempts to bring activities closer to residents so that they can fulfil their needs by using low carbon options such as active travel or public transport. The problem exists in maintaining quality of life and space in high population density environments, and what options for built environment interventions are available where high-density living is not feasible. Aditjandra *et al.* [6] concluded that the specific layout of towns and cities in a low carbon world was as yet unclear.

Santos *et al.* [33] considered a wide range of low carbon transport policy options, which they grouped into three categories:

- *physical policies*, concerning infrastructure and service provision,
- *soft policies*, aimed at bringing about behavioural change, and
- *knowledge policies*, emphasising the role of investment in research and development for future sustainable mobility.

They argued that policy integration was the key and that optimal blending of mutually reinforcing policies was essential. Banister [34] discussed the concept of the sustainable mobility paradigm, which again required integrated, mutually supporting sets of policies. He then suggested a possible policy framework for low carbon mobility systems (see also [5]). However, this still required determination of relevant policy sets. A major question is just what policies should be considered and how can policy settings be optimised? Nakamura and Hayashi [35] reviewed international developments on low carbon mobility policies and strategies. They concluded that the specific development processes of individual cities significantly affected the feasibility and effectiveness of different policies.

Taeihagh *et al.* [36] developed a virtual environment system for the exploration and analysis of different configurations of policy measures in order to build and assess alternative policy packages. Packages such as these and decision support systems are required to assist policy makers to develop effective policies with respect to low carbon mobility. Good planning practices will allow low carbon mobility to become a reality.

One useful approach to developing policy packages for low carbon mobility is the 'avoid-shift-improve' (ASI) framework [35,37,38]. ASI has found wide recognition in Europe and Japan. The framework provides a blueprint for policy packages aimed at promoting alternative mobility solutions and sustainable transport systems.

As was discussed in Chapter 3, the ASI framework is extended in this book to include the recent developments in car-sharing and ride-sharing services. The extended framework '*avoid, shift, share, improve*' is described below (Chapter 3):

- *Avoid*, considering measures and policy instruments that reduce or avoid the need to travel and thus improve the efficiency of the transport system. Planning for LUTI and management of travel demand can reduce the overall need for physical travel activity and, in particular, seek reductions in trip length
- *Shift* policy instruments, which seek to improve trip efficiency, generally through shifts from high-energy consuming modes (e.g. the private car) to more environment-friendly modes such as active transport (walking and cycling) and public transport

- *Share*, enables and promotes a shift from car ownership models and private vehicle use towards car-sharing, ride-sharing, bike-sharing (and even vehicle parking space sharing in resource restrained urban centres)
- *Improve*, which considers measures that focus on technology and infrastructure use, especially with regard to vehicles, propulsion systems and energy and fuel efficiency, including the potential of alternative energy sources for transport.

The reader is referred to Chapter 3 for further coverage and analysis of these policy dimensions.

5.7 Case studies

A useful way to envisage the concepts and applications of LUTI is through examining case study implementations of integrated planning and design ideas. Three case studies are offered here. These are a greenfields development (Mawson Lakes in northern Adelaide, Australia), a brownfields redevelopment in an inner city suburb (Subiaco, Perth, Australia), and metropolitan Portland, Oregon, USA. The case studies are presented in summary form, with information on how to gather more information about each of them.

5.7.1 Mawson Lakes

The Mawson Lakes development is a large mixed use, mixed density suburban development in the northern suburbs of Adelaide, about 12 km from the CBD. The development commenced on a greenfields site in the late 1990s. It comprises mixed density residential development, along with a town centre next to a railway station on the Gawler suburban line, and employment centres, notably Technology Park. It is adjacent to the Mawson Lakes campus of the University of South Australia. The area of the development is 6.37 km^2. The present residential population of Mawson Lakes is about 12,000, and there are about 5,000 jobs in the development, Technology Park and the university. The overall residential density is about 1,750 persons per km^2.

The transport network in the development was designed to provide separate access for pedestrians and cyclists throughout the area, focussing on the town centre and the railway station and utilising green spaces and fingers throughout the area. Transit-oriented development principles were used in the design of the residential parts and the town centre, but the development is a ‘hybrid’ TOD in that the railway station is designed as a transport hub for the surrounding suburbs, including feeder bus access, ‘kiss and ride’ facilities, and substantial parking as a ‘park and ride’ venue. This latter development has led to some degree of separation between the station and the main development, especially the town centre.

More details on Mawson Lakes are available in [39].

5.7.2 Subiaco

Subiaco is an inner suburb of Perth, about 3-km west of the CBD on the Fremantle suburban rail line. The revitalised town centre, ‘Subi Centro’, was developed around the railway station, beginning by the undergrounding of the station and rail line to provide substantial space in the town centre and removing the historical separation between north and south Subiaco due to the railway. The redevelopment is a strong example of urban regeneration sympathetic to the historical and heritage aspects of the area. The redevelopment has strengthened the nearby, popular Rokeby Street retail development, incorporated almost 1,500 new dwellings at medium density compatible with historical developments in the area, and added substantial additional commercial floorspace with more than 300 new jobs. Provision of high-quality public space through best practice urban design principles is a major feature of the new Subiaco. More details about this brownfields development can be found in [40].

5.7.3 Portland

Portland, Oregon is widely acclaimed as 'an international pioneer in transit-orientated developments', and is seen by many as the premier city in North America for sustainable transport planning based on LUTI principles. Its 'TriMet' public transport system based around light rail transit, buses and some suburban rail is one of the most successful public transport operations in the United States of America. The public transport system serves a residential population of about 1,400,000. Transport policy in the Portland metropolitan area has favoured transit development since the late 1970s, and these policies are integrated with strong land-use and urban development policies that adhere to a strict 'urban growth boundary' around the metropolis and seek to encourage TOD. This is done through a 'TOD and Centres' programme which seeks to create public–private partnerships for TOD projects, involving vibrant urban centres that favour travel by transit, walking and cycling and create a sense of strong local community identity.

See [41] for more information about Portland, its transit system and its LUTI-based planning philosophy and system implementation.

5.8 Conclusions

Attaining sustainable transport and low carbon mobility is vital for the future success of our cities. To be blunt, for transport systems this means focusing on moving people and goods instead of cars and trucks. Land-use planning and transport planning have key roles in shaping our cities and in delivering social, economic and environmental sustainability. The successful delivery of these outcomes requires strong integration of land-use and transport planning. Only then can we achieve a sustainable urban land-use and transport system featuring low carbon mobility. LUTI can facilitate a sustainable city by

- reducing the need to travel
- reducing the length of trips
- making it easier and safer for people to access services and facilities
- reducing transport impacts on communities

- providing for the efficient distribution of goods and services to businesses and the community
- providing choices in travel modes and
- ensuring flexibility now and in the future to address changes in demand in the economy and in society.

On the basis of the discussion in this chapter, the most important contribution that LUTI could make to achieving low carbon mobility would be the reorganisation of facilities and services in an urban area so that trip destinations are generally closer to trip origins, reducing travel distances and making use of the active transport modes more feasible and attractive. Arranging residential areas clustered around activity nodes and with good access – again by the active modes – to line haul public transport is also important for longer distance travel, but the proximity of destinations is a key feature of the integrated land-use and transport system.

Amongst the many issues facing planners are environmental impacts, both local (e.g. noise and air quality) and global (climate change and GHG emissions), economic productivity, conservation of resources (especially the use of land), traffic congestion, social severance and public health. Following the established principles and practices of LUTI – while these may not always coincide with what has been past practices in our cities – offers a systematic path to long-term resolution of these issues.

References

[1] V. Rauland and P. Newman, "Decarbonising Australian cities: a new model for creating low carbon, resilient cities", in *MODSIM2011: 19th International Congress on Modelling and Simulation*, F. Chan, D. Marinova and R.S. Anderssen, Eds. Sydney: Modelling and Simulation Society of Australia and New Zealand, 2011, pp. 3073–3079.

[2] M. Philp and M.A.P. Taylor, "A research agenda for low carbon mobility: issues for 'new world' cities", *International Journal of Sustainable Transportation*, 2016, http://dx.doi.org/10.1080/15568318.2015.1106261.

[3] A. Chavez and A. Ramaswami, "Progress toward low carbon cities: approaches for transboundary GHG emissions' footprinting", *Carbon Management*, vol. 2 (4), 2011, pp. 471–482.

[4] P. Donoso, F. Martinez, and C. Zegras, "The Kyoto Protocol and sustainable cities – potential use of clean-development mechanism in structuring cities for carbon-efficient transportation", *Transportation Research Record*, vol. 1983, 2006, pp. 158–166.

[5] D. Banister, "City transport in a post carbon society", in *Moving Towards Low Carbon Mobility*, M. Givoni and D. Banister, Eds. Cheltenham, UK: Edward Elgar Publishing, 2013, pp. 255–266.

[6] P. Aditjandra, C. Mulley, and J.D. Nelson, "The influence of neighbourhood design on travel behaviour: empirical evidence from North East England", *Transport Policy*, vol. 26, 2013, pp. 54–65.

[7] DCCEE, *Australian National Greenhouse Accounts: National Inventory Report 2010*, vol 1, Department of Climate Change and Energy Efficiency, Canberra, 2012, www.climatechange.gov.au/emissions.

[8] P.W. Newton and P. Newman, "The geography of solar photovoltaics (PV) and a new low carbon urban transition theory", *Sustainability*, vol. 5, 2013, pp. 2537–2556.

[9] H. Minken, "A sustainability objective function for local transport policy evaluation", in *World Transport Research*. Selected Proceedings from the Eighth World Conference on Transport Research, vol. 4: Transport Policy, H. Meersman, E. Van De Voorde and W. Winkelmans, Eds. Oxford: Elsevier-Pergamon, 1999, pp. 269–279.

[10] KonSULT, *Knowledgebase ON Sustainable Urban Land use and Transport*. Institute for Transport Studies, University of Leeds, 2016, www.konsult.leeds.ac.uk.

[11] C. Hale and P. Charles, "Making the most of transit oriented development opportunities", Papers of the 29th Australasian Transport Research Forum, 2006, www.atrf.info/papers.

[12] L. Meng, M.A.P. Taylor and D. Scrafton, "Combining latent class models and GIS models for integrated transport and land use planning", *Urban Policy and Research*, 2016, http://dx.doi.org/10.1080/08111146.2015.1118372.

[13] P.J. Hills, "What is induced traffic?", *Transportation*, vol. 23 (1), 1996, pp. 5–16.

[14] D.A. Hensher, J.M. Rose and W.H. Greene, *Applied Choice Analysis: A Primer*. Cambridge: Cambridge University Press, 2005.

[15] D. Niemeier, "Accessibility: an evaluation using social welfare", *Transportation*, vol. 24, 1997, pp. 377–396.

[16] F. Primerano and M.A.P. Taylor, "An accessibility framework for evaluating transport policies", in *Access to Destinations*, D.M. Levinson and K.J. Krizek, Eds. Oxford: Elsevier, 2005, pp. 325–346.

[17] F. Primerano and M.A.P. Taylor, "Increasing accessibility to work opportunities in metropolitan Adelaide", *Journal of the Eastern Asia Society for Transportation Studies*, vol. 6, 2005, pp. 4097–4112.

[18] M. Wegener, "Reduction of CO_2 emissions of transport by reorganisation of urban activities", In *Transport, Land-Use and the Environment*, Y. Hayashi and J.R. Roy, Eds. Dordrecht: Kluwer Academic Publishers, 1996, pp. 103–124.

[19] J.M. Morris, P.L. Dumble and M.R. Wigan, "Accessibility indicators for transport planning", *Transportation Research A*, vol. 13, 1979, pp. 91–109.

[20] W.G. Hansen, "How accessibility shapes land use", *Journal of the American Institute of Planners*, vol. 25, 1959, pp. 73–76.

[21] M.A.P. Taylor, S.V.C. Sekhar and G.M. D'Este, "Application of accessibility based methods for vulnerability analysis of strategic road networks", *Networks and Spatial Economics*, vol. 6 (3–4), 2006, pp. 267–291.

[22] C. Curtis and J. Scheurer, "Benchmarking public transport accessibility on Australasian cities", Papers of the 35th Australasian Transport Research Forum, 2012, www.atrf.info/papers.
[23] A. Reggiani, P.J. Nijkamp and D. Lanzi, "Transport resilience and vulnerability: the role of connectivity", *Transportation Research Part A: Policy and Practice*, vol. 81, 2015, pp. 4–15.
[24] K.T. Geurs and J. Östh, "Advances in the measurement of transport impedance in accessibility modelling", *European Journal of Transport and Infrastructure Research*, vol. 16 (2), 2016, pp. 294–299.
[25] M.E. Ben-Akiva and S.R. Lerman, *Discrete Choice Analysis: Theory and Application to Travel Demand*. Boston, MA: MIT Press, 1985.
[26] ABS, Canberra, *ACT: Australian Bureau of Statistics*, 2016, www.abs.gov.au/websitedbs/D3310114.nsf/Home/Geography.
[27] K. Train, *Discrete Choice Methods with Simulation*, 2nd ed. Cambridge, UK: Cambridge University Press, 2009.
[28] BTS, *NSW Bureau of Transport Statistics*, Sydney, 2016, www.bts.nsw.gov.au.
[29] P.W. Newton, A. Pears, J. Whiteman and R. Astle, "The energy and carbon footprints of urban housing and transport: current trends and future prospects", in *Australia's Unintended Cities: The Impact of Housing on Urban Development*, R. Tomlinson, Ed. Melbourne: CSIRO Publishing, 2012, pp. 153–189.
[30] R. Hickman and D. Banister, "Looking over the horizon: transport and reduced CO_2 emissions in the UK by 2030", *Transport Policy*, vol. 14, 2007, pp. 377–387.
[31] R. Hickman, "Urbanization and future mobility", in *Moving Towards Low Carbon Mobility*, M. Givoni and D. Banister, Eds. Cheltenham, UK: Edward Elgar Publishing, 2013, pp. 60–74.
[32] P. Newman, J. Kenworthy and G. Glazebrook, "Peak car use and the rise of global rail: why this is happening and what it means for large and small cities", *Journal of Transportation Technologies*, vol. 3, 2013, pp. 272–287.
[33] G. Santos, H. Behrendt and A. Teytelboym, "Part II: policy instruments for sustainable road transport", *Research in Transportation Economics*, vol. 28 (1), 2010, pp. 46–91.
[34] D. Banister, "Cities, mobility and climate change", *Journal of Transport Geography*, vol. 19 (6), 2011, pp. 1538–1546.
[35] K. Nakamura and Y. Hayashi, "Strategies and instruments for low-carbon urban transport: an international review on trends and effects", *Transport Policy*, vol. 29, 2013, pp. 264–274.
[36] A. Taeihagh, R. Banares-Alcantra and M. Givoni, "A virtual environment for the formulation of policy packages", *Transportation Research Part A: Policy and Practice*, vol. 60, 2013, pp. 53–68.
[37] H. Nakamura, Y. Hayashi and A.D. May, Eds. *Urban Transport and the Environment: An International Perspective*. Oxford: Elsevier, 2004.

[38] Deutsche Gesellschaft für Internationale Zusammenarbeit (GIZ) *Sustainable urban transport: avoid-shift-improve (A-S-I)*, 2016, www.giz.de/transport.
[39] Mawson Lakes Living, 2016, http://mawsonlakesliving.info
[40] Heart Foundation of Australia, 2016, www.healthyplaces.org.au/site/case-studies.php.
[41] Center for Transit-Oriented Development, 2016, www.ctod.org/portal/Portland-Metros-TOD-Strategic-Plan.

Glossary

ABS	Australian Bureau of Statistics
ASI	'avoid-shift-improve' paradigm for policy packages aimed at promoting alternative mobility solutions and sustainable transport systems
CBD	central business district
CO2-e	carbon dioxide equivalents, a metric of total greenhouse gas (GHG) emissions
CS	consumer surplus
GHG	greenhouse gas
IV	inclusive value, a parameter extracted from a set of discrete choice models
LUTI	land-use-transport integration
MNL	multinomial logit model, a form of discrete choice model
NGGI	National Greenhouse Gas Inventory
SA1	Statistical Areas Level 1, the smallest area of output for the ABS Census of Population and Housing
SA2	Statistical Areas Level 2, a medium-sized general purpose area built up from whole Statistical Areas Level 1 (SA1s). They are used by the ABS to represent a community that interacts together socially and economically
SA3	Statistical Areas Level 3, geographical areas used by the ABS for the output of regional data The SA3s are used to create a standard framework for the analysis of ABS data at the regional level through clustering groups of SA2s that have similar regional characteristics
TOD	transit-oriented development

Further Reading

[1] B. Armstrong, G. Davison, J. de Vos Malan, B. Gleeson and B. Godfrey, *Delivering sustainable urban mobility. Report for the Australian Council of Learned Academies*, Canberrra ACT, 2015, www.acola.org.au.
[2] F. Creutzig, "Evolving narratives of low-carbon futures in transportation", *Transport Reviews*, 2015, http://dx.doi.org/10.1080/01441647.2015.1079277.

[3] C. Curtis, "Integrating land use with public transport: the use of a discursive accessibility tool to inform metropolitan spatial planning in Perth", *Transport Reviews*, vol. 31 (2), 2011, pp. 179–197.

[4] M. Givoni and D. Banister, Eds., *Moving Towards Low Carbon Mobility*. Cheltenham, UK: Edward Elgar Publishing, 2013.

[5] International Energy Agency, "A tale of renewed cities: a policy guide on how to transform cities by improving energy efficiency in urban transport systems". Policy Pathway Report, International Energy Agency, Paris, 2013, www.iea.org.

[6] D.M. Levinson and K.J. Krizek, Eds., *Access to Destinations*. Oxford, UK: Elsevier, 2005.

[7] R.L. Mackett, A.D. May, M. Kii and H. Pan, Eds., *Sustainable Transport for Chinese Cities*. Bingley, UK: Emerald Group Publishing, 2013.

[8] OECD, *Greening household behaviour: overview from the 2011 survey* – revised edition. OECD Studies on Environmental Policy and Household Behaviour, OECD Publishing, Paris: Organisation for Economic Cooperation and Development, 2014, http://dx.doi.org/10.1787/9789264214651-en.

[9] J. Stanley and P. Brain, *Urban land use transport integration and the vital role for Australia's forgotten inner/middle suburbs*. Bus and Coach Industry Policy Paper 5, Canberra ACT: Bus Industry Confederation, 2015, www.ozebus.com.au.

Chapter 6

Decarbonising suburban mobility

Peter Newton[1], Michael A.P. Taylor[2], Peter Newman[3], John Stanley[4], Chris Rissel[5], Billie Giles-Corti[6] and Rocco Zito[7]

Abstract

Greening suburban mobility in low-density cities will require a step change in both the demand and supply sides of urban travel. Supply-side innovation is critical in order that the opportunities and alternatives are established for car-dependent suburban residents to embrace more effective forms of active, low carbon travel. This chapter provides a high-level roadmap of the innovations and planning strategies required for a low carbon transition in each of the key supply-side arenas of urban transport.

Keywords

Low carbon mobility; suburban transport; public transport; active transport; walking; cycling; urban development

6.1 Introduction

Many studies have shown that as an economy grows so does its vehicle kilometres of travel (VKT) [1–3]. This directly increases the amount of greenhouse gas (GHG) emissions produced by the transport sector. In addition, the rate of increase of associated emissions produced from the growing VKT is higher than other

[1]The Centre for Urban Transitions, Swinburne University of Technology, Melbourne, Australia
[2]School of Natural and Built Environments, University of South Australia, Adelaide, Australia
[3]Curtin University Sustainability Policy Institute, Curtin University, Perth, Australia
[4]Institute of Transport and Logistics Studies, University of Sydney, Sydney, Australia
[5]Sydney School of Public Health, University of Sydney, Sydney, Australia
[6]Melbourne School of Population and Global Health, University of Melbourne, Melbourne, Australia
[7]School of Computer Science, Engineering and Mathematics, Flinders University of South Australia, Adelaide, Australia

industrial sectors [4]. If left unchecked, emissions from transport, in which suburban travel is a significant proportion, would become the highest contributing sector of the economy to global warming and climate change. We need to find ways of implementing a low carbon transition for transport that breaks the link between economic prosperity and increasing transport emissions while simultaneously providing safe, efficient and equitable urban mobility.

A global analysis of cities [5] has revealed comparable levels of liveability between Western European cities and their counterparts in North America and Australasia but a significant divide reflecting differing levels of resource consumption and GHG emissions associated with their contrasting forms of built environment and urban living. Much of that variability in ecological footprint can be associated with the contrasting types of housing (floor area, density) and modes of transport that have characterised the post-WW2 era of city development. The low-density car-oriented suburban development that occurred during this 70-year period, in countries such as Australia, now presents governments with major planning challenges in re-directing urban development towards delivery of more sustainable, resilient, productive, equitable and liveable cities. This is a significant challenge, requiring a major transition in land-use and transport (Chapter 5) capable of creating smart, accessible cities that deliver high levels of low carbon mobility to their resident populations.

Three kinds of transport-related urban fabric have evolved over time via the walking city, the transit city and the automobile city to a stage where all cities can be seen to have the three urban fabrics represented – albeit in varying proportions – expressed together in the form of different elements, functions and qualities [6]. In the Australian context, there is a stark divide between the pre-war inner suburbs (established during the walking and transit city era) and the post-war outer suburbs. Melbourne is illustrative of the spatial variability in public transport access levels (PTALs, [7]) that exist in Australian cities (Figure 6.1): there are 15 municipalities in the inner region, all with PTAL scores above the metro average, and 16 in the outer suburbs, all with below average scores. The challenge of delivering more public transport (PT) and active transport options in the outer region is clear, given that the area to be serviced is currently 810,000 ha – ten times that of the inner suburbs. Continued car-dependent greenfield sprawl (where most of the new housing in Australian cities is being constructed) exacerbates the transport-mobility-carbon problem.

An analysis of daily personal travel in Melbourne using the Victorian Integrated Survey of Travel and Activity database [8] reveals that outer region residents drove 67% more kilometres per household as driver and 60% more as passenger, but walked 40% less and cycled 200% less compared with inner city residents (Table 6.1). School bus was an important mode of travel for outer suburban households, as was the train – for those who had access from the eight corridors (again see Figure 6.1) and where person-kilometres of travel was 30% longer – reflecting the concentration of jobs in the inner city and central business district (CBD). Overall, individuals living in the outer suburbs of Melbourne spent 10% more time in total daily urban travel activities than those in the inner suburbs. There

Figure 6.1 Melbourne's inner and outer regions based on PTAL; showing urban transit lines (suburban rail and tram) (Source: Peter Newton and Stephen Glackin – unpublished)

is a clear inner–outer divide that has environmental, social and productivity implications. Transport and access are at their heart. For example, how time is used in cities has a critical impact on economic productivity and social well-being. The extra 10% of a daily activity budget spent in travel could be allocated to more valuable and creative activities. The $16.5 billion congestion bill for Australia's cities in 2015 is close to a 30% increase on the 2010 cost of $12.8 billion and is made up by about $6 billion in private time costs, $8 billion in business time costs, $1.5 billion in extra vehicle operating costs and $1 billion in extra air pollution costs [3]. The target of a 30-min city is aligned to Marchetti's anthropogenic constant of the average time workers spent travelling from home to work throughout the history of urban settlement – a figure now exceeded in Australia's largest cities [9]. It tends to represent a tipping point for human settlements around which city structures as well as household locational choices and travel-time budgets are re-shaped. For Australia, this is resulting in a trend to more compact city living, a focus on urban infill rather than continued low-density expansion on the periphery, and a transition from what have been archetypal 'suburban cities' up until the end of the twentieth century to more urban cities.

The social divide between inner and outer suburbs in Australia's cities has also been widening, driven by a major economic restructuring of cities that has seen the displacement of inner city manufacturing jobs in favour of information, knowledge and

Table 6.1 Daily travel rates in Melbourne by region of residence

Indicator	Inner region		Outer region	
	Per capita	**Per household**	**Per capita**	**Per household**
Person-kilometres of travel (PKT) by mode				
Vehicle driver	15.80	38.63	21.70	64.57
Vehicle passenger	6.90	16.86	8.99	26.76
Motorcycle	0.14	0.33	0.10	0.28
Walk	0.90	2.19	0.44	1.30
Bicycle	0.51	1.24	0.14	0.41
Taxi	0.20	0.50	0.05	0.15
Train	2.28	5.56	2.44	7.26
Tram	0.58	1.43	0.05	0.15
School bus	0.07	0.18	0.40	1.18
Public bus	0.40	1.64	0.38	1.12
Total	28.06	68.62	34.68	103.21
Person-hours of travel (PHT) by mode				
Vehicle driver	0.58	1.41	0.63	1.73
Vehicle passenger	0.24	0.58	0.26	0.73
Motorcycle	0.01	0.01	0.00	0.01
Walk	0.21	0.51	0.10	0.27
Bicycle	0.04	0.10	0.01	0.03
Taxi	0.01	0.02	0.00	0.01
Train	0.07	0.18	0.07	0.18
Tram	0.04	0.11	0.00	0.01
School bus	0.00	0.01	0.01	0.04
Public bus	0.03	0.07	0.02	0.06
Total	1.22	2.99	1.11	3.06
Vehicle-kilometres of travel (VKT/day)	16.14	39.46	21.85	65.01
Vehicle-hours of travel (VHT/day)	0.59	1.44	0.63	1.75
Total person hours of travel/day	1,957,586		2,151,746	
Personal income (2012–13) p.a.	$68,400		$51,700	

Source: original analysis of Victorian Integrated Survey of Travel and Activity database [8].

creative industries and their higher paid workforces [10,11]. Manufacturing has either decentralised or shifted off-shore to lower cost locations. This has helped drive a re-urbanisation process within Australian cities (and many North American cities) that is now reflected in an inner city housing market that is increasingly unaffordable to those on below average incomes [12]. As a consequence of this economic restructuring, there has been an associated socio-spatial restructuring evident via a decentralisation of social disadvantage in Australia's major cities with three inter-related dimensions: increasing concentration of lower income households (Table 6.1; average incomes are one-third lower in the outer region); poor access to jobs, tertiary education and specialist health services [12]; and a concentration of social problems [11].

Table 6.2 Annual greenhouse gas emissions from personal travel in Melbourne by region of residence

Indicator	Inner region	Outer region	Metropolitan wide
Percentage of GHG emissions by mode			
Vehicle driver	34.1%	57.0%	91.1%
Vehicle passenger	0.0%	0.0%	0.0%
Motorcycle	0.1%	0.1%	0.3%
Walk	0.0%	0.0%	0.0%
Bicycle	0.0%	0.0%	0.0%
Taxi	0.4%	0.1%	0.6%
Train	3.2%	4.2%	7.5%
Tram	1.0%	0.1%	1.1%
School bus	0.1%	0.4%	0.4%
Public bus	0.5%	0.4%	0.9%
Total	38.7%	61.3%	100.0%
GHG emission rates (for personal travel)			
CO_2-e tonnes per person per annum	1.83	2.39	2.14
CO_2-e tonnes per household per annum	4.48	6.59	5.58
CO_2-e grams per person-km of travel	65.3	68.8	67.4
Private vehicle CO_2-e g/person-km of travel	194.9	198.0	196.8
Private vehicle occupancy rate (persons/veh)	1.44	1.41	1.42
Estimated population (persons)	1,598,456	1,945,253	3,543,709
Estimated population (% of people)	45.1%	54.9%	100.0%
Estimated population (households)	653,676	703,890	1,357,566
Estimated population (% of households)	48.2%	51.8%	100.0%

Source: original analysis of Victorian Integrated Survey of Travel and Activity database [8].

The environmental consequences of low-density car-oriented urban development characteristic of Australian and North American cities have been established for a number of decades [13] and are clearly reflected in the statistics on GHG emissions in Table 6.2. The automobile is responsible for 91.1% of GHG emissions from personal transport. Outer region residents use cars more than inner area residents, generating 62.3% of all VKT. This private vehicle traffic load is responsible for 57.2% of travel-related GHG emissions by Melbourne residents, noting that the outer area population of about 1,945,000 (in 2010) was 54% of the metropolitan total (about 3,454,000). Private vehicle use by inner area residents generates about 34.6% of travel-related GHG emissions, whilst comprising 45.1% of the metropolitan population. Travel by households in the outer region generates almost 50% more CO_2-e per annum on average (6.59 t/hhld) than the inner region (4.48 t/hhld).

The urban fabric laid down during previous eras of urban development provides opportunities as well as constraints for the decarbonisation of housing and transport in cities (Figure 6.2 and [14]). In the outer suburbs, decarbonisation of housing is being driven by the rapid penetration of solar photo-voltaic (PV) on the roof-tops of increasingly energy-efficient detached dwellings – a form of residential development that has not been conducive to date to low carbon transport interventions.

Built environment fabric	Low/zero carbon technologies: Housing	Low/zero carbon technologies: Transport
Suburban	• Renewable energy technologies for individual buildings, e.g., solar PV • Precinct scale technologies	• EVs, hybrid, hydrogen vehicles and associated infrastructures • Active transport (walk, cycle) • Smart buses • New TODs (heavy, light rail)
Urban	• Precinct scale low emission energy technologies, e.g., co-generation; tri-generation	• Public transit • Active transport (walk, cycle) • EVs, hybrid, hydrogen vehicles and associated infrastructures

Figure 6.2 Model framework for low carbon housing and transport interventions in urban and suburban forms of the built environment (Source: Newton and Newman [14]. Reprinted with permission from [14]) [Permission to re-produce obtained and attached with this submission]

After 70 years of almost total reliance on roads and automobiles to support the outer expansion of Australasian and North American cities, a level of transformational change beyond that required for the transition from walking city to transit city and transit city to auto city is required for sustainable low carbon living in the twenty-first century. The transition will require integrated long-term planning and implementation of a type largely absent in the last half century in Australian city development. Greening suburban mobility in low-density cities will require a step change in both the demand and supply sides of urban travel. Supply-side innovation is critical in establishing the opportunities and alternatives for car-dependent suburban residents to embrace more effective forms of active, low carbon travel. Indeed, there are significant co-benefits to be gained from decarbonising suburban transport in addition to environmental: increasing accessibility and mobility of residents to jobs and services; increasing productivity by reduced road congestion and travel times; and enhancing human health by introducing more opportunities for walking and cycling [15].

This chapter provides a high-level roadmap of the innovations and planning strategies required for a low carbon transition in each of the key supply-side areas of transport in the outer suburbs. The roadmap will need to accommodate a city that better develops and integrates the multiple transport modes that already exist as well as those that are emerging (e.g. autonomous electric vehicles (AEVs) capable of providing new ways to service the mobility needs of city residents) by harnessing 'smart' information and communications technologies systems within evolving spatial structures and forms that can deliver better cities across all the key dimensions of performance. In this context, as cities grow to a mega-metropolitan scale, ability to plan a multi-nodal city based around activity centres of different scale (employing concepts such as the '20 min city' [16]), interlinked by different modes and hierarchies of transport within the city (and extending beyond city boundaries to include nearby provincial cities or towns in what is effectively a functional urban region) will determine whether twenty-first century urbanisation is sustainable.

6.2 Next wave developments in suburban transport

As data from the introduction clearly demonstrate, Australian cities require new PT systems and services and active transport initiatives, especially in the suburbs, if universal low carbon mobility is to be achieved. In terms of PT systems, bus systems provide wide-area coverage and can support rail systems, which of themselves provide for fast, high capacity services and facilitate suitable land-use developments. Encouragement of, if not a new reliance on, the active transport modes – cycling and walking – must also feature in plans for low carbon urban mobility.

6.2.1 Buses

Buses are typically the workhorse mode for PT in a suburban setting, providing both direct trunk services and local services and playing a feeder role to rail trunk services (Chapter 7). The opportunity for buses to contribute effectively to decarbonising suburban mobility depends on both (1) reducing the carbon footprint of the mode and (2) increasing its modal share at the expense of cars. For the road transport sector more broadly, Stanley *et al.* [17] have shown that, in urban Australia, changing to low carbon fuel sources is more important than behaviour change for lowering GHG emissions, but both have roles to play.

Decreasing the carbon footprint of buses in urban operation is about shifting from fossil-fuelled energy to electric, which will often happen via a hybrid pathway, as in London. A focus on the behaviour change side, mainly concerned with achieving a modal shift from car to bus in suburban operation, can support lower GHG emissions. However, the main societal benefits from such an outcome will not come from lower GHG emissions but from co-benefits that flow from reducing car use. For example, in Melbourne, a low-density city where buses are the dominant form of suburban PT, Stanley and Hensher [18] found that social inclusion benefits and congestion cost savings were the dominant benefits from bus, accounting for 75% of total measured benefit, together with user benefits (a further 22%). The concept of co-benefits is thus vital to bus planning [15,18].

6.2.1.1 Trunk bus services

These can be provided as a high service level by bus rapid transit (BRT) operating in its own right-of-way, in which system capacities can match rail. For example, Brisbane's BRT system runs 200 buses/h, with peak loadings of 9,000 persons/h (pph), but one corridor has achieved loadings over 14,000 pph. Hensher [19] points out that BRT operating on an exclusive lane has a demonstrated capacity of 25,000 pph but that the Bogota BRT has achieved loadings approaching 50,000 pph! More usually, however, bus has to compete with other road users on shared road space, with congestion a major barrier to high capacities and service levels. On-road, bus priority treatments are a common response to speed up bus services and improve their reliability. The main benefits from trunk bus services are user benefits and congestion reduction, with carbon emission reduction more significant for this than for local services.

The mode share that bus can capture (particularly from car users) for trunk services depends on service characteristics such as [16]

- coverage – the proportion of population within about 400-m distance of the bus route, although distances from BRT (and rail) can typically be further, up to about 800 m, because of the higher service levels
- frequency – where 15-min peak frequencies would be considered only fair and cities like London plan around 10 min for a wide span of hours
- span of hours – the time period over which services operate, which need to be sufficient to complete return journeys on a given day. Eighteen or so hours on weekdays seems to be common practice in cities serious about achieving mode share
- speed of service (including wait times)
- reliability – the extent to which the service meets its timetable (which has high value)
- need for transfers – transfers are a significant penalty to users, as evidenced by studies on time valuation.

The effect of the first three of these factors is sometimes summarised in an aggregate short-hand measure called the 'service elasticity'. This measures the relative responsiveness of patronage levels to service kilometres provided. When Melbourne introduced its SmartBus services, bus patronage had been stable at around 75 million passengers p.a. over the preceding two decades but jumped by over 30% between 2006/07 and 2011/12 [16] following the development of a set of high-quality SmartBus trunk routes, with 15-min service frequencies over extended service hours, and provision of real-time passenger information. Service elasticities of two were found. Most of these routes operated circumferentially through Melbourne's middle suburbs, complemented by upgraded local bus services in outer suburban areas, to progressively achieve a (very modest) minimum frequency of an hourly service across all routes. This was a very strong result, exceeding expectations. Fare levels are usually a relatively insignificant influence on patronage, with elasticity values commonly estimated at around −0.3 (i.e. a 10% increase in fare levels leads to about a 3% reduction in patronage).

Ewing and Cervero [20] have shown that land-use characteristics are important for PT use, with key influencing factors for trunk services being:

- destination density – particularly the density of the central area, which is usually the single most important destination for trunk PT services. Densities of lower order nodes in a polycentric city are important for the contribution bus can make to decarbonising suburban mobility

- distance from the CBD – the closer the CBD, the higher the PT mode share
- diversity of land uses – greater mixed-use development encourages PT use. This includes development of mixed use transit corridors, as is the practice in Vancouver [21]
- design – the more direct the walk to a bus (or rail) stop/station, the higher the service use
- distance from transit – proximity supports use.

Some of these factors can be used to promote use of buses in suburban settings. For example, zoning for higher density mixed use suburban transit (bus) corridors can help build patronage growth, as in Vancouver, particularly if parking supply is tightened [21]. Similarly, land-use strategies to promote growth of knowledge-based suburban hubs are supportive of higher urban productivity levels, more equitable sharing of the flow-on benefits and increased PT use, a strategy Melbourne is currently pursuing.

6.2.1.2 Local services

Most people live most of their daily lives locally, not city-wide. Their well-being is therefore at least as much tied up in how well their local neighbourhood functions as it is in how the wider city functions [16]. The concept of a city being comprising a series of constituent '20-min cities' brings together ideas about integrated land-use transport for (inter alia) a low carbon footprint, with the mobility dimension spanning both local and trunk bus services and rail for the trunk component – see also Chapter 5. A 20-min city needs to enable most people to undertake most of the activities they need for a good life within 20-min travel time by foot, cycle or PT. It refers to a larger area than a neighbourhood, perhaps an area with a 5-km radius, and could encompass 200,000 to 300,000 people at modest densities. It is primarily an urban design principle advanced for sub-metropolitan district planning. Local bus services are an essential element in the 20-min city, providing access to local opportunities and trunk interchanges. In contrast to trunk bus services, the major benefits from suburban local bus services are user benefits, with social inclusion benefits being particularly high for bus in an Australian suburban setting [11,22]. Patronage drivers for local services tend to concern service coverage, frequency and operating spans rather than speed.

In Australian cities, outer urban areas and parts of the middle suburbs are where most focus is needed to support achievement of a series of 20-min cities. Supportive local bus service levels are likely to require a 30-min minimum service frequency on local services for about 15–18 h a day, with increased peak frequencies being justified if loadings suffice: Vancouver's local services typically operate at frequencies of between 30 and 60 min, depending on demand.

Demand responsive bus services are being trialled in low-density settings, as a means of providing timely service. Opportunities to integrate across route bus services, community transport/para-transit and school bus services, where these exist, in a demand-responsive 'total transport' model and perhaps utilising AEV technology, provide a promising new way forward.

6.2.2 Rail

Australian cities need and want new rail projects, especially in the suburbs. There is a need to create activity centres that are efficient at concentrating local jobs and services; and it is clear that railways create the land value uplift that brings investment and developers. A new model is required for extending rail into the suburbs developed historically for the car.

6.2.2.1 Principles and innovations for extension of rail into suburban areas

The role of value capture [23,24] is well established but now needs to be taken further to solve the problem by tapping private investment. There is no shortage of private capital, especially superannuation funds looking for good, safe investments. But how can you make money out of urban rail when governments have to subsidise them? The answer lies in land development around stations. If enough land can be found to enable redevelopment by the private sector to sell and lease buildings around stations, or to redevelop jointly with private owners or government for mutual benefit, then this can create the capital for developers to build the rail line, to own it and to operate it. This is the entrepreneur rail model (ERM) [25].

The result is not only to have a new rail line that opens up the opportunity for suburbs to gain a sustainable transport option but also transit-oriented development in which residents can reach many services locally. These strong local activity centres are the planners' goal, though hard to achieve in practice. Transit-land-use integration is hardly happening compared to problematic urban fringe development. Under the present greenfield, development model if transit is planned it is rare to get the associated land use until decades afterwards, or if outer suburban centres are developed at higher densities, the transit is lacking.

The secret to achieving this is a new governance instrument that integrates transit, land use and finance in the ERM. This reverses the traditional approach to transit planning by turning the process on its head. Rather than use estimates of transit patronage to seek funding for transit system development and then see what land use is possible with that system, the ERM looks at what land use is possible, seeks the funding required for this, and then estimates transit patronage.

Instead of government planning the rail system, the private sector suggests the most important opportunities for creating viable redevelopment projects and therefore how much private investment can be attracted to build a rail line. This is how tram and train lines were first built and how they are now built in Japan and Hong Kong. It uses private sector development skills – as cities are built by the private sector – aligned to a government metropolitan planning strategy.

Rail and land development add value to a city. Rail projects raise the value of land around stations through: travel time savings, agglomeration economies in activity centres, land development efficiencies, and environmental gains due to reduced automobile dependence. Studies in Perth indicate that a new rail line raised land values in station precincts by 42% in 5 years, above that of the general metropolitan-wide trend, and for commercial land even higher [23]. Thus significant redevelopment potential is unlocked by new rail lines. The traditional land value capture approach is to calculate such potential value uplift and see how much of it flows through into taxes; government can then use this money to finance the rail project if the risk is worthwhile. However, the private and government land value increase can be tapped directly by engaging the private sector to estimate the redevelopment potential and use this to raise their own finance. The various sources of land value-based funding are set out in Figure 6.3 and are briefly summarized below [25].

How to deliver such a rail project is suggested through three approaches:

- *Unsolicited bids* – a consortium of land developer, train builder, train operator and financier provide government with a bid that makes a rail project proceed to an evaluation phase.
- *Government calls for bids* – with a general consensus that a particular corridor could have the required land development potential and fulfil transport needs then government can request bids from consortia before evaluating the best one.
- *Government controls internally* – a new government agency (or revamped land agency) creates a rail project through land development as in Hong Kong. This could be a semi-private enterprise.

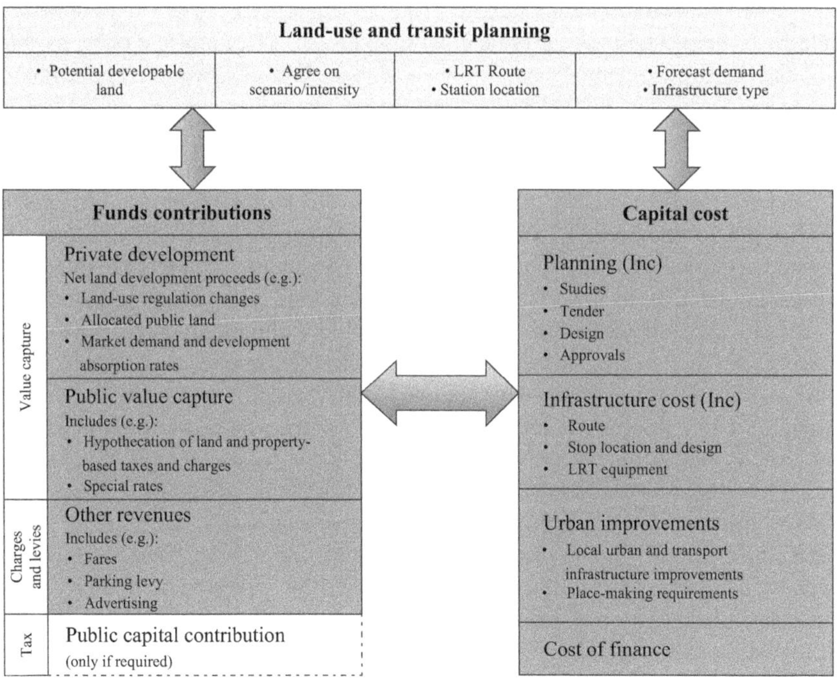

Figure 6.3 Land value-based funding framework (Source: Peter Newman – unpublished)

There are also three ways of funding and financing such projects:

- *Totally private capital.* Government's role is kept to in-kind activity to ensure land assembly and land acquisition, zoning and other transport planning integration is fully covered. This depends on sufficient land being available to generate the capital and enabling necessary mechanisms to generate private investment.
- *Substantial private and some public capital.* Substantial private capital can be supplemented by some government capital. Government's expected land value-based tax flow-on could be hypothecated to cover its contribution. This approach ensures that the rail project is generating all the capital required although some is from public sources (potentially across the three levels of government).
- *Some private and substantial public capital.* This seeks help from private sources through land development but primarily raises government capital through a mixture of sources such as parking levies, tolls on associated private traffic, developer contributions, an increase in registration fees or some other form of tax hypothecated to the rail project.

Governments can seek combinations of these approaches and funding/financing. The preferred option is to seek a process of Bids to Government based on 100% private capital as the goal of the ERM. If some small contribution from public capital is needed, then this would be the next level to be sought. A Federal Government role could be to help fund bids for critical nation-building urban projects (such as those prioritised by Infrastructure Australia on a cost–benefit basis). However, governments responsible for metropolitan planning must have a comprehensive long-term vision of development so cities know what infrastructure products to demand and where.

It is important that a government bidding process is controlled by Treasury as the central agency required to ensure that private sector funds are attracted to achieve public-good goals. Treasury would ensure consortia are evaluated by financial criteria, land development criteria and transit criteria in an integrated way. This cannot be done by a transit agency as its emphasis is on choosing the routes in detail first. This will not optimise land development opportunities, so the rail will not get built. A transit agency's only task in the ERM model is to ensure transit system compatibility with any new rail corridors. The delivery process requires the powers of a redevelopment agency to provide government's role in land acquisition, zoning and land assembly to unlock the latent value in land development around the stations.

The importance of enabling private sector investment is the critical step in unleashing the new governance instrument. Under the ERM, land development becomes the cornerstone of its funding, just as the integration of private land development entrepreneurial skill unlocks access to private capital. The power of this model is that the unlocking of private development in new activity centres could not occur unless it is completely integrated with the amenity-creating, value-creating power of a new urban rail service. A full discussion of the ERM and its implementation is given in [25].

6.2.3 Cycling

Cycling is a genuine solution for increasing low carbon urban mobility. It offers multiple benefits, especially the personal and societal health outcomes (Chapter 9) directly related to the frequency and intensity of cycling [26,27]. There are congestion benefits in which personal motor vehicle travel is replaced by bicycle, and bicycles contribute zero GHG emissions. Studies of the benefits (e.g. health) and costs (e.g. injury and pollution exposure) associated with cycling consistently conclude that the benefits significantly outweigh the costs [28].

Cycling is a popular recreational activity; the fourth most popular sport or recreational activity (after walking, swimming and aerobics) in Australia [29]. Men cycle more than women in Australia, both for sport and transport, but this is not necessarily the case in other countries with good cycling infrastructure [30]. A 2015 national survey [31] found that about 36% of Australians cycled at least once in the previous year, and about 17% in the past week, mainly for recreation rather than transport. Cycling mode share is low in Australia at about 1%–2% of trips, although this does vary by region (as indicated in Table 6.1).

There is also variation in cycling behaviour by socio-demographic characteristics, with cycling for transport more frequent among commuters with higher levels of education and higher social status in high income countries [32,33], but more frequent among poorer people in low-income countries. Perceived status associated with transport choices influences cycling participation, with aspirational communities often viewing bicycles as low status [34–36].

6.2.3.1 Principles and innovations required for further extension of cycling into suburban areas

The actions required to increase urban cycling for transport are reasonably clear, based on communities that have successfully increased cycling transport mode share [37–40]. These principles include political leadership, strategic policies and planning, infrastructure provision and behavioural programmes and encouragement. Political leadership is essential for setting the tone and direction of the public dialogue about the role of cycling in society. This is particularly important in shifting the focus of cycling from recreation to cycling as a legitimate form of everyday transport available to most people.

Strategic policy direction and frameworks are needed for government to align resources to achieve stated targets for increased cycling mode share. It requires a meaningful prioritization of travel modes such that a focus on walking is the first priority, then cycling followed by PT and then private motor vehicles. Australia has a national cycling plan [41], and most states and local government authorities have cycling plans, but these plans need to be resourced if they are to be effective. Policy directions in non-transport areas can also influence cycling. For example, integrated transport and land-use planning can create destinations of interest within distances able to be cycled and populations likely to use cycling infrastructure.

Bicycle infrastructure, such as dedicated bicycle paths and lanes, bicycle parking, and end of trip facilities (showers and lockers), is essential for attracting new riders and increasing safety. Changes to the physical environmental are relatively inexpensive compared to building roads, and even small urban design changes (such as ramps or signage) can facilitate much larger benefits across the population [42]. However, connectivity of a bicycle network is fundamental.

Behavioural programmes that encourage people to start and continue to cycle contribute to changing social norms about cycling participation. Programmes that increase cycling skills and confidence as well as programmes that encourage trials of cycling (e.g. cycling to work) are effective at increasing cycling activity [43,44]. Building a 'culture' of cycling, in which cycling is seen as acceptable and fashionable, also contributes to supporting cycling participation. The community bicycle sharing systems common in European cities (and found in Melbourne and Brisbane) also offer an avenue to encourage cycling [27].

In 2007, the City of Sydney released a 10-year strategic plan to increase cycling [45]. The plan had a focus on infrastructure and facilities, specifically the building of a coordinated trunk network of cycling routes to interconnect the City's villages, including 55 km of dedicated bike paths, plus a series of social initiatives required to encourage the behavioural change required to ensure that cycling became a regular and normal mode choice [46]. A specific target was to increase the number of bicycle trips in the City of Sydney, as a percentage of total trips, from less than 2% in 2006 to 10% by 2016.

In the Australian context, there needs to be much stronger national political leadership for cycling, with appropriate funding. State governments are variable in their support of cycling. At the local government level, there is perhaps the strongest recognition of the benefits of increasing cycling. Local government often builds the cycling infrastructure but needs state government support for strategic and regional connections. Better connectivity across local government areas also needs state government direction.

To facilitate more cycling in sprawling urban Australia where distances to work or shopping can be a barrier to cycling, better integration with PT is needed. Businesses and work places can also contribute by consistently providing better end of trip facilities. Creating a positive social norm that cycling is a personal and social 'good' creates a supportive environment for people considering cycling.

6.2.4 Walkability

Despite over two decades of research showing that people living in more walkable neighbourhoods are more likely to walk (and thus to reduce their risk of chronic disease), low-density housing developments poorly served by PT, social infrastructure, shops and services continue to be built on our urban fringes [47]. Unsuspecting residents seeking 'affordable' housing, buy into these fringe developments, often waiting years (if not decades) for promised infrastructure to arrive [48]. With limited PT, nowhere to walk to, and inadequate cycling infrastructure residents quickly realise their housing choice may not provide affordable 'living': lengthy and costly commutes to employment are the norm, with driving the only reasonable option [49]. Time-spent driving increases the risk of mortgage vulnerability [50], as well as weight gain [51], obesity [52] and poorer mental health outcomes [53] due to decreasing opportunities for physical and social activity. The way we design communities can therefore have a profound impact on physical and mental health (Chapters 8 and 9).

The development industry regularly argues that 'the market chooses' these fringe developments, and the industry delivers what 'the market wants'. However, for those with few choices, low-density development poorly served by shops, services and PT is what they are given: car dependency [54]. People living in communities designed for walking, walk more. Moreover, there is a large unmet latent demand for better designed communities: for example 62% of people living in low-walkable neighbourhoods in Perth would prefer neighbourhoods where they could walk to local shops and services [55].

In Australian cities, levels of walkability vary markedly. For example, Giles-Corti *et al.* [56] found that inner city neighbourhoods of Melbourne (established in the late 1800s with their regular grid pattern street network, higher dwelling densities and main street activity centres) were uniformly ranked 'highly walkable'. Although there were small pockets of highly walkable areas in middle and outer suburbs, particularly those around train stations, most were ranked as having low or moderate walkability, with very low levels of local walking.

The evidence is clear: walking is more likely in walkable communities with connected street networks, higher density housing, mixed land uses with a diversity of local destinations, footpaths and well-managed traffic. Yet there is a huge gap between what we know and what we deliver on-the-ground. There are major health benefits from getting more people walking, and the OECD argues that planning, transport and health ministers have a responsibility to put in place the 'legal, administrative and technical frameworks' to encourage walking [57].

6.2.4.1 Principles and innovations for enhancing walkability in suburban areas

In the twenty-first century, we should not be building neighbourhoods unless they are walkable and accessible, with the basic local services required for daily living and a multi-modal transport system linking residents to regional and metropolitan employment and services:

- Outer suburban development requires consistent urban design guidelines that mandate higher dwelling densities, the provision of local mixed use development and connected street network design; and that neighbouring housing developments be connected to each other.
- Key elements for successful accessible local communities include the early delivery of high-quality PT linking walkable communities with local destinations to metropolitan and regional employment and service hubs, and the provision of safe connected cycling infrastructure linking local communities to regional PT and activity hubs [58].

- Achieving high-quality PT and local shops and services requires higher dwelling density in urban fringe developments. Minimum dwelling densities of around 35 dwellings per hectare (dwellings/ha) is required to deliver better PT [59]. Density guidelines in fringe developments vary across Australian cities ranging from 20 dwellings/ha in Perth [48] to 15 dwellings/ha in Melbourne [60]. There is also a substantial gap between design recommendations and what is delivered on-the-ground [48], further undermining the delivery of walkable neighbourhoods. Higher levels of density facilitate the delivery of more efficient PT, and the viability of local shops and services. In addition, fixed infrastructure costs for utilities, footpaths and roads are virtually halved if density is doubled, and more people make use of the available infrastructure [61]. The delivery of higher densities requires the provision of more diversified housing stock in outer suburban developments. Barriers to building high-density buildings must therefore be overcome.
- Construction of outer suburban development needs to be sequenced to ensure that the housing is built at the same time as the infrastructure is delivered. A Victorian Auditor-General report identified an $18 billion underspend in infrastructure on Melbourne's growth areas: government cannot keep up with the pace of new development. Disincentives are required for developers who build out-of-sequence. For example, developers could be required to build basic levels of social infrastructure to support residents to live locally if they choose to develop out of sequence.

There is also an urgent need to retrofit infrastructure-rich established suburbs – particularly within and around transport nodes and activity centres – to create accessible higher density, mixed-use developments with affordable housing (Chapter 5). These developments could provide alternative housing options for young people buying their first home, and for older adults who wish to age-in-place. As well as creating walkable areas, these developments should aim to reduce motor vehicle dependency for residents.

These ideas are not new. As indicated in Chapter 5, in the 1970s Portland, Oregon adopted a successful city plan with integrated land-use and transport accepted and built upon by successive governments for over 30 years. Underpinned by a fine grain connected urban grid, inner-city Melbourne provides many examples of industrial areas successfully retrofitted to create highly sought after, higher density, mixed-use neighbourhoods well served by PT. A comprehensive evaluation of the Western Australian state government's *Liveable Neighbourhood* [55] demonstrated that subdivision design codes in greenfield sites can help produce better outcomes for residents. Yet the study highlighted the importance of evaluation, showing that overall the policy was only 47% implemented. Nevertheless, for each 10% increase in implementation, the odds of people walking locally increased by 50% [48].

If we build it, they will come. Walkable neighbourhoods are desirable, health-promoting and more sustainable, and there is a latent demand for their delivery.

6.3 Smart mobility

'Smart Mobility' encompasses 'Smart Users' making informed decisions about their travel choices, 'Smart Vehicles' that can drive themselves in an optimal way using renewable energy sources, and 'Smart Infrastructure' that can communicate with vehicles and users and inform them in real time about changes in travel conditions. It attempts to integrate all of these new technology systems and applications in ways that meet user needs, while providing safe, efficient and low or zero emissions mobility (Chapters 10–13). This section discusses some of the smart infrastructure, vehicle and user systems currently being implemented, with the aims of increasing safety and efficiency and decreasing emissions from transport sources.

6.3.1 Principles and innovations required for smart mobility in suburban areas

Intelligent transport systems originated in the early 1990s [62], employing new information and communication technologies to increase the capacity and safety of existing transport infrastructure while lowering environmental impact. Building more roads at every increasing cost is not effective in managing congestion. New technologies can provide information about travel conditions in real time. Traffic data previously only used for traffic control purposes can now be processed and disseminated in real time via variable message signs and dedicated Radio Data System Traffic Message Channels. Today most Global Positioning Systems enabled navigation aids have access to this real-time traffic data and can use it to determine the real-time best path to a destination [63]. More recently Bluetooth data collection systems have become widespread [64,65], and together with video detection can provide real-time data on travel times across selected routes within a metropolitan area. Real-time information for travellers can lead to more efficient use of existing road capacity and hence decreases in congestion, lower travel times and lower fuel consumption and emissions. Suburban travel directly benefits from this information through informed decisions on when to travel and the optimal route to take.

Just as road infrastructure has seen new technologies allow for greater efficiencies and the lowering of emissions, vehicles are also undergoing disruptive changes towards similar outcomes. In the passenger vehicle market these changes are primarily driven by safety outcomes rather than environmental gains. The advent of anti-lock braking systems and airbags has seen a significant drop in serious injury and fatal road crashes [66]. Manufacturers are looking at new technologies that can prevent or reduce the severity of crashes. These include autonomous vehicle (AV) technologies that automatically avoid frontal collisions and automated freeway driving and lane changing. Although there is debate as to whether or when fully autonomous or driverless vehicles will be generally available [67,68], there is no doubt that they will arrive soon for specific applications. The main issues concern public acceptance and legislative changes to allow AVs on the road. South Australia is the first Australian state to legislate for AVs to drive on public roads under trial conditions [69]. These 'Smart Vehicles' have the ability to communicate with other 'Smart Vehicles' and the 'Smart Infrastructure'. This allows for enhanced safety and for real-time information to be used to determine optimal routes to destinations. These efficiency gains made through the use of 'Smart vehicles' on 'Smart Infrastructure' should see emissions from transport decrease. However in order to have a significant impact on suburban travel emissions, the vehicles will need to be fuelled using renewable energy. Although vehicle manufacturers can proclaim the environmental credentials of electric vehicles to buyers, they have little influence on the source of the energy used to recharge their batteries. Since the national electricity grid in Australia is still dominated by coal-based power stations, the most rapid route to decarbonisation of private cars will be via the uptake of vehicles that can be recharged at home with solar PV and battery storage, and the suburbs are where these prospects are perhaps the greatest [14].

Another application for 'Smart Vehicles' is in suburban demand-responsive PT applications. Users could request an autonomous electric shuttle to provide door-to-door service to local destinations such as shops, medical centres and schools, or to trunk transit services [64]. These vehicles would reduce personal travel. Since these vehicle fleets are likely to be owned and operated by government or commercial organisations it is likely that renewable energy would be used for recharging, significantly reducing emissions. A Western Australian trial of this technology is commencing in 2016 [70].

The final piece of the puzzle are 'Smart Users'. No matter how 'Smart' the road infrastructure or the vehicles are, if they are not being used appropriately, then any potential safety, efficiency or environmental benefits will not be realised. Providing users with relevant and timely travel information in order to make informed 'Smart' decisions about their travel choices is a difficult challenge. Suburban travel is a good example of this, with the journey to work becoming more complex and changing over time [71]. Generally, suburban mobility needs are becoming more complex than just finding a fast and reliable way to the CBD. Suburban users are requiring access to services in their local areas, cross town and within the CBD, at all times of the day and for multiple purposes. Technologies and Apps, coupled with smart pricing systems that allow users to become informed of their travel options and their costs and consequences, could assist 'Smart Users' to utilise 'Smart Infrastructure' and 'Smart Vehicles' appropriately and hence provide 'Smart Mobility' options for travellers, thereby improving safety, efficiency and environmental outcomes [72]. Active transport coupled with PT linking the suburbs and the CBD – and augmented where necessary by personal transport providing mobility to less accessible locations requiring cross suburb travel – can deliver a more efficient, healthier and more environmentally friendly framework for suburban travel.

6.4 Conclusions

Achieving the vision outlined in this chapter requires bi-partisan political leadership to deliver city plans, rather than government plans. A 'planning deficit' has been evident for most of the past 50 years in relation to long-term planning of Australia's major cities – an increasing catalogue of strategic plans that have either been abandoned with a change of government or at best inadequately implemented. A major casualty of these failures is the mobility available within cities, and especially the outer suburbs [73].

There is a continuing need to halt automobile-dependent urban sprawl, a process that is compounding now well-known negative externalities such as loss of agricultural land in addition to those catalogued in this chapter. A combination of settlement strategies is required that can harness centripetal as well as centrifugal forces of population geography: disperse population growth to regional cities as well as redirect population inwards to the established suburbs via infill and urban intensification. In both cases, sound strategic planning by government that can activate market investment is critical. The failed decentralisation policies of an earlier era need to be replaced by those that recognise the agglomeration economy benefits that accrue to large cities and to ensure regional cities can become part of a functional mega-metropolitan region – a 'system of cities' – via fast rail. For example, a 350-km/h high speed service between Melbourne and provincial cities such as Geelong, Ballarat, Bendigo and Warragul would transform these centres into the equivalent of present day middle ring suburbs of Melbourne where 30-min commutes to work are typically the norm – see Figure 6.4 and [74].

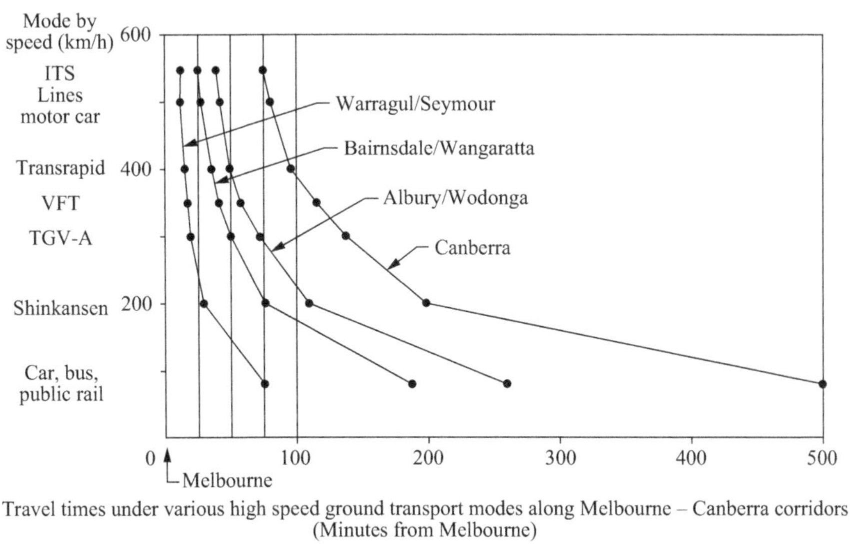

Figure 6.4 Space shrinking impacts of high speed ground transportation (Source: Newton et al. [74]) [Peter Newton holds copyright to this figure]

Urban regeneration in the established suburbs of cities is also required to deliver more sustainable outcomes for future populations. Here significant transformation in transport and land use is required. Intensification of housing has been a major focus in recent years via consolidation and infill policies, which still have significant scope for innovation [75]. Urban transport planning has witnessed much greater inertia. The set of innovations and guiding principles capable of injecting greater low carbon mobility, with all the attendant co-benefits, into the established middle and outer suburbs of Australia's cities (and those cities internationally that share low density car dependent urban forms, see [76]) outlined in this chapter represent a way forward for the next generation of strategic metropolitan planning.

References

[1] RAND Corporation, Ecola, L, and Wachs, M (2012). *Exploring the Relationship between Travel Demand and Economic Growth*, Prepared for US Federal Highway Administration, December 2012. [online] Available at: http://www.fhwa.dot.gov/policy/otps/pubs/vmt_gdp/.

[2] Eurostat (2016). *Passenger transport statistics*. [online] Available at: http://ec.europa.eu/eurostat/statistics-explained/index.php/Passenger_transport_statistics#Relative_growth_of_passenger_transport_and_the_economy [Accessed 4 April 2016].

[3] Bureau of Infrastructure, Transport and Regional Economics (2015). *Traffic and Congestion Costs for Australian Capital Cities, Information Sheet 74*, BITRE, Canberra. [online] Available at: www.bitre.gov.au.

[4] Department of the Environment (2015). *National Inventory Report 2013 Volume 1*, Commonwealth of Australia.

[5] Newton, P W (2012). Liveable and sustainable? Socio-technical challenges for 21st century cities. *Journal of Urban Technology*, 19(1), 81–102.

[6] Thomson, G, Newton, P and Newman, P (2016). Urban regeneration and urban fabrics in Australian cities. *Journal of Urban Regeneration and Renewal*, 10(2), 1–22.

[7] Szafraniec, J, Rawnsley, T and Maloney, J (2011). *Public Transport Access Levels*, SGS Planning, Melbourne.

[8] Victorian Department of Economic Development, Jobs, Transport and Resources (2016). *Melbourne, Victorian Department of Economic Development, Jobs, Transport and Resources*. Available at: http://economicdevelopment.vic.gov.au/transport/research-and-data/vista.

[9] BITRE (2016). 'Lengthy Commutes in Australia', Research Report 144, Bureau of Infrastructure, Transport and Regional Economics, Canberra.

[10] Gipps, P G, Brotchie, J, Hensher, D, Newton, P and O'Connor, K (1997). *The Journey to Work, Employment, and the Changing Structure of Australian Cities*, Research Monograph No. 3, Australian Housing and Urban Research Institute, Melbourne.

[11] Stanley, J K, Hensher, D, Stanley, J R and Vella-Brodrick, D (2011). Mobility, social exclusion and well-being: Exploring the links. *Transportation Research A*, 45(8), 789–801.

[12] Rawnsley, T and Spiller, M (2012). 'Housing and Urban form: A New Productivity Agenda', in *The Unintended City*, R Tomlinson (ed), CSIRO Publishing: Melbourne.

[13] Newton, P W (2000). 'Urban Form and Environmental Performance', in *Sustainable Urban Form*, K Williams, E Burton and M Jenks (eds.), Achieving, E&FN Spon: London.

[14] Newton, P and Newman, P (2013). The geography of solar photovoltaics (PV) and a new low carbon urban transition theory. *Sustainability*, 5, 2537–2556.

[15] Taylor, M A P and Thompson, S M (2015). 'RP – 2015 Carbon Reductions and Co-benefits: Literature and Practice Review of Australian Policies Relating Urban Planning and Public Health'. Final Report – Part II: An Analysis of Current Levels of Active Transport Usage in Australia – Towards a Measure of Baseline Activity, Research Report, CRC for Low Carbon Living, Sydney.

[16] Stanley, J K, Stanley, J R and Davis, S (2015). *Connecting Neighbourhoods: The 20 Minute City*, Bus and Coach Industry Policy Paper 4, Canberra: Bus Industry Confederation.

[17] Stanley, J, Hensher, D A and Loader, C (2011). Road transport and climate change: Stepping off the greenhouse gas. *Transportation Research Part A: Policy and Practice*, 45(10), 1020–1030.

[18] Stanley, J and Hensher, D (2011). 'Economic Modelling, in New Perspectives and Methods' in *Transport and Social Exclusion Research*, G Currie (ed.), Emerald: Bingley.

[19] Hensher, D (2016). Why is light rail starting to dominate bus rapid transit yet again? *Transport Reviews*, 36(3), 289–292.

[20] Ewing, R and Cervero, R (2010). Travel and the built environment. *Journal of the American Planning Association*, 76(3), 265–294.

[21] Chatman, D (2013). Does TOD need the T? *Journal of the American Planning Association*, 79(1), 17–31.

[22] Stanley, J K, Stanley, J R and Hensher, D (2012). Mobility, social capital and sense of community: What value? *Urban Studies*, 49(16), 3595–3609.

[23] McIntosh, J, Trubka, R and Newman, P (2015). Tax increment financing framework for integrated transit and urban renewal projects in car dependent cities. *Urban Policy and Research*, 33(1), 37–60. 3 December, DOI: 10.1080/08111146.2014.968246.

[24] Newman, P and Kenworthy, J (2015). *The End of Automobile Dependence: How Cities Are Moving Beyond Car-based Planning*, Island Press: Washington, DC.

[25] Newman, P, Jones, E, Green, J and Davies-Slate, S (2016). *The Entrepreneur Rail Model. [online] Curtin University Sustainability Policy Institute, Perth.* Available at: http://www.curtin.edu.au/research/cusp/publications.cfm.

[26] Garrard, J, Rissel, C and Bauman, A (2012). 'Health Benefits of Cycling', in *City Cycling*, J Pucher and R Buehler (eds.), MIT Press: Cambridge, Massachusetts. p. 31–55.
[27] Philp, M, Taylor, M A P and Thompson, S (2015). *RP-2015 Carbon Reductions and Co-benefits: Final Report – Part I, Literature and Practice Review of Australian Policies Relating Urban Planning and Public Health*. Research Report. Cooperative Research Centre for Low Carbon Living, Sydney. [online] Available at: www.lowcarbonlivingcrc.com.au.
[28] Rissel C (2015). 'Health Benefits of Cycling', in *Cycling Futures*, J Bonham and M Johnson (eds.), University of Adelaide Press: Adelaide. p. 43–62.
[29] Australian Bureau of Statistics. *Participation in Sport and Physical Recreation*, Australia, 2011–2012 (2013). [online] Available at: http://www.abs.gov.au/ausstats/abs@.nsf/mf/4177.0. [Accessed 14 February 2016].
[30] Garrard, J (2003). Healthy revolutions: Promoting cycling among women. *Health Promotion Journal of Australia*, 14(3), 213–215.
[31] Austroads, National Cycling Participation Survey (2015). *National Results*, Sydney: Austroads.
[32] Rissel, C, Greenaway, M, Wen, L M and Bauman, A (2014). Active travel to work in New South Wales 2005–2010 in NSW, individual characteristics of people walking and cycling to work and association with body mass index. *Australian and New Zealand Journal of Public Health*, 38(1), 25–29.
[33] Adams, E J, Goodman, A, Sahlqvist, S, Bull, F C, Ogilvie, D, iConnect Consortium (2013). Correlates of walking and cycling for transport and recreation: Factor structure, reliability and behavioural associations of the perceptions of the environment in the neighbourhood scale (PENS). *The International Journal of Behavioral Nutrition and Physical Activity*, 10, 87.
[34] Green, J, Steinbach, R and Datta, J (2012). The travelling citizen: Emergent discourses of moral mobility in a study of cycling in London. *Sociology*, 24(2), 189–218.
[35] Haixiao P (2012). 'Evolution of Urban Bicycle Transport in China', in *Cycling and Sustainability*, J Parkin (ed.), Emerald: Bingley. p. 161–180.
[36] Steinbach, R, Green, J, Datta, J and Edwards, P (2011). Cycling and the city: A case study of how gendered, ethnic and class identities can shape healthy transport choices. *Social Science Medicine*, 72(7), 1123–1130.
[37] Pucher, J and Buehler, R (2008). Making cycling irresistible: Lessons from the Netherlands, Denmark and Germany. *Transport Reviews*, 28(4), 495–528.
[38] Pucher, J, Dill, J and Handy, S (2010). Infrastructure, programs, and policies to increase bicycling: An international review. *Preventive Medicine*, 50(Suppl 1), S106–S125.
[39] Goodman, A, Panter, J, Sharp, S J, and Ogilvie, D (2013). Effectiveness and equity impacts of town-wide cycling initiatives in England: A longitudinal, controlled natural experimental study. *Social Science & Medicine*, 97, 228–237.
[40] Pucher, J and Buehler, R (2012). 'Promoting Cycling for Daily Travel: Conclusions and Lessons from Across the Globe', in *City Cycling*, J Pucher and R Buehler (eds.), The MIT Press: Cambridge, MA. p. 347–363.

[41] Australian Bicycle Council (2010). *The Australian National Cycling Strategy 2011–2016*. Austroads, Sydney.

[42] Stewart, G, Anokye, N K and Pokhrel, S (2015). What interventions increase commuter cycling? A systematic review. *BMJ Open*, 5(8), e007945.

[43] Rissel, C and Watkins, G (2013). *AustCycle – Evaluation of the Healthy Communities Initiative 2010–13*, Cycling Australia: Alexandria.

[44] Rose, G and Marfut, H (2007). Travel behaviour change impacts of a major ride to work day event. *Transportation Research Part A*, 41, 351–364.

[45] City of Sydney (2007). *Cycle Strategy and Action Plan 2007–2017*, City of Sydney: Sydney.

[46] City of Sydney (2010). *Enabling Cycling Strategy*, City of Sydney: Sydney.

[47] Allender, S, Cavill, N, Parker, M and Foster, C (2009). Tell us something we don't already know or do! – The response of planning and transport professionals to public health guidance on the built environment and physical activity. *Journal of Public Health Policy*, 30(1), 102–116.

[48] Hooper, P, Giles-Corti, B and Knuiman, M (2014). Evaluating the implementation and active living impacts of a state government planning policy designed to create walkable neighborhoods in Perth, Western Australia. *American Journal of Health Promotion*, 28(3 Suppl), S5–S18.

[49] Falconer, R, Newman, P and Giles-Corti, B (2010). Is practice aligned with the principles? Implementing New Urbanism in Perth, Western Australia. *Transport Policy*, 17(5), 287–294.

[50] Dodson, J and Sipe, N (2008). Shocking the suburbs: Urban location, home ownership and oil vulnerability in the Australian city. *Housing Studies*, 23(3), 377–401.

[51] Sugiyama, T, Merom, D, van der Ploeg, H P, Corpuz, G, Bauman, A and Owen, N (2012). Prolonged sitting in cars: Prevalence, socio-demographic variations, and trends. *Preventive Medicine*, 55(4), 315–318.

[52] Frank, L D, Andresen, M A and Schmid, T L (2004). Obesity relationships with community design, physical activity, and time spent in cars. *American Journal of Preventive Medicine*, 27(2), 87–96.

[53] Putnam, R (2000). *Bowling Alone: The Collapse and Revival of American Community*, Simon & Schuster: New York.

[54] Giles-Corti, B, Knuiman, M, Timperio, A, *et al*. (2008). Evaluation of the implementation of a state government community design policy aimed at increasing local walking: Design issues and baseline results from RESIDE, Perth Western Australia. *Preventive Medicine*, 46(1), 46–54.

[55] Bull, F, Hooper, P, Foster, S, Giles-Corti, B (2015). *Living Liveable: The Impact of the Liveable Neighbourhoods Policy on the Health and Wellbeing of Perth Residents*, Perth, Western Australia: The University of Western Australia.

[56] Giles-Corti, B, Mavoa, S, Eagleson, S, Davern, M, Roberts, R and Badland, H (2014). *How Walkable is Melbourne? The Development of a Transport Walkability Index for Metropolitan Melbourne*. Melbourne: University of Melbourne.

[57] OECD (2011). *International Transport Forum 2011, Pedestrian Survey*, Urban Space and Health: Summary Document.
[58] Giles-Corti, B, Eagleson, S and Lowe, M (2014). *Securing Australia's Future – Sustainable Urban Mobility. The Public Health Impacts of Transportation Decisions*, Melbourne: Consultancy Services.
[59] Giles-Corti, B, Hooper, P, Foster, S, Koohsari, M and Francis, J (2014). *Low Density Development: Impacts on Physical Activity and Associated Health Outcomes*, Melbourne: Heart Foundation (Victorian Division).
[60] State of Victoria (2014). *Plan Melbourne – Metropolitan Planning Strategy, in Department of Transport Planning, and Local Infrastructure* (eds.), Victorian Government: Treasury Place, Melbourne.
[61] Gunn, L D, Lee, Y, Geelhoed, E, Shiell, A and Giles-Corti, B (2014). The cost-effectiveness of installing sidewalks to increase levels of transport-walking and health. *Preventive Medicine*, 67, 322–329.
[62] ITS America (2016). *Intelligent Transportation Society of America*. [online] Available at: http://www.itsa.org/. [Accessed 12 April 2016].
[63] TomTom. 2016. *TomTom Traffic*. [online] Available at: https://www.tomtom.com/en_gb/licensing/products/traffic/real-time-traffic/hd-traffic/ [Accessed 6 April 2016].
[64] South Australian Department of Planning Transport and Infrastructure (2015). *Journey decisions made easier as travel time signs switched on*. [online] Available at: http://dpti.sa.gov.au/news/?a=173180/ [Accessed 12 April 2016].
[65] Queensland Government Department of Transport and Main Roads (2014). *Dealing with traffic congestion in south east Queensland*. [online] Available at: http://www.tmr.qld.gov.au/Community-and-environment/Planning-for-the-future/Congestion.aspx [Accessed 6 April 2016].
[66] National Roads and Motorists Association (2016). *The history of car safety*. [online] Available at: http://www.mynrma.com.au/motoring-services/safety-advice/safer-driving/history.htm [Accessed 5 April 2016].
[67] Driverless car market watch. 2016. *Autonomous car forecasts*. [online] Available at: http://www.driverless-future.com/?page_id=384 [Accessed 6 April 2016].
[68] Newman, P (2015). *Going down the same old road: Driverless cars aren't a fix for our transport woes, The Conversation*, 1 December. Available at: https://theconversation.com/going-down-the-same-old-road-driverless-cars-arent-a-fix-for-our-transport-woes-50912 [Accessed 3 December 2016].
[69] Government of South Australia (2016). *SA becomes first Australian jurisdiction to allow on-road driverless car trials*. [online] Available at: http://www.premier.sa.gov.au/index.php/stephen-mullighan-news-releases/337-sa-becomes-first-australian-jurisdiction-to-allow-on-road-driverless-car-trials [Accessed 4 April 2016].
[70] NAVYA (2016). *Royal Automobile Club: NAVYA ARMA is set to arrive in Western Australia*. [online] Available at: http://navya.tech/2016/02/royal-automobile-club-navya-arma-australia/?lang=en [Accessed 7 April 2016].

[71] Xu, B and Milthorpe, F (2010). *Analysis of Journey to Work Travel Patterns in Sydney*, Papers of the 2010 Australian Transport Research Forum, Canberra. [online] Available at: www.atrf.info/papers.
[72] CISCO (2016). *Smart Transportation Pricing (STP): An Innovative Connected and Sustainable Mobility Pilot. Cisco Internet Business Solutions Group (IBSG) and the Seoul Metropolitan Government*. [online] Available at: http://www.cisco.com/c/dam/en_us/about/ac79/docs/cud/STP_Fact_Sheet_051309_SM_TKv2_FINAL2.pdf.
[73] ACOLA (2015). *Delivering Sustainable Urban Mobility*, Australian Council of Learned Academies, Canberra.
[74] Newton, P W, Brotchie, J F and Gipps, P G (1997) *Cities in Transition: Changing Economic and Technological Processes and Australia's Settlement System, State of the Environment Technical Paper Series* (Human Settlements), Environment Australia, Canberra.
[75] Newton, P (2017). 'Framing New Retrofit Models for Regenerating Australia's Fast Growing Cities', in *Retrofitting Cities for Tomorrow's World*, M Eames *et al*. (eds.), Wiley-Blackwell: London.
[76] Philp, M and Taylor, M A P (2017). A research agenda for low carbon mobility: Issues for 'new world' cities. *International Journal of Sustainable Transportation*, 11(1), 49–58. [online] Available at: http://dx.doi.org/10.1080/15568318.2015.1106261

Glossary

AEV autonomous electric vehicle
AV autonomous vehicle
BRT bus rapid transit
CBD central business district
ERM entrepreneur rail model
EV electric vehicle
GHG greenhouse gases
ICT information and communications technologies
ITS intelligent transport system
LRT light rail transport
PTAL public transport access level
PV photo-voltaic (solar)
TOD transport oriented development
VFT very fast train
VKT vehicle kilometres travelled
VMS variable message signs

Further Reading

[1] Bento, A M, Cropper, M L, Mobarak, A M and Vinha, K (2005). The effects of urban spatial structure on travel demand in the United States. *The Review of Economics and Statistics*, 87(3), 466–478.

[2] Fagnant, D J and Kockelman, K (2015). Preparing a nation for autonomous vehicles: Opportunities, barriers and policy recommendations. *Transportation Research Part A: Policy and Practice*, 77(July), 167–181.

[3] Giles-Corti, B, Foster, S, Koohsari, J and Hooper, P (2015). 'The Influence of Urban Design and Planning on Physical Activity', in *The Routledge Handbook of Planning for Health and Wellbeing*, H. Barton, S. Thompson, S. Burgess and M. Grant (eds.), Routledge: London.

[4] Givoni, M and Banister, D (eds.) (2013). *Moving Towards Low Carbon Mobility*. Edward Elgar Publishing: Cheltenham.

[5] Loader, C and Stanley, J (2009). Growing bus patronage and addressing transport disadvantage – The Melbourne experience. *Transport Policy*, 16, 106–114.

[6] Nakamura, K and Hayashi, Y (2013). Strategies and instruments for low-carbon urban transport: An international review on trends and effects. *Transport Policy*, 29, 264–274.

[7] Public Health England (2016). *Working Together to Promote Active Travel: A Briefing for Local Authorities*. Public Health England: London (PHE publications gateway number: 2016070).

[8] Rajé, F and Saffrey, A (2016). *The Value of Cycling*. Department for Transport and University of Birmingham: London.

[9] Sallis J F, *et al.* (2016). Physical activity in relation to urban environments in 14 cities worldwide: A cross-sectional study. *The Lancet*, 387(10034): 2207–2217.

[10] Schoettle, B and Sivak, M (2015). *Motorists' Preferences for Different Levels of Vehicle Automation*, The University of Michigan, Transportation Research Institute: Ann Arbor, MI, Report No. UMTRI-2015-22, July 2015.

Chapter 7

The 'disruption' we really need: public transport for the urban millennium

John Stone[1] and Yvonne Kirk[1]

Abstract

Mass transit is the only form of motorised transport that can move large numbers of people to the same destination at the same time without either alienating the space needed for social and economic interaction or allowing cities to encroach further upon vital natural environments or agricultural land. Despite the rapid development and deployment of 'disruptive' technologies in urban transport, mass transit will still have a vital role to play in the transport systems of the world's great urban regions in coming decades. This is largely because cities of the future will face increasing competition over space. No urban region has been entirely successful in creating mass transit networks that offer speed and convenience approaching that of the private car, but some have done much better than others. This chapter presents case studies of relative success in the creation of space-efficient transport systems in the urban regions of Vienna, Zurich and Vancouver. It gives an overview of transport system performance including operating costs, infrastructure investments and mode share, together with contextual demographic data. In each case, this is accompanied by a short outline of the political and institutional processes that have enabled these outcomes to be achieved. Common features in all the three cities include consistent and skilful engagement in local political processes by transit advocates and planners and coherent use of transport planning practices that give primacy to meet the needs of mass transit users at the lowest possible cost.

Keywords

Transit; public transport; urban transport

[1]University of Melbourne, Melbourne, Australia

7.1 Introduction

Transport systems, which now typically alienate 30% or more of the land area of the developed cities, will face increasing competition over space for essential social and economic interaction and pressures to restrict further encroachment into vital natural environments and agricultural land. The demands of a carbon-constrained world will impose very strict transport energy budgets; and, the health costs of sedentary life-styles will force a re-assessment of the value of active travel.

So-called 'disruptive' information and communications technology (ICT) innovations in urban transport appear to be advancing in development and deployment at an increasingly rapid rate. However, traditional approaches to transport engineering and urban planning, which encourage greater use of mass transit and active travel, will have a vital role to play in the transport systems of the world's great urban regions in the coming decades. This is because such approaches will offer cities of the future better solutions to a series of complex and inter-related challenges forced upon them by population growth and climate change.

This chapter argues that the real 'disruption' needed in transport planning and engineering is to overcome the policy and political obstacles that prevent most cities from vastly improving networks for rapid transit and active travel, rather than further entrenching the inefficiencies of road-based transport through an uncritical adoption of new technologies.

No urban region has been entirely successful in creating the conditions in which public and active transport systems offer access to urban life with speed and convenience approaching that of the private car, but some have done much better than others.

This chapter presents case studies of relative success in the creation of space-efficient transport systems in the urban regions of Vienna, Zurich and Vancouver. It gives an overview of transport system performance, together with contextual demographic data. In each case, this is accompanied by a short outline of the political and institutional processes that have enabled these outcomes to be achieved. Common features in all the three cities include consistent and skilful engagement in local political processes by public and active transport advocates and planners and coherent use of transport planning practices that give primacy to meeting the needs of public transport users at the lowest possible cost. The success of public transport systems in large cities of East Asia – such as Singapore, Seoul, Hong Kong and Tokyo – and European mega-cities including Paris and London also offer important lessons, but we have chosen these case studies because, independent of issues of scale and density, they show how civic leaders can tackle the vital mobility and access challenges facing other cities of comparable size in the Global North.

7.2 Challenges of the urban millennium: the role for 'traditional' public transport

The challenges for mobility and access in the cities of the twenty-first century are well documented (Chapter 2): how in the face of population growth, climate change and social instability can we improve access to employment, education and social and cultural activity? Any solutions to these questions will need to recognise the complex interactions between land-use decisions and investments in transport systems. Although it is outside the scope of this chapter, there will clearly need to be greater success in planning and implementing a more equitable spatial distribution of urban jobs and services than we have managed in the era of neo-liberalism. The benefits of future transport system investments will need to be measured for their ability to reduce the need for alienation of increasingly contested urban space and the occurrence of illnesses caused by sedentary lifestyles, as well as the more traditional measures of travel time.

From this perspective, it is important that the excitement surrounding 'disruptive' technologies of autonomous vehicles and other ICT innovations in transport do not blind us to the reality that cities of the future will need effective and efficient public and active transport systems.

Expected benefits of autonomous vehicles are in travel time and convenience, but how will the wide adoption of autonomous vehicle technologies measure up on the key indicators of transport system performance: the efficient use of constrained urban space and the promotion of physical activity to meet vital health objectives? Many of the anticipated benefits, particularly those related to efficient use of road space and the potential to reduce congestion, will only come with full deployment of autonomous vehicles. Some analysts estimate that the technological, regulatory, financial and social conditions for full deployment will not be reached until 2050 or 2060 [1].

Widespread deployment of autonomous vehicle technology (Chapter 11) will be necessary before congestion relief is possible because the principal gains will only be made when vehicle separation distances and acceleration at intersections are markedly improved. It is also unclear to what extent circulation of empty autonomous vehicles in space-constrained inner-city locations will add to congestion. Further, the political dynamics, particularly in increasingly gentrified city cores, will tend to favour reclamation of road space for active travel and civic amenity purposes. What restrictions on pedestrian movement in a busy shopping street might Google demand for the operation of its new technology?

Optimistically, autonomous vehicles may eventually allow traffic systems to operate with less road space per trip, but population growth and induced demand mean that we will gain at best a short reprieve before transport again demands the alienation of more of our urban space. Autonomous vehicles, through removal of driver stress and by allowing 'drivers' to use travel time for other purposes, may encourage longer average travel distances with consequent increasing pressure on suburban expansion and destruction of land vital to agriculture and bio-diversity.

The RAND Corporation, a technologically optimistic think-tank, is guarded in its predictions of the impact of autonomous vehicles on urban traffic congestion. They recognise that: a decreased cost of driving may lead to an increase in vehicle miles travelled (even though) the technology can enable increased throughput .... Thus, the overall effect of autonomous vehicle technology on congestion is uncertain [2].

Even under relatively optimistic assumptions contained in recent modelling done for the Organisation for Economic Co-operation and Development, the International Transport Forum suggests that autonomous vehicle technology in the form of self-driving taxis will only succeed in large cities if it is implemented as a complement to mass transit services rather than as a substitute [3]. This is principally because rapid mass transit is the only form of motorised transport that can move large numbers of people over long distances to the same destination at the same time without alienating the space needed for the social and economic interactions, which are the fundamental purpose of cities.

So, benefits from autonomous vehicles will only accrue if urban rapid transit is maintained and extended.

Autonomous vehicles and other ICT innovations (Chapters 11 and 13) may well revolutionise the way vehicles are owned and operated, and 'disrupt' current models for the distribution of profits from the mobility economy. It remains to be seen whether this disruption will be an economic benefit to ordinary citizens or to public institutions responsible for investment in cities, or whether it will result in further concentration of capital [4].

In any case, there is a great danger that unreflective enthusiasm for new forms of motorised transport will entrench rather than disrupt many of the inefficient and costly features of transport systems already struggling to meet demands of urban population growth and space constraints. Without sufficient thought, we may also be blinded to the health imperatives that should drive transport planners and engineers to find new ways to build physical activity into routine daily travel patterns.

The true 'disruption', for most of the cities, will be to see autonomous vehicles as a new means to better provide 'first-and-last-mile' connections in suburban locations as part of urban transport systems in which the bulk of long-distance movements are by rapid transit and active modes dominate in space-constrained higher density nodes and corridors.

7.3 Can priority for public and active transport succeed?

Very few urban regions in the developed world have successfully established a balance of mode shares in which motor vehicles play a residual role. However, some have made significant moves in the right direction. This 'real world' evidence of the potential for moving away from automobile-dependence (Chapter 4), supported by investigation of politics, policies, and practices of transport and urban planning in these cities, allowed us to identify the elements of a theory of planning for effective and efficient public and active transport systems. Such approaches need to be kept firmly in mind when shaping policies for the implementation of new transport technologies.

While acknowledging the importance of good planning policies and practices to encourage growth in active transport modes, this chapter will now focus on public transport planning.

7.3.1 Principles of success

In cities in all parts of the world, the automobile and the provision of roads and parking spaces have dominated transport planning since the 1960s. Billions are spent annually by governments in road construction that aims to enable increasing volumes of traffic to go anywhere at any time, but which almost inevitably creates new locations for congestion. For example, in Melbourne, where more lane-kilometres of freeway have been built since the 1970s than in any other Australian city [5], congestion levels for long periods during weekends are now reaching levels higher than in afternoon weekday rush hours [6].

The endless repetition of these consistently unsuccessful attempts to build a way out of congestion is justified by the assertion that there is no alternative: many argue that low population density and dispersed trip patterns in the suburbs where most people live are insurmountable obstacles to competitive public transport at an affordable cost. Fortunately, however, this is not the case. In several important works, the late Mees [7] demonstrated the fragility of many arguments for the 'impossibility' of suburban public transport (and 'A Very Public Solution'). For a summary of Mees' analysis of the use and abuse of population density data as evidence to support the 'impossibility' argument, see the chapters by Burke and Dodson and Petersen in *The Public City: Essays in honour of Paul Mees* [8].

The theoretical insight that allows us to see the possibilities for creating effective and efficient public transport systems even in dispersed suburbia is the recognition of the existence of the 'network effect'.

In *A Very Public Solution*, Mees [9] constructed the simple, theoretical town of 'Squareville' (see Figure 7.1), in which demand for travel is evenly spread between all possible origin–destination pairs. He used this to compare the impacts on potential ridership of different ways to distribute an increase in public transport service supply from a base of a series of disconnected lines chosen to represent typical conditions in a dispersed suburb. With this pattern of service supply, the choice of public transport by one-third of all those for whom such an option is feasible would produce a market share of only 3%.

Improvements to public transport are usually made by investing in higher frequency services along existing lines. This response is useful and can be welcomed for people whose destinations lie along their local line. However, in the Squareville example, if we assume, as do traditional transport economists, that the elasticity of demand is 0.5, then the new modes share across the town will grow to only 4.5%. The cost of the new service will not be covered by increased revenue from new passengers. Such patterns of service supply, and the associated law of diminishing returns, are built into many demand-based transport models, guaranteeing an unfavourable benefit cost analysis. If, however, that same level of investment in higher frequencies is used to supply services that cross the existing routes in a regular pattern, then, in 'Squareville' at least, any destination can be reached from any other point in the town with a trip requiring at most a single transfer. Compared with the original collection of disconnected lines, the new network provides a huge increase in the potential public transport ridership. Even if

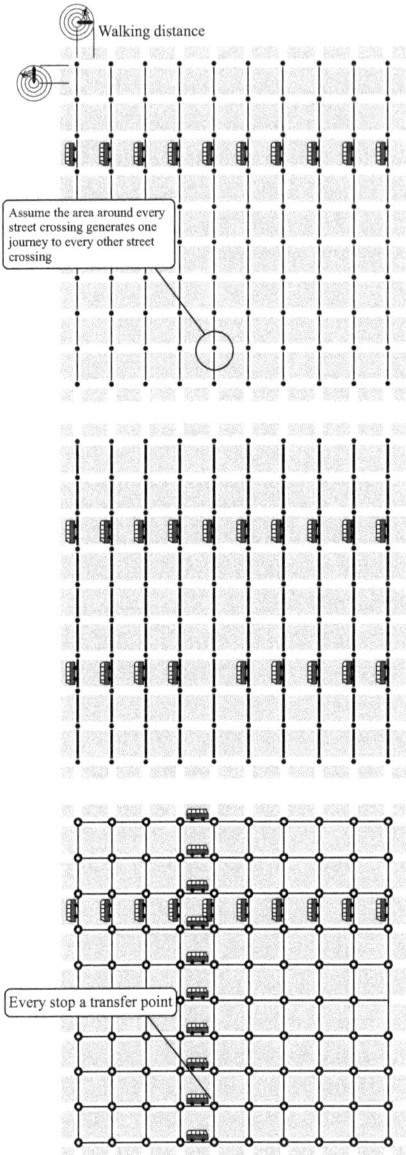

The network effect: the "Squareville" example

❶ "Squareville" with ten bus lines running north–south

The hypothetical city of "Squareville" has a grid-iron street pattern. The streets are well suited for a bus service since they are 800 meters apart. "Squareville" is a homogeneous city with a travel demand that is entirely dispersed. Assume the area around each of the city's street crossings generates on journey to every other street crossing; 9900 trips per day in total.

For the whole of "Squareville", the ten bus lines can only serve 900 trips in the city, which is less than 10 percent of the total trips of 9900. Assume that the public transport service presently attracts one-third of the journeys it can theoretically serve. This gives 300 trips per day by public transport, which is a modal share across the whole city of only 3 percent.

❷ "Squareville" with ten bus lines running north–south, double frequencies on all lines

Imagine that services on the existing bus lines are doubled in order to make more people in "Squareville" to use public transport. According to traditional transport demand modelling the elasticity of demand might be assumed to be some 0.5. This means that a 100 percent increase in service will produce a 50 percent increase in demand. The result will be 450 public transport trips per day and a modal share of 4.5 percent. Since the operational costs are likely to increase by more than 50 percent, the cost-recovery through fares is likely to fall.

❸ "Squareville" with twenty bus lines running north–south and east–west

Imagine that the extra operating resources instead were used to run ten new bus lines in the east–west direction. This would create a grid network of twenty lines. The number of trips that are directly served would double to 1800; the 900 initial north–south journeys and the 900 new east-west journeys that can be made without transferring between lines. But if passengers are willing to transfer, then all 9900 trips between all blocks can be served by this network; 1800 directly and 8100 by transferring. Assume that the modal share for journeys involving a transfer is half of that for direct journeys, i.e. one-sixth of these trips that can be attracted to public transport. This gives a total number of 1950 public transport trips per day (1800/3 + 8100/6). The modal share has increased dramatically from 3 to 20 percent.

This gives an elasticity of demand that is 5.5, rather than the traditional figure of 0.5. Increased revenue from the fares should more than cover the extra costs of operation and vehicle occupation would rise.

(Adapted from Paul Mees, 2000).

Figure 7.1 Squareville: the 'network effect' explained [10] [printed with permission from Truls Lange]

public transport attracts only one-third of potential non-transfer trips and one-sixth of trips requiring a transfer, the new mode share for public transport grows to 20%, with an elasticity of demand of 5.5.

To sum up, the 'network effect' enables 'services to be added in ways that increase demand at faster rates than those at which resources are added' [11].

This challenges the traditional 'diminishing returns' argument used by transport economists and is fundamental to understanding how public transport planners in both developed and developing cities can, by creating the 'network effect, do much more with their available resources than most transport economists predict.

Traditional public transport planning approaches take users' dislike of transfers as evidence that it is necessary to eliminate them wherever possible, thus creating service patterns in which each individual line is used to connect a multiplicity of origins and destinations. As we have seen, this is self-defeating unless huge financial resources are available.

In contrast, a public transport network must be designed and operated in a hierarchy of interconnected multi-directional nodes. Transfers are essential to the ability of such system to operate both efficiently and effectively.

Typical obstacles to 'seamless' transfer include: poor physical design, uncoordinated timetables, and financial penalties imposed by fare systems that, for example, charge separately for bus and train travel within a single journey.

The key principles and practices that underlie the creation of the network effect in a metropolitan setting are summarised in Table 7.1. Here, 'principles' guide the actions of decision-makers, and the 'practices' identify the means required to achieve the desired outcome.

The fundamental performance slogan for an effective network is: frequency, speed and connections.

As we will see below, transport planners in different cities have responded in quite different ways, consciously or not, to apply these principles in their own local specific conditions of demography, geography and political structure.

Table 7.1 Principles and practices to create the 'network effect'

Principles and key practices of Public Transport Network Planning	
Principles	
(1) Provide throughout the day (and into the night) stable inter-connected public transport through a structured network and easy to learn timetable	
(2) Recognise and plan for the need of users to transfer between different modes of public transport to arrive at intended destinations	
Key practices	
(a) Realise a line/route structure (individually and within the network) that is understandable and direct for the user	(c) Aim to match the convenience of the car. Transfers are essential and must be made to work – physically easy, minimal time penalty and no extra cost (integrated ticketing)
(b) Install a hierarchy of lines feeding into the network that can provide for high speed and capacity cross-city links, inter-suburban links and local 'feeders'	(d) Install clear and understandable way-finding mechanisms (e.g. signage describing routes, connections and station arrival times) and simple ticketing structures and technology

Source: From [12] based on [10,11,13].

7.3.2 Delivering the network effect

There is empirical evidence for Mees' theory of the 'network effect'. This can be found by benchmarking cities with similar demographics, geography and urban form: cities that adopted a network planning approach significantly outperformed those that adopted a radial/direct-route approach: the 'network' cities recorded higher patronage increases and lower rises in subsidies [14].

Looking to Europe, Buehler and Pucher [15] argued that public transport success in Germany is the result of the 'tight coordination of transit services, fares and schedules'. In Switzerland, Austria and Germany, such coordination is achieved through regional governing 'alliances' (in German – *Verkehrsverbünde*) that bring together the many operators and government authorities to deliver and market coordinated services [16].

In the United States, researchers have also found good supporting data. Thompson and Matoff [17] suggest that effective multi-destination transit (i.e. network planning) is not dependent on transport-related factors such as the strength of a city's Central Business District (CBD) (i.e. its population density and/or employment base), surrounding land use(s), rapid population growth and density of populations (or lack of) in city areas. More recently, Brown and Thompson [18] provided further evidence for this in a study of 45 US cities. As a note of caution, this Florida researcher group has also found that creating too sparse a network can lead to decline in patronage compared with 'traditional' radially focussed systems [19].

The success of a network also relies on maintaining a universally high level of service through cross-subsidies from busy radial trunk lines to the less patronised, but essential, feeder and cross-town services. This observation supports the notion that the creation and maintenance of an effective network requires planning and management by a public agency. Mees [7] argued that the responsibilities of this agency must include planning and coordination of the network design and the oversight of the distribution of fare revenues and subsidies to operators.

A further requirement for success is to ensure that coordination of a public transport network is done within a framework for land-use planning that recognises the relationship between transport policy choices and objectives for economic development, population health, the maintenance of good air quality and the protection of the environment [20,21]. With such relationships established, often through community engagement to identify common values and visions, it is possible to set clear targets for desired changes in mode share.

Many cities struggle to find the right balance between transport and land-use policies that favour the individual convenience of the private car and those that encourage shifts to alternatives that bring greater collective benefits. Vuchic has recommended that policies should be developed that to support complementary measures that provide disincentives for car-use and incentives for public transport: a balance between the two modes can be achieved with moderate disutility changes in each, because the two measures work 'toward' each other. A conceptual balance is achieved by a smaller effort on each mode than would be required if measures were applied to only one of the modes [20].

7.3.3 How have some cities 'won' the politics?

In a seminal work on the question of how some cities have achieved some measure of success in reducing the negative consequences of over-dependence on the private car, Bratzel [22] examined the trajectories of transport policy and outcomes in six European urban regions during the 1980s and 1990s. The cities were Amsterdam and Groningen in Holland, Freiburg and Karlsruhe in Germany, and Zurich and Basel in Switzerland. He identified three common stages in the political process:

1. well-organised popular opposition to existing car-based transport policies;
2. a radical change in political leadership and in incumbency in important political positions relating to transport policy;
3. institutionalisation of new operational practice – this required the new political leaders to maintain and extend the mandate for new transport policies over sufficient time during which new practices could be established.

A similar model has been used to explore reasons for success and failure in exploiting political 'windows of opportunity' for change in the direction of transport policy in cities in Australia and Canada [23]. This work is based on theoretical approaches to studying policy process described by Sabatier [24].

Road-based planning agencies, with their strong economic and political power bases, have long demonstrated their ability to maintain transport policies and programs that support the continued dominance of the car. In part, this has been done by maintaining the myths about the 'impossibility' of greater suburban public transport use that were described earlier. However, many public transport managers share these beliefs in the limits to public transport growth. For them, public transport in the car-dominated city has a role only in limited markets such as CBD/downtown work travel, major sporting or cultural events, or the local needs of pensioners and schoolchildren.

In addition, the long-established institutions responsible for rail operations in many cities have a culture of attention to the industrial hardware at the expense of the needs of the users. Vuchic [20] provides some telling descriptions of the 'old' public transport planning mindset that gives low priority to the user.

In urban regions where public transport mode share is growing, transport planning is being transformed from the old operationally focussed practices to new 'user-oriented' networks, and public transport is being positioned at the centre of 'city-shaping' planning and development agendas. The transformation from an industrial producer to a service provider is complex and requires the creation of new operational and planning practices and performance measures. Even in the Swiss railways, the world leaders in user-oriented operations, the process of invention and adaptation is continuing after three decades of progress [25,26].

Having now set the theoretical framework for transport planning at the broad scale of urban policy, and at the specific level of how to create a public transport system that can effectively compete with the private car at an affordable cost, we can now turn to an examination of attempts to put this into practice across an entire urban region.

7.4 Case studies of relative success

In looking for real world expression of the theoretical concepts described above, it is necessary to define success in relatively narrow terms. Here, 'success' is taken to be reversal of growth in private car use in the urban core and, at least, a slowing of growth across the whole urban region over a period of two decades or more.

For each of the three case studies, we provide a short description of the characteristics of the urban region and the structure of transport networks. This is followed by presentation of data and analysis of the performance of these networks, with particular attention given to the public transport networks. To conclude, we discuss the place of public transport in the context of the long-term trajectory of urban planning policies and the dynamics of urban politics.

Due to differences in the way demographic and transport-system performance data is collected and reported in each city – and because the three urban regions vary in size, form and structure – the data categories are not uniform. Even so, the broad trends in performance and practice can be observed across the three case studies.

7.4.1 Vancouver

7.4.1.1 Transit networks and patterns of urban development

Vancouver's early transit infrastructure shaped the region's distribution of commercial and residential development in the brutally regular grid of streets set out by the colonial surveyors. Its streetcars were closed down in the post-war years and new rapid transit infrastructure was not built until 1986. There are now three automated light rail lines (Skytrain), which operate at headways as short as 90 s in both elevated and underground rights of way. The rest of the urban region is served by a grid pattern of trolley and conventional buses that run in a legible pattern of reasonably straight lines in mixed traffic, but at much lower frequencies than the Skytrain. There are several higher-frequency express bus routes (B-Line), which are geared to carry very large passenger loads (see www.translink.ca).

Together, this pattern of transit supply creates a reasonably effective network on the models described earlier. The high Skytrain frequencies provide the basis for successful transfers. Daytime bus headways between 10 and 15 min in suburban areas do allow some passengers to choose to transfer where routes intersect, but, based on international experience, these frequencies are close to the upper limit for public acceptance. This home-grown 'network' philosophy is well-understood by local transit planners and is codified in internal technical documents [27].

Much international interest in urban development in Vancouver focusses on the new high-rise neighbourhoods built on the downtown peninsula and in the regional town centres (see e.g. [28]). However, during the period of greatest growth

Table 7.2 Population numbers and densities in Greater Vancouver planning zones [29]

	Urban land area (ha)	Population			Population density (persons/ha residential)		
		1996	2001	2006	1996	2001	2006
Eight regional town centres	2,356	102,560	122,115	162,560	44	52	n/a
Vancouver metro core	926	72,985	88,410	104,750	79	95	n/a
Growth concentration area	37,967	1,176,560	1,289,288	1,356,290	31	34	n/a
Vancouver metro (statistical district)	75,979	1,831,665	1,986,965	2,116,580	24	26	n/a
Within 500 m of SkyTrain	4,034	105,785	120,853	n/a	26	30	n/a

in transit use, over 90% of residents in the region lived in dispersed suburban dwellings (see Table 7.2). In addition, the CBD employment market is small: 50% of jobs in 2005 lay outside preferred 'growth concentration zones' (Greater Vancouver Regional District Planning Department, pers. comm.), with growth in logistics and transport associated with expansion of port facilities along the Fraser River, south of established suburban centres.

7.4.1.2 Performance

Trip making by public transport in Greater Vancouver is reported here as annual per capita linked trips (i.e. a one-way journey, using one or more transfers, counts as a single trip). Changes in this measure from 1971 to 2015 can be seen in Figure 7.2[1].

Even in the 1970s, operators were able to use an injection of funds to reverse an earlier large decline in transit tripmaking associated with the post-war availability of automobiles. This growth peaked again in the late 1980s with the opening of the first Skytrain line and stagnated in the 1990s as municipal and provincial governments with pro-suburban development policies cut back funding for public transport services. A reversal of political fortunes for transit in the 1990s led to a trend of modest but steady growth, exceeding 100 trips per capita in 2011 following significant investment in new transit infrastructure.

[1] As with many transit agencies, boardings and transfer rates are calculated from ticket sales and head counts. Methodologies for these calculations are not provided and change over time. Changes typically produce a higher ridership number from the same raw data. This inflationary tendency has been taken into account to allow meaningful comparisons to be made across different years. Patronage spikes during the 2010 Winter Olympics, and troughs during extended bus strikes in 1985 and 2001, have also been discounted.

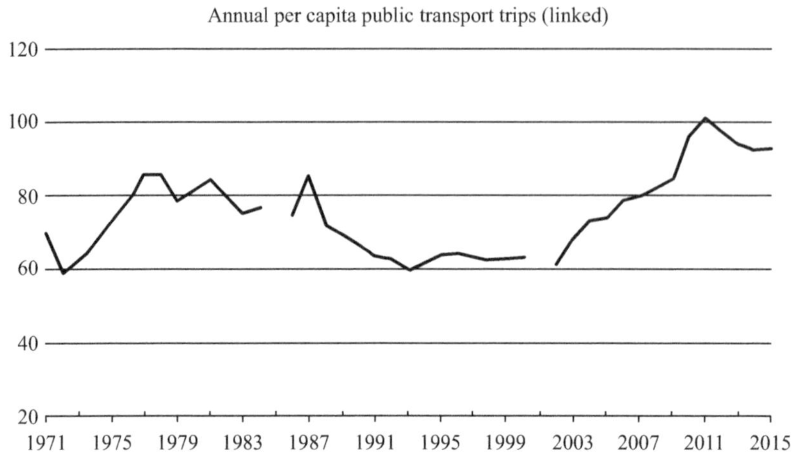

Figure 7.2 Annual per capita trip-making in Vancouver (1971–2015) (Source: operator Annual Reports)

Political contention over directions for urban transport and planning has returned, and the operator has had to make significant service cuts in response to constrained budgets. Although there is no historical data, recent annual reports show bus punctuality at around 85% (with a relatively tight definition of 'late' as over 3 min outside the timetable). As buses operate in mixed traffic with almost no lane or intersection priority, it is likely that this performance has worsened over recent years in the face of increased traffic due to population growth. Together, these pressures have resulted in a decline in tripmaking from 2011 to 2015. A measure of the declining quality of the network can be seen in falling transfer rates (Figure 7.3).

Public and active transport have each performed better relative to the private car. This is demonstrated in the changing mode splits for the journey to work (JTW) from 2001 to 2011 (see Table 7.3).

Within the City of Vancouver, which comprises 25% of the land area and population and around 40% of the wealth of the urban region, many targets for transit mode share, set for different market segments in regional plans in the 1990s, are now being met (see Table 7.4).

In this data, it is interesting to note how rapidly the desired changes in transport behaviour were achieved. In fact, the transit targets for downtown and Central Broadway were reached after the opening of the Canada Line Skytrain to the international airport in 2010. Also significant is that the active transport targets were achieved through a dramatic increase in walking and not cycling as the planners had anticipated. The growth in transit use to the hard-to-reach University of British Columbia (UBC) Campus can be attributed to a well-designed student transit ticket (the U-Pass) which covers all public transport use at a

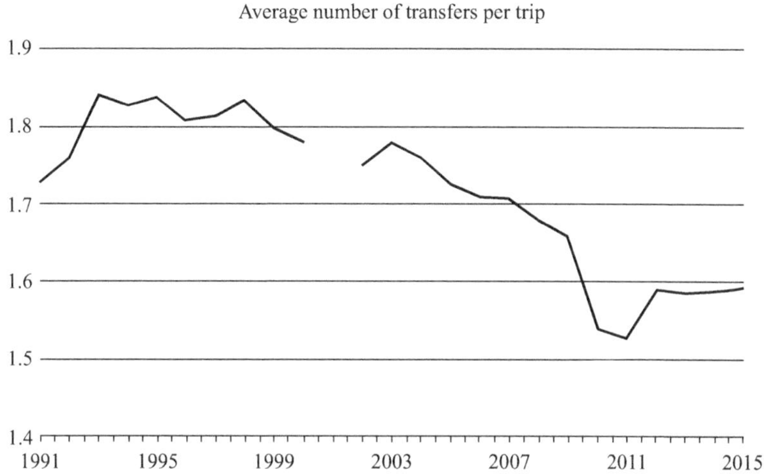

Figure 7.3 Average transfer per trip: Greater Vancouver: 1991–2015 (Source: operator Annual Reports)

Table 7.3 Mode splits for the journey to work in Greater Vancouver [30–32].

	Car	Public transport	Walked	Bicycle	Other
2001	79.2	11.5	6.4	1.9	1.0
2006	74.4	16.5	6.3	1.7	1.2
2011	70.8	19.7	6.3	1.8	1.4

Table 7.4 Progress towards mode share targets for 2021 in the City of Vancouver [33]

<table>
<tr><th rowspan="2">Mode</th><th colspan="2">Downtown</th><th colspan="2">Central Broadway</th><th colspan="2">Vancouver</th><th colspan="2">UBC</th></tr>
<tr><th>Target</th><th>Actual (2004)</th><th>Target</th><th>Actual (2004)</th><th>Target</th><th>Actual (2004)</th><th>Target</th><th>Actual (2004)</th></tr>
<tr><td>Driver</td><td>36%</td><td>30%</td><td>45%</td><td>54%</td><td>44%</td><td>50%</td><td>41%</td><td>37%</td></tr>
<tr><td>Passenger</td><td>12%</td><td>9%</td><td>15%</td><td>10%</td><td>15%</td><td>12%</td><td>16%</td><td>19%</td></tr>
<tr><td>Transit</td><td>34%</td><td>30%</td><td>25%</td><td>20%</td><td>23%</td><td>17%</td><td>33%</td><td>42%</td></tr>
<tr><td>Bike</td><td rowspan="2">18%</td><td>3%</td><td rowspan="2">15%</td><td>3%</td><td rowspan="2">18%</td><td>3%</td><td rowspan="2">10%</td><td>1%</td></tr>
<tr><td>Walk</td><td>27%</td><td>12%</td><td>17%</td><td>1%</td></tr>
</table>

Source: Targets were set for 2021 in [34]; 2004 data from [35].

Table 7.5 Average travel time for the journey to work and return [36,37]

	1992	1998	2005	2010
Toronto	68	76	79	81
Montreal	62	65	76	76
Vancouver	70	68	67	74

much-discounted cost and is paid for as part of student-services fees: guaranteed revenue for the transit operator and more students housing on redundant carparks at UBC.

A final measure of transport system performance reflects the relative success of Vancouver's transport policies and its supportive land-use policies of concentration of commercial and residential development. The average time for the round trip to and from work fell from 70 to 67 min between 1992 and 2005 [36]. More recent data is less encouraging – the average round trip in 2010 took 74 min [37] – but the long-term trend shows that Vancouver has made more progress than larger Canadian cities towards the planners' holy grail of 'urban containment' (Table 7.5).

A number of transport and planning policies have contributed to this relative success. Investment has continued in locally proven transit technology, and service patterns have been explicitly designed to create an efficient intermodal network [38]. The region, since 1993, has been explicitly used congestion as a travel-demand tool [39]. There is only one radial commuter freeway – the Trans-Canada Highway – but even this comes no closer than 5 km to the downtown district. Road capacity expansion, until recently, has been confined to freight routes in the outer suburbs.

The efficacy of this policy mix is amply demonstrated by comparison with Auckland: a city with many demographic and geographic similarities to Vancouver but which has followed a car-based transport and land-use policy mix for more than half a century (Table 7.6).

Of note in this table is the evidence of the network effect: Vancouver achieves more than four times the annual per capita boardings with less than twice the per capita level of service supply. This is reflected in the lower subsidy required per trip. This low subsidy is also evidence of the ability of Vancouver transit planners to do a great deal within constrained budgets. The B-Line express bus services are a classic example of a sweated asset: the no. 99 B-Line to UBC, for example, carries 60,000 people per day on old diesel bendy-buses. This is more than most light rail transit systems can achieve.

As an aside, Auckland is now completely redesigning its bus system into a 'network' which aims to improve efficiency through effective transfers [40]. The design of this new network is based on the work of Walker [41].

A final piece of data on the performance of Vancouver's transport system is the evidence of continuing growth in car ownership (Figure 7.4). The de-coupling of vehicle ownership from vehicle use is shown by the fact that mode share of car travel for the JTW fell by nearly ten percentage points over this time (Table 7.3).

Table 7.6 Urban form and transport performance in Auckland and Vancouver in 2009 (see [42] for details on sources for this data)

	Vancouver	Auckland
Total population (millions)	2.1	1.3
Density (persons/ha in the urban region)	17.1	18.9
Urbanised area (km^2)[a]	1,230	690
Jobs in CBD (% of total jobs in the urban region)	12.6	13.5
Passenger cars/1,000 residents	519	523
Car travel cost index[b]	2.3	4.2
Share of work trips:		
Public transport	16.5	7.0
Walk and cycle	8.0	5.6
Car	74.4	87.4
Public transport boardings (millions of 'unlinked' trips)	283	52
Annual boardings/capita	135	40
Public transport service (km) (bus- and tram-km + train-km: millions)	116.2	42.3
Service (km/capita)	55	32
Subsidy per boarding[c]	$1.10	$2.54

[a]Areas are defined by the boundaries of contiguous urban development.
[b]This index measures 2009 average petrol prices relative to the median individual wage.
[c]Equivalent in $NZ at October 2009 exchange rates.

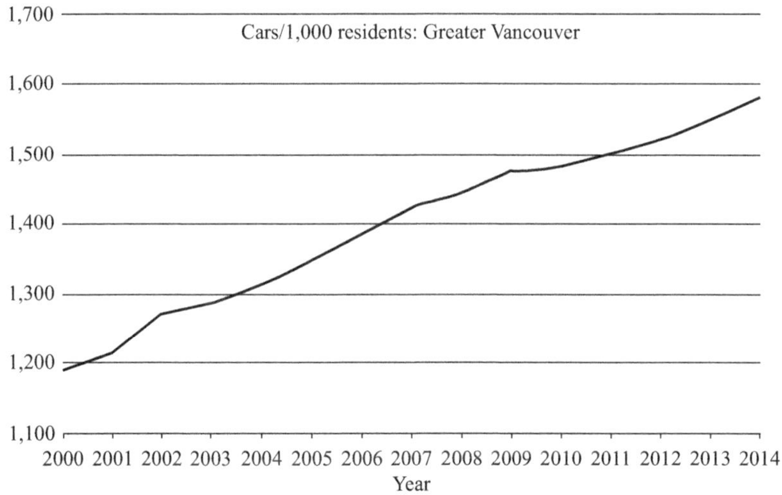

Figure 7.4 Per capita car ownership in Greater Vancouver (2000–14) [43]

7.4.1.3 Long-term planning framework

At the municipal level, the City of Vancouver's building codes 'impose significant constraints on developers' [44] and have won international praise for their achievement of 'liveable' outcomes [28]. At a regional level, a strong urban growth boundary has been maintained since the 1970s: the constraints of mountains, sea and the US border provide a compelling case. The competition between the car and its favoured alternatives has chiefly been regulated through the fostering of a differential in time and convenience with road congestion used as a conscious travel-demand strategy.

Vancouver's regional government – a peculiarly local form of cooperation among municipal councils [45] – has, until recently, maintained a political consensus around this approach in the face of significant opposition from supporters of untrammelled development and greater investment in roads and bridges for suburban passenger cars. The legitimacy for the 'compact city' vision was established through the actions of citizens groups and political entrepreneurs in the 1970s and re-established in the 1990s. This was achieved against strong political opposition, particularly from the province, which typically draws its political power from the extractive industries in regional British Columbia. The establishment of political legitimacy has required regional and local politicians and planners to adopt public planning processes expressly designed to facilitate strong community engagement. Detailed accounts of this can be found here [23,46].

The erosion of the effectiveness of Vancouver's transit system documented here – and of the 'compact city' agenda more widely – can be seen as a failure to re-establish the legitimacy of this vision in face of the changing economic and demographic structures in the region and the growing 'democratic deficit' common within neo-liberal regimes.

The defeat, in a 2015 referendum, of a proposal to fund transit and road infrastructure through a small increase in sales tax revealed the growing polarisation between city and suburb. This setback has thrown out a challenge to supporters of transit-oriented planning to find new ways to re-establish their legitimacy and power. This task is made all the more difficult because of an apparent hollowing out the capacity of independent civil-society organisations to engage effectively in political processes. This has many causes: from neo-liberal attacks on funding and legitimacy to the complexities of a larger and more culturally diverse population. In addition, the institutionalisation of public engagement, through processes such as the 'visioning' exercises used by regional government planners to build support for the planning strategies of the 1990s, may also have unwittingly contributed to the weakening of civil-society capacity.

The robust contention that has characterised transport politics in Vancouver for many years is set to continue between proponents of new roads and bridges and suburban industrial development, and the keepers of the vision of Vancouver as a 'cities in a sea of green'. It remains to be seen which side will develop the best tactics and mobilise most skilfully in coming years.

Despite this uncertain future, the trajectory of transport and land-use planning in Greater Vancouver since the 1970s shows that it is possible to create effective alternatives to private car use in the physical and cultural context of a dispersed North American city. The two following shorter case studies reveal what is possible when the political legitimacy of planning for alternatives to the car is much more firmly established.

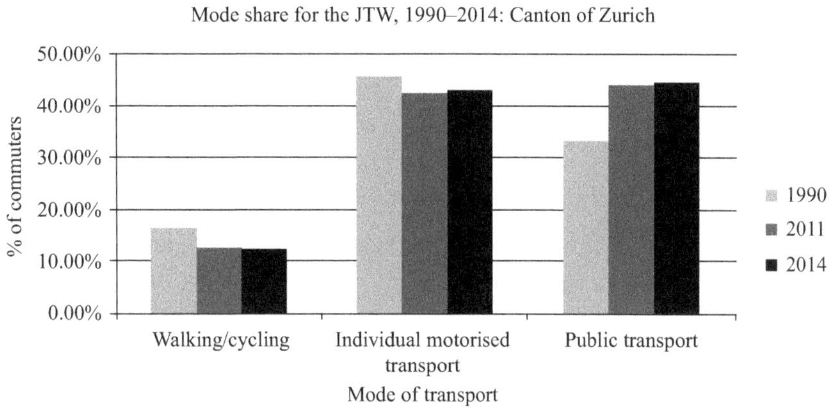

Figure 7.5 Mode share changes for the JTW in the Canton of Zurich: 1990–2014 [47–49]

7.4.2 Zurich

The City of Zurich has long been recognised as an international exemplar for its achievements in planning and delivering a transport system that allows car-use to be optional for almost all trip purposes. This success is based on a long political and social consensus around transport policies that provide disincentives for drivers and considerable incentives in service quality and price for users of alternative modes. What is less well recognised is the success of public transport in the wider Canton of Zurich (see Figure 7.5), where policies of incentives and disincentives have been less clearly applied: road construction has gone hand-in-hand with the creation of an effective public transport network.

The Canton has a population of 1.3 million people. The urbanised area of the Canton, only 350 km^2, is dispersed within rural land and forest over a total area of 1,840 km^2.

Zurich has met the challenge of economically operating its large number of rail lines in areas with very low populations by integrating them with feeder and cross-suburban bus services to serve a large variety of trip purposes and destinations.

The basis for Zurich's levels of both operational efficiency and user choice is the time-pulse network that is built around carefully crafted national rail timetables. The creation of these timetables – based on the principal of travel speeds that are 'as fast as necessary' rather than 'as fast as possible' – allows regular pendulum of regional trains arriving and departing key nodes on the hour and the half-hour. Using this strong 'scaffolding', timetables and on-road priority requirements for local trains, buses and trams can be established, and the 'Verkehrsangebot' (the 'service offer') can be created using what is called a 'planning triangle' (see Figure 7.6).

This style of planning leads to clear proposals for new infrastructure and service funding that can be debated through the unique Swiss system of cooperative federalism that relies on a systematically engaged population making decisions through frequent referenda.

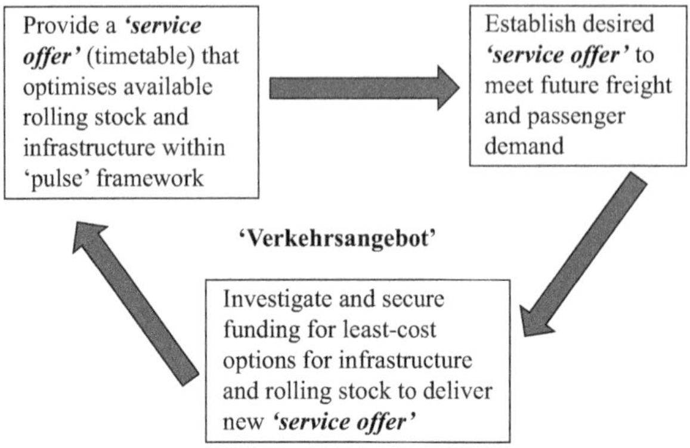

Figure 7.6 The planning framework used by Swiss railway engineers [26]

Within the City of Zurich, planning for public transport has allowed trams to operate without interference from car traffic, which is kept away from tram routes by a unique 'metering' system of traffic signals that keep traffic queues on the arterial roads [50]. A system of parking restrictions that support tram priority, pedestrian access and civic amenity is maintained politically through an agreement of 20 years standing still referred to as the 'Historic Compromise'. This agreement requires on-road parking spaces to be removed on a one-to-one basis whenever new off-street parking is created.

The tram and bus system in the City of Zurich is a very effective 'go anywhere anytime' network, but in contrast to the theoretical 'Squareville' network model, the Zurich network geometry is the based on the triangle [51]. This observation should be very valuable for planners and engineers seeking to create new public transport networks in cities with travel patterns focussed on nodes and with imbalances in the demand between radial and circumferential directions.

7.4.3 Vienna

A key indicator of the performance of the Vienna regional public transport system is the public transport 'accessibility index' derived from the Scheurer and Curtis Spatial Network Analysis for Multi-Modal Urban Transport Systems (SNAMUTS) model [52]. On this index, Greater Vienna is one of the world's most 'transit-connected' cities. The effectiveness of public transport in Vienna can also be seen from the network coverage map (Figure 7.7) which shows that 79.7% of residents and jobs lie within walking distance of an effective public transport service. Use of motorised modes has fallen continuously since the early 1990s (see Figure 7.8).

Wiener Linien – the city's public transport operator – uses the slogan *Die Stadt gehört Dir* ('the city belongs to you'). Evidence that this is not just empty marketing can be found in the fact that one million annual transit passes, at a subsidised price of €356, were sold in the first year of its availability to a resident population of 1.8 million in the city and further 800,000 in the surrounding metropolitan region (Wiener Linien senior planners, pers. comm., 2015).

Figure 7.7 Network coverage of public transport in Greater Vienna from SNAMUTS

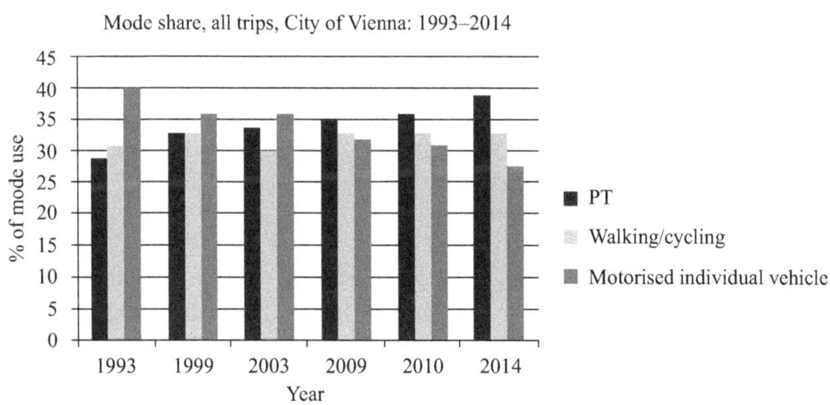

Figure 7.8 Mode share, all trips, City of Vienna: 1993–2014 [53]

Performance standards for transit supply reflect a clear understanding of the importance of networks and access to a range of destinations. The measure of adequate transit supply for any residential location in the suburbs is access to a rapid transit line within 15-min travel-time (on foot or by bus). This compares favourably with Melbourne, for example, where the measure of adequate supply is simply the availability of a public transport stop within walking distance, regardless of the quality of the service available at that location.

These two examples of planning practice are embedded in a planning culture that uses housing affordability to unify policy and planning practices across all city functions.

7.5 Conclusions: overcoming obstacles to implementing effective and efficient public transport systems

The case studies of Vienna, Zurich and Vancouver demonstrate that transport planners have been able to achieve something that has often been considered impossible. The public transport systems in these urban regions have consistently increased their mode share at the expense of the private car in their inner and middle suburbs. And, this has been done at a cost that was affordable to the wider urban economy.

These results have been achieved in cities in both North America and Europe, but the element that is common to all is a particular approach to transport planning. Policy principles underpinning this approach can be drawn from this chapter and from the texts suggested for further reading. These include:

- A coordinated commitment to giving priority for public and active transport over the private car in investment and in the allocation of road space.
- Integrated land-use and transport planning to encourage the development of destinations (employment, services, schools, colleges, etc.) at nodes that can be serviced by multi-directional transit networks.
- Planning and development of effective conditions for intermodal transfers. This includes not just stations and their connections with surrounding neighbourhoods, but also fare structures that do not penalise passengers making transfers and which allow for cross-subsidies from highly patronised lines to support necessary frequencies on feeder and circumferential lines.
- Creation of institutions for the planning and operation of effective public transport networks: these are typically public agencies that are given some measure of budgetary independence.

By understanding and employ these principles, the planners and engineers who will design and operate future urban transport systems – even those employing new ICT technologies – have a strong practical and theoretical foundation from which they can provide sustainable and healthy mobility and access for the cities of the future.

References

[1] Litman T. *Autonomous Vehicle Implementation Predictions: Implications for Transport Planning*. Victoria, BC: Victoria Transport Policy Institute; 2015.

[2] Anderson J, Kalra N, Stanley K, Sorensen P, Samaras C, Oluwatola O. *Autonomous Vehicle Technology: A Guide for Policymakers*. Santa Monica: RAND Corporation; 2014.

[3] International Transport Forum. *Urban Mobility System Upgrade: How Shared Self-driving Cars Could Change City Traffic (Corporate Partnership Board Report)*. Paris: ITF; 2015.

[4] Urry J. *Sociology Beyond Societies: Mobilities for the Twenty-first Century*. London: Routledge; 2000.

[5] Stone J. Contrasts in Reform: How the Cain and Burke Years Shaped Public Transport in Melbourne and Perth. *Urban Policy and Research*. 2009; 27 (4): 419–34.

[6] Armstrong B, Davison G, de Vos Malan J, Gleeson B, Godfrey B. *Delivering Sustainable Urban Mobility*. Canberra: Australian Council of Learned Academies; 2015.

[7] Mees P. *Transport for Suburbia*. London: Routledge; 2010.

[8] Gleeson B, Beza B. *The Public City*. Melbourne: Melbourne University Press; 2014.

[9] Mees P. *A Very Public Solution*. Melbourne: Melbourne University Press; 1996.

[10] Nielsen G, Lange T. *HiTrans Best Practice Guide No. 2, Public Transport: Planning the Networks*. Stavanger, Norway: HiTrans, EU Interreg IIIB North Sea Programme; 2005.

[11] Mees P, Stone J, Imran M, Nielson G. *Public Transport Network Planning: A Guide to Best Practice in NZ Cities*. Wellington: New Zealand Transport Agency 2010.

[12] Stone J, Beza B. Public transport: elements for success in a car-oriented city. In: Gleeson B, Beza B, eds. *The Public City: Essays in Honour of Paul Mees*. Melbourne: Melbourne University Press; 2014.

[13] Mees P, Dodson J. *Public Transport Network Planning in Australia: Assessing Current Practice in Australia's Five Largest Cities*. Research Paper 34. Brisbane: Griffith Urban Research Program; 2011.

[14] Mees P, Stone J, Imran M, Neilson G. 'Public transport network planning: a guide to best practice in New Zealand cities'. NZTA Research Report No. 396. Wellington: New Zealand Transport Agency; 2010.

[15] Buehler R, Pucher J. *Making public transport financially sustainable*. Transport Policy. 2011; 18: 126–38.

[16] Stone J. *Can European Models of Public Transport Governance Help to Save Australian Cities? State of Australian Cities*. Melbourne: Australian Sustainable Cities and Regions Network; 2011.

[17] Thompson G, Matoff T. Keeping Up with the Joneses: Radial vs Multi-destinational Transit in Decentralizing Regions. *Journal of the American Planning Association*. 2003; 69 (3): 296–312.
[18] Brown J, Thompson G. Service Orientation, Bus—Rail Service Integration, and Transit Performance: Examination of 45 U.S. Metropolitan Areas. *Transport Research Record: Journal of the Transportation Research Board*. 2008; 2024 (1): 82–9.
[19] Jaroszynski M, Brown J, Bhattacharya T. An Examination of the Relationship between Urban Decentralisation and Transit Decentralisation in a Small-sized US Metropolitan Area. *Urban Studies*. 2016; 1–19, DOI: 10.1177/0042098015626687.
[20] Vuchic V. *Transportation for Livable Cities*. New Brunswick, NJ: Rutgers University Press; 1999.
[21] Stevenson M, Thompson J, De Sá TH, *et al.* Land use, transport, and population health: estimating the health benefits of compact cities. *The Lancet*, 2016; Series: Urban design, transport, and health.
[22] Bratzel S. Conditions of Success in Sustainable Urban Transport Policy—Policy Change in 'Relatively Successful' European Cities. *Transport Reviews*. 1999; 19 (2): 177–90.
[23] Stone J. Continuity and Change in Urban Transport Policy: Politics, Institutions and Actors in Melbourne and Vancouver Since 1970. *Planning Practice and Research*. 2014; 29 (4): 388–404.
[24] Sabatier P. *Theories of the Policy Process*. Boulder, CO: Westview Press; 1999.
[25] Laube F, Mahadevan V. *Bringing Customer Focus into Every Nut and Bolt of the Railway: Swiss Federal Railways Path into the Future*. Paris: UIC, The Worldwide Railway Organisation; 2008.
[26] Stone J. *Planning for Affordable Transit Infrastructure and Service Expansion: Two European Case Studies*. 36th Australasian Transport Research Forum; 2013; Brisbane: ATRF.
[27] GVTA. *Transit Service Guidelines: Technical Report*. Vancouver: Greater Vancouver Transportation Authority (TransLink); 2004.
[28] Punter J. *The Vancouver Achievement: Urban Planning and Design*. Vancouver: UBC Press; 2003.
[29] GVRD. *2001 and 2006 Census Bulletin #1 – Population and Dwelling Counts*. Vancouver: Greater Vancouver Regional District Policy and Planning Department; 2002 and 2007.
[30] City of Vancouver. *Open Data Catalogue: Census Local Area Profiles—2006*. Vancouver; 2006.
[31] City of Vancouver. *Open Data Catalogue: Census Local Area Profiles 2011*. 2011. [7/5/2016]. Available from: http://data.vancouver.ca/datacatalogue/censusLocalAreaProfiles2011.htm.
[32] City of Vancouver. *Open Data Catalogue: Census Local Area Profiles 2001*. 2001. [7/5/2016]. Available from: http://data.vancouver.ca/datacatalogue/censusLocalAreaProfiles2001.htm.

[33] City of Vancouver. *Vancouver Transportation Plan, Progress Report*. Vancouver: City of Vancouver; 2006.

[34] City of Vancouver. *Transportation Plan Progress Report*. Vancouver: City of Vancouver; 2006.

[35] Translink. *Vancouver-UBC Transit Plan, Ridership Statistics Appendix*. Vancouver: Translink; 2005.

[36] Turcotte M. *The Time it Takes to Get to Work and Back*. Ottawa: Statistics Canada; 2005.

[37] Turcotte M. *Commuting to Work: Results of the 2010 General Social Survey*. Ottawa: Statistics Canada; 2011.

[38] Stone J, Mees P, Imran M. Benchmarking the Efficiency and Effectiveness of Public Transport in New Zealand Cities. *Urban Policy and Research*. 2012; 30 (2): 207–24.

[39] GVRD. *A Long-Range Transportation Plan for Greater Vancouver*. Vancouver: Greater Vancouver Regional District; 1993.

[40] Auckland Transport. *New Public Transport Network*. 2016. [11/5/2016]. Available from: www.at.govt.nz/projects-roadworks/new-public-transport-network/#purpose.

[41] Walker J. *Human Transit: How Clearer Thinking about Public Transit Can Enrich Our Communities and Our Lives*. Washington, DC: Island Press; 2011.

[42] Stone J, Mees P, Imran M. *Network planning for more effective public transport in New Zealand cities*. Selected Proceedings, 12th World Conference on Transport Research; 2010; Lisbon.

[43] Metro Vancouver. *Annual Planning Data*. 2015. [7/5/2016]. Available from: http://www.metrovancouver.org/services/regional-planning/data-statistics/annual-planning-data/Pages/default.aspx.

[44] Leo C. The urban economy and the power of the local state: the politics of planning in Edmonton and Vancouver. In: Frisken F, ed. *The Changing Canadian Metropolis: A Public Policy Perspective*, Vol 2. Berkeley: Institute for Governmental Studies Press; 1994.

[45] Sancton A. *Governing Canada's City-Regions: Adapting Form to Function*. Montreal: Institute for Research on Public Policy; 1994.

[46] Taylor Z, Burchfield M. *Growing Cities: Comparing Urban Growth Patterns and Regional Growth Policies in Calgary, Toronto, and Vancouver*. Toronto: Neptis Foundation; 2014.

[47] Statistique suisse. *Relevé structurel du recensement fédéral de la population 2011: pendulaires*. 2011. [7/5/16]. Available from: http://www.bfs.admin.ch/bfs/portal/fr/index/themen/11/01/new/nip_detail.html?gnpID=2060-186.

[48] Statistique suisse. *Relevé structurel du recensement fédéral de la population 2014: pendularité*. 2014. [7/5/16]. Available from: http://www.bfs.admin.ch/bfs/portal/fr/index/themen/11/01/new/nip_detail.html?gnpID=2016-009.

[49] Steinmetz R, Pola M, *Ernst Basler+Partner AG*. Eidgenössische Volkszählung 1990: Pendlermobilität in der Schweiz (Federal Census: Commuter Travel in Switzerland). Bern: Bundesamt für Statistik; 1997.

[50] Cervero R. *The Transit Metropolis: A Global Inquiry*. Washington, DC: Island Press; 1998.
[51] Bell K. From Squaresville to Triangle Town: Geometries for Public Transport Network Planning. *State of Australian Cities 2015*. Australia: Gold Coast; 2015.
[52] Curtis C, Scheurer J. *Planning for Public Transport Accessibility*. Farnham: Ashgate; 2016.
[53] Wiene Srtadtwerke. *Modal Split*. 2015. [8/5/2016]. Available from: http://www.nachhaltigkeit.wienerstadtwerke.at/daseinsvorsorge/oepnv/modal-split.html.

Glossary

CBD	Central Business District
ICT	Information and communications technology
JTW	Journey to work
LRT	Light rail transit
OECD	Organisation for Economic Co-operation and Development
SBB	Swiss Federal Railways
SNAMUTS	Spatial Network Analysis for Multi-Modal Urban Transport Systems
UBC	University of British Columbia

Further Reading

[1] Litman T. *Autonomous Vehicle Implementation Predictions: Implications for Transport Planning*. Victoria, BC: Victoria Transport Policy Institute; 2014.
[2] Mees P. *Transport for Suburbia: Beyond the Automobile Age*. London: Earthscan; 2010.
[3] Millard-Ball A, Schipper L. Are We Reaching Peak Travel? Trends in Passenger Transport in Eight Industrialized Countries. *Transport Reviews*. 2011; 31 (3): 357–78.
[4] Nielsen G, Lange T. *HiTrans Best Practice Guide No. 2, Public Transport: Planning the Networks*. Stavanger, Norway: HiTrans, EU Interreg IIIB North Sea Programme; 2005.
[5] Skinner I, van Essen H, Smokers R, Hill N. *EU Transport GHG: Routes to 2050?—Towards the Decarbonisation of the EU's Transport Sector by 2050*. Brussels: European Commission Directorate-General Environment; 2010
[6] Vuchic V. *Transportation for Livable Cities*. New Brunswick, NJ: Rutgers University Press; 1999.

Chapter 8

Urban transport and impacts on public health

Ghassan Abu-Lebdeh[1]

Abstract

The focus of this chapter is the impacts of urban transport on health: physical, mental, well-being, and livability. Specifically, and besides the traditionally obvious and well-studied subjects of transport-induced crash injuries and pollution health impacts, the chapter presents the modern day non-contagious ailments such as stress, cardiovascular diseases, physical inactivity, and obesity and highlights their links and interactions with urban transport systems. These health impacts are mostly significant in the long term. They are subtle, slow, and pervasive. They have traditionally not been addressed or even identified as transport related; only recently and just qualitatively, these impacts came to the fore of the interrelated subjects of urban form, transport, and modern day health problems. The chapter will identify the needs for research to both further the understanding of the dynamics between transport and urban form on one hand and the identified health impacts on the other. The chapter will also address the question of how to objectively measure and quantify these health impacts with a particular view to evaluating the benefits that can be realized through low carbon mobility solutions.

Keywords

Urban transport; public health; active transport; noise; traffic

8.1 Introduction

Directly or indirectly—and sometimes both—the contributions of transport systems in our lives is omnipresent. Equally pervasive, but far less understood, is the influence of transport systems on public health. Transport systems impact the economy, quality of life, livability of a community, and they impact public health.

[1]Department of Civil Engineering, American University of Sharjah, Sharjah, UAE

There are so many positives that transportation brings to the socio-economic and strategic goals of individuals, communities, and countries. Transportation is also at the heart of prosperity of urban life. There are, however, downsides to transportation. These are beside the obvious health impacts in terms of crashes and the resulting deaths and or injuries.

The worldwide statistics on transport health issues are staggering. Motorized road transport's toll on health includes 1.5 million deaths and 79.6 million healthy years of lost lives due to road crashes and air pollution annually (Chapter 2). Roadway crashes account for majority of the impact: 1.3 million deaths and 78.2 million injuries, conservative estimates, most likely, since in many parts of the world official government statistics underreport road injuries. Pollution from vehicles causes the death of 184,000 globally. Of the globally top 10 causes of death, six are due to injuries and pollution from vehicles [1]. But deaths and injuries alone do not tell the whole story of the true health impact of transport. The impacts on the mental health and well-being of individuals and communities are not included. Nor are they easy to measure or quantify. Perhaps, that is why they are often missing form most studies and statistics on transport health impacts.

But what is health/healthy? Per the World Health Organization, "Health is a state of complete physical, mental and social well-being and not merely the absence of disease or infirmity." This is a more holistic definition; here health and well-being as being the result of a combination of physical, social, intellectual, and emotional factors. Four elements of health living are identified as shown in Figure 8.1.

There is a negative definition of health and well-being: the absence of physical illness, disease, and mental distress. And there is a positive one: the achievement and maintenance of physical fitness and mental stability.

In presenting health impacts of transportation in this chapter, the holistic definition is adopted so that all positive and negative heath impacts are addressed; the ones that are already known and understood as well as those that emerged more recently and/or were known to exist but were not well understood and which may have been either neglected or overlooked. The traditional well-researched and debated topic of traffic road crashes, which is more consistent with the negative definition of health and well-being, will be presented, but it is not the main focus of the chapter.

Figure 8.1 Elements of healthy living

8.1.1 Context

The health state of an individual is a function of three factors: genetics, diet, and life style (Figure 8.2). Transport choices and decisions are part of the life style. There is a reciprocal relationship between diet and life style, but not between those two and genetics, although recent findings suggest that healthier life style and diet can alter the shape and functioning of our genes thus our health [2]. Transportation also contributes to health indirectly by enabling certain behaviors and accessibilities thus enforcing or eliminating key influencing food consumption behaviors [3,4]. Diet (and choices of foods) can be strongly connected to availability and choice of transport/mobility options. A walking-only option combined within certain socio-economic conditions could well mean access to only not-so-healthy food outlets thus offsetting, perhaps entirely, the positive health impact of walking. Directly, transport choices contribute to health through crash injuries, pollution-induced illnesses, and auto-oriented more sedentary less active lifestyle (Chapter 9). As such, and in so far as health is concerned, transportation can be doubly influential.

The depiction in Figure 8.2 however does not portray the relative importance of each factor. Transportation is a key element and maker of the built environment. Its direct and indirect influence on health is significant but has largely been underestimated. The well-known Frieden's health pyramid, from which Figure 8.3 is adapted, shows that the built environment, of which transport is a key element, is the second most influential factor after socio-economic factors [5]. Low carbon mobility options (LCMOs) are part of the built environment and all fit within "changing the context to make individual's default 'transport' decisions healthy". LCMOs are potentially significant good health enablers.

8.1.2 Why health impacts of transport now?

The impacts of transport options on health are as old as the existence of transportation. What is new is the realization and understanding of the cause and effect between transportation and public health; we are far ahead now as what aspects of transportation choices impact what aspects of human health. Simply, we know more. And we can measure and observe and track more easily and precisely. Of course, some of those impacts were obvious from day one: vehicle crashes and the resulting death and or/injuries. What was not as clear and well understood were

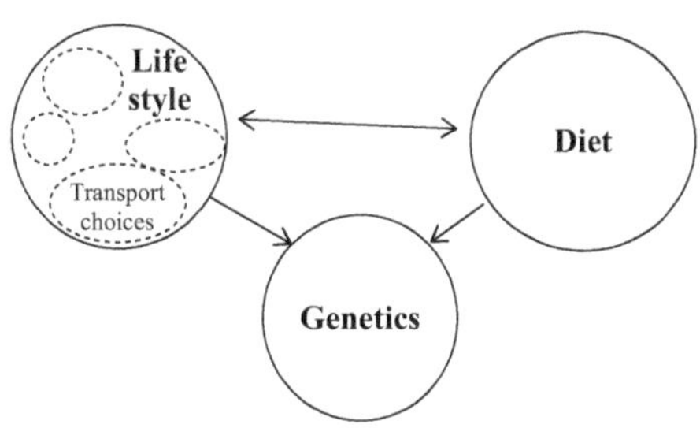

Figure 8.2 Health factors

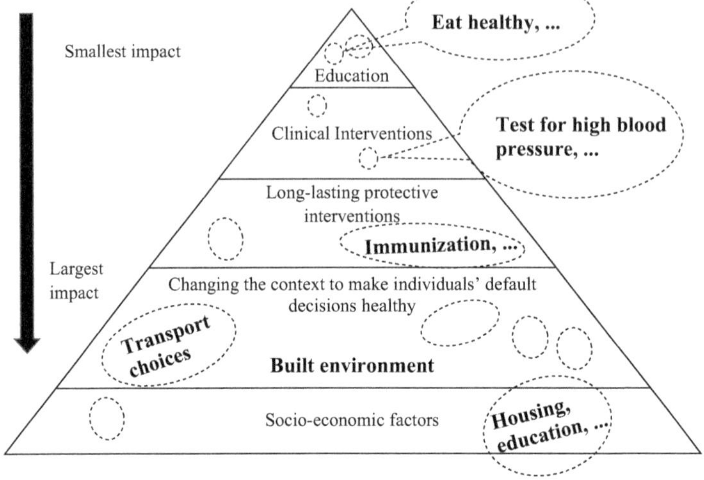

Figure 8.3 Factors that affect health (adapted from [5])

both the impact of transport pollutants and the psychological impact of other aspects of transport as long commute and traffic noise. With time, our understanding of those impacts has improved although it is far from perfect. It is good enough, though, that we can quantify, with a reasonable level of accuracy, the amount of harm a particular transportation option, change, or decision will have on the health of those impacted by that change.

But there are still challenges, some of which are basic. There is no one standard way to present the health impacts of transport. Those impacts can be presented based on desirability (i.e. positive or negative), causes (crashes vs non-crash), specific health issues caused (physical ailments vs psychological), or reach of impact (individual, community). Neither of these approaches fits the purpose of this chapter, thus a different approach and division of topics is used. The reason is simple: (1) there are too many confounding factors; isolating the transportation systems-related ones is not easy, (2) some of the health issues and/or their causes are overlapping; in some cases, the interdependences or correlations and the nature of the cause-and-effect process are known; in others, they are not. A mixed approach is used in this chapter: The material on specific health impacts of transport systems is presented within the following topical areas: accidents and crashes, air pollution, noise (as an ambient stressor), active transport, community health and well-being, and soil and water pollution. Regardless of cause(s) or areas, the health outcomes of transport are classified according to the four elements of healthy living presented earlier in Figure 8.1. Some of the health outcomes are linked to multiple of the aforementioned topics. Those health outcomes are not independent of each other; there are strong interactions between them. Physical health of an individual, for example, contributes to their state of mental health and social well-being; community health and livability are dependent on the health and social well-being of its population. In all elements, a transportation system is a player: traffic crash injuries and accessibility to health care facilities impacts physical health; traffic noise and stress from long commute and congestions impact mental health; ease of mobility accessibility to educational, health, and recreational facilities impacts community health and quality of life; individual's abilities to meet, connect, and interact with other individuals and groups impact social well-being.

8.1.3 This chapter

The chapter attempts to address the full spectrum of urban transport health impacts. More emphasis will be placed on less understood and documented non-crash-related health impacts. Where crash health-related impacts are discussed, the discussion will be limited to the part less researched and considered by the transport community. These include the long-term, mainly psychological, impacts of the resulting injuries and/or disabilities. These can often be significant to both crash victim, their families and to the larger society. If for nothing else, relevant statistics on crashes and crash-related injuries will be reported for completion of the picture on the health impacts of transportation. Even then, there likely are additional health impacts that at some level are related and or caused, perhaps in part, by transportation. For example, parts of vehicle emissions negatively impact ozone levels in earth's surface, which in turn, and indirectly, can lead to health-related issues through changes in weather pattern, droughts, flooding, mass migration, loss of habitat, etc. This is only partly understood in so far as its impacts on health. Although transportation is a contributing factor, it is one among many.

The remainder of the chapter is presented as follows. Section 8.2 presents a brief history of modern land use and development and, by extension, urban form and travel choices. The connections to public health are noted. A brief presentation of LCMOs is the subject of Section 8.3. Section 8.4 is the focal part of this chapter. It presents the health impacts of urban transport systems, but it does not follow a traditional order where different types of health impacts are presented. The material on the specific health impacts of transport systems is presented within the following topical areas: (1) accidents and crashes, (2) air pollution, (3) noise, (4) active transport, (5) community health and well-being, and (6) soil and water pollution. Section 8.5 discusses the relevance and means to formally measure, assess, and quantify health impacts of transportation policies and/or projects, a necessary step that would inform how we plan, design, operate, and maintain transport systems to support and further public health goals. A discussion of the salient points and current and future needs is presented in Section 8.6.

8.2 History and background of urban form, transportation, and public health

In the late nineteenth and early twentieth centuries, cities were far more compact and land use was mixed; worker lived, worked, and gotten other necessities all within walking distance. Motor transport was not a choice. Breakouts of illnesses at the time were mistakenly blamed on overcrowding and high densities of urban areas. Calls were abound for change and spreading out, lowering the density of communities, and segregation of land uses. Supporting policies and zoning ordinances followed, and that was the origins of the current suburbanization and urban scrawl. It took time, but it turned out that overcrowding at the time was not to blame for the diseases breakouts; the cause was more to do with hygiene and waterborne bacteria. But by then, it was too late to turn the momentum around. A new life style has emerged: suburbia, urban scrawl, and, with it, high carbon mobility options (HCMOs).

The unintended consequences of such ill-founded notions and policies soon followed in the form of auto-oriented low-density developments, long commutes, traffic congestion, and increased air pollution. Although apparent, the extent and severity of the negative public health impacts associated with these "transport" issues would take time to fully mature. Only recently, we began to understand, measure, and quantify the negative health impacts associated with these common urban transport problems. And what we are finding is not pleasant: the new urban form that began to take shape at the dawn of industrialization, with its transport problems is now a serious health burden. In this context, many of today's sustainable traffic and transport solutions and trends, including LCMOs, are also public health tools. But how to undo or halt the harm that has been done is not an easy fate. As Holden puts it "how to untie the knot constituting mobility's positive and negative effects" is no easy undertaking [6].

8.3 Low carbon mobility options and health

Low carbon mobility is defined as "mobility that results in substantially lower levels of carbon" [7]. In this chapter, LCMO refers to: low carbon transport modes and vehicle technologies, land development strategies, and enabling policies for both. LCMOs, besides their obvious positive impacts on the environment, and to making cities more livable—thus improving social well-being—are also positive to public health. All LCMOs are positive to health. When a negative impact is there, it is insignificant. Transit and transit-oriented developments, cycling, walking, among others, are all associated with positive health outcomes. The benefits of these, through more physical activity and lower carbon footprint, are obvious. Their negative impacts in the form of increased exposure to crashes and/or pollution levels are not, especially when supportive policies are in place.

The LCMOs discussed here are presented for the purpose of, later, assessing their qualitative contribution to the specific health issues that are presented in Section 8.4. Their presentation here, therefore, is not a comprehensive one.

- Walking: increased physical activity means better health as walker are more likely to get all or part of their recommended daily exercise through "transporting" themselves. More walking trips means fewer use of HCMOs thus even more health benefits due to reduced air pollution.

- Cycling: similar to walking but fewer calories burnt per unit distance or time. It is nonetheless a healthy mode of transportation. Reduction in air pollution is an added health benefit.
- Mass transit (bus, metro) and transit-oriented developments: public transport and transit-orient development are both lower carbon mobility options and at the same time associated with positive public health impacts, both in physical and psychological terms. Transit-oriented development is a land-use solution that makes provision and use of public transit viable.
- High density mixed use land developments: reduces number and length of trips thus a reduction in energy consumption and pollution. It increases the shares of walking, cycling, transit and social interactions, and consumes less land.
- Flexible work hours: reduce congestion and, hence, the number of hours a driver is experiencing congestion and delay, and thus less stress and anxiety. These also mean fewer hours in sedentary situations (driving). The less of all of this is the better.
- Ride sharing: means less time at the wheel navigating and maneuvering through traffic, which translates into lower stress levels. It also means lower vehicle ownership and less land dedicated to vehicles. Long term, this is a definite health gain. Ridesharing means fewer vehicle trips for the same person trips. That means less fuel consumed, less pollution, which in translates into an even more health perks.
- Telecommuting: besides the all clear positive impacts on health of telecommuting—through reduction of pollution, crashes, stress, etc., Novaco and Gonzales noted that telecommuting in a way is also a mean to cope with stress as technological, socio-economic, and psychological factors have made work less place-dependent [8].
- Vehicle technology: this includes hybrid and electric vehicles and small and/or aerodynamic vehicles, and new designs as the active-lean Toyota I-Road. Those vehicles reduce energy consumption and pollution and use less for travel and parking. Reduced congestion, thus shorter commute duration, means less stress and shorter sedentary time.

8.4 Transport health impacts

Transportation systems, or travel within them, is viewed here as one of four-linked domains or environments (Figure 8.4). The health impacts of travel are thus not isolated from the influences of factors at home, work, or leisure places. Factors from home, work, or leisure domains can trigger, accentuate, or dampen health impacts/outcomes from transport systems. Factors from one or all of the four environments can—and are actually used—to offset or cope with transport system health impacts. In so far as health is concerned, LCMOs can be seen as a coping or mitigation tool—an effective one at that, especially if viewed in the long term. The dashed lines in Figure 8.4 reflect the notion that these major life domains and environmental settings are naturally/objectively connected in both space and time

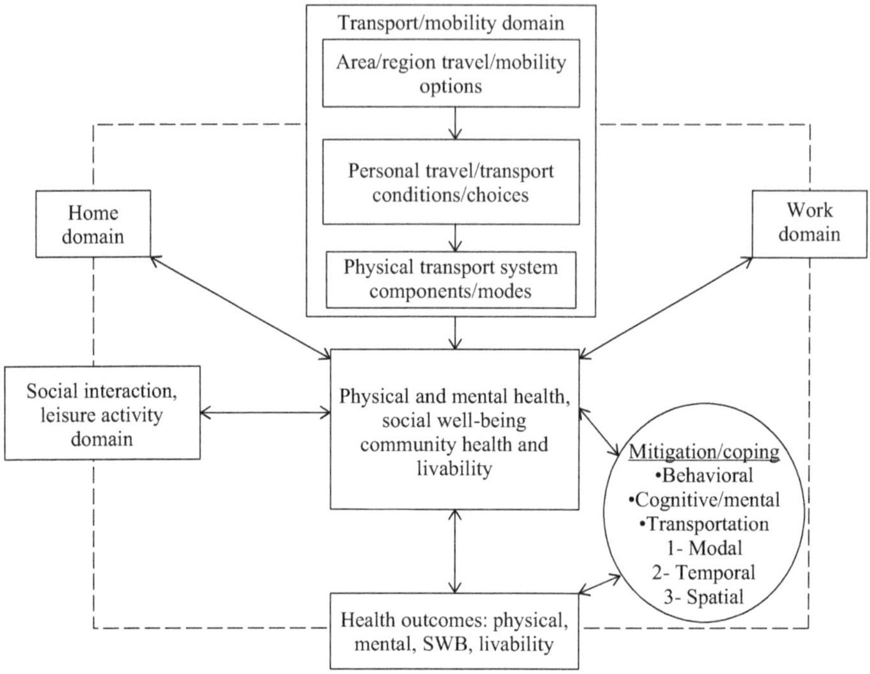

Figure 8.4 Ecology of transport and health

as part of the individual daily routines. Studying transport system health issues in isolation from factors in other environments is an incomplete portrayal of the issues.

Health impacts of transport systems range from immediate, clear, and easily countable or measureable (injuries in road crashes) to subtle, delayed, and hard to measure. Some of the impacts are positive; others are negative. Some of the impacts are acute; others are gradual. In some causes, the cause-and-effect relationship is clear; in others, a correlation is clear but little understood causality. Figure 8.4 depicts a proposed ecology of health impacts of transport systems. The transportation system is seen as one of multiple interconnected environments that all contribute to the health of the individual and the community. A mitigation set of options is part of the proposed framework. As with all health issues and concerns, individuals and communities can tackle transport system-related health issues through different coping strategies and means. When no modal, temporal, or spatial options are available, individuals can simply use their own mental and cognitive abilities and skills to deal with health issues.

8.4.1 Transport crashes and injuries

This is the most obvious, well-studied, and easiest way to quantify aspect of transport health impacts. The size of the impact is staggering. The numbers are numbing: road traffic injuries are the leading cause of death among people aged 15–29 years [9]. Worldwide, 1.2 million deaths occur due to road crashes, and 2 to 50 million sustain non-fatal injuries per year. The share of pedestrian, cyclist, and motorcyclists is disproportionately high and overrepresented. Low- and middle-income countries are at far higher risk. For the same countries, 5% of gross domestic product is lost due to road traffic deaths and injuries [9]. Troubling these numbers as they may be, they tell only part of the story: there are post-crash physical and mental health issues related to crash victims and family caregivers. The social well-being of both groups and their communities usually suffer well beyond what the crash and economic statistics. In so far as social well-being, community health, and livability, those post-crash issues are serious but not easy to quantify.

Although all these statistics and facts vary widely with countries and socio-economic factors, they would all improve by LCMOs and suitable policies and regulations. That around 65% of crashes happen in built-up areas reinforces that fact the LCMOs for futures cities is a health necessity. However, as speed and vehicle mass play significant roles in risk, proper policies and measure must accompany LCMOs. For example, speed reductions and management must accompany LCMOs lest the increased volume of higher crash risk walk- and cycle-based trips negates the health gains associated with those two modes.

8.4.2 Air pollution

The impacts of air pollution on health are serious, but the exact dynamics involving the different pollutants and the health outcomes are not well understood because, among other reasons, there are different types of pollutants, and their interaction with other confounding factors as traffic mix and technology, fuel types, weather, and topology. Regardless of exact source or health outcome, health impacts due to air pollution, to which transportation contributes a significant share, are serious. In some cases, the health effects are immediate; in others, they are incremental and occur over a lifetime.

Some of the immediate health issues attributed to air pollution include obvious and easy to see respiratory issues and eye irritation. Longer term subtle impacts include serious health ones include cognitive and developmental issues (especially in infants and youngsters), developmental and mental effects/impairment due to Lead emissions. Boothe *et al.* [10] noted that there is evidence that childhood leukemia is associated with residential traffic exposure during the postnatal period. Recent research conducted in the Los Angeles area in California, USA, found connections between people living near freeways and asthma, reduced lung function, cardiovascular disease, autism, and other health effects [11]. This is directly related to pollution. Levy *et al.* [12] evaluated the premature deaths results from people breathing primary PM2.5 (fine Particulate Matter 2.5 μm in diameter or smaller) and secondarily formed particles specifically during periods of traffic congestion for in the USA. They estimated that congestion-related PM2.5, NO_x, sulfur dioxide (SO_2) emissions cause around 3,000 premature deaths at a cost of $24 billion in 2005.

In most cases and health impacts, the exact cause and effect is not understood; in most studies, it was difficult to untangle the contributions of the many contributing factors and pollutants. For the purpose of this presentation, different pollutants are noted here, and their associated health impacts are mentioned. The more common pollutants of interest are: carbon monoxide (CO), nitrogen oxides (NO_x), volatile organic compounds (VOCs), particulate matter (PMx), lead, and SO_2. The exact shares from transportation of those pollutants vary. For example, in the USA, the share of all transportation is 53%, of which 34% is due to highway vehicles. Highway traffic contributions to specific pollutants are as follows (all transportation shares shown in parenthesis): CO 51% (77%), NO_x 34% (56%), VOCs 29% (47%), PM10 1.3% (3%), lead 0.5% (12%), SO_2 2.1% (7%) (Transportation Research Board, 2002). These shares may vary significantly based on distribution of population in urban and rural areas, thus the contribution of roadway traffic. The more significant pollutants, in terms of health effects, are:

- Carbon oxides: ozone, CO, carbon dioxide (CO_2) are also associated with some serious health impacts. In some cases, the negative health impacts are not through direct exposure. For example, the health impacts of CO_2 are the result of the climate change caused by this gas. Those indirect health impacts appear in the form of deaths due heat waves, floods and droughts, disturbance to complex physical and ecological processes. Potentially, other health impacts will appear but only in the long run [13]. When viewed through this compounded and long-term perspective, carbon oxides are a primary source of health concern.

  Besides its known role in global warming, ozone is also associated in reductions in lung functions, increased bronchial reactivity and admissions to hospitals [13]. Further to its indirect contribution to greenhouse gases though oxidation, independent health effect of CO include contribution to cardiovascular diseases and related hospital admissions [14]. Exposure to CO reduces blood's oxygen carrying capacity, which in extreme cases can lead to death.
- Particulate matter (PM): PM is "a complex mixture of extremely small particles and liquid droplets. Particle pollution is made up of a number of components, including acids (such as nitrates and sulfates), organic chemicals, metals, and soil or dust particles" [15]. There are different types based on size of PM (Figure 8.5): (1) Inhalable coarse particle (PM2.5-PM10), found near roadways and dusty industries and (2) fine PM (PM-2.5 and smaller), found in smoke and haze. These are either directly emitted from sources or they can form when gases from automobiles react in the air. Fine PM (PM2.5) is respirable (can be breathed in); however, there is evidence that even these smaller matters might be more relevant to health than larger ones. Even smaller, ultra-fine particles (PM0.1) have been linked to stronger effects on lung function and symptoms in individuals with asthma than PM2.5 and PM10. Those ultra-fine PM can enter the blood stream, which transports them to organs as the liver, kidneys, and brain (VOC, 2013).

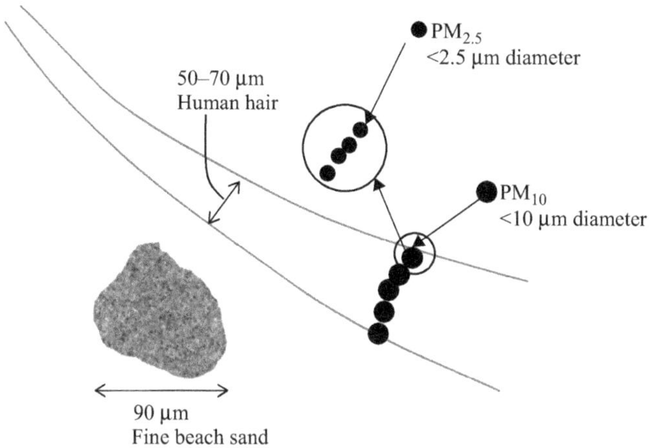

Figure 8.5 Illustration of different sizes of PMs (adapted from [15])

In many cities, even in the developed world, road traffic is the main source of $PM_{2.5}$. PM affects the heart and lungs and cause serious health effects. Quantifying the health effects of PM is not as easy in part because we still do not understand the exact cause and effect. In most situations, those PM are present with many other pollutants; isolating their role and effects from other pollutants is difficult. Studies have documented clear correlation between increased levels of PM and increased hospital admission of respiratory health cases. In Vienna, Austria, the daily number of hospital patients admitted for respiratory disorders increased by 5.5% for each 10-µg increments of PM2.5 per cubic meter [16].

- Sulfur and nitrogen dioxides: The health effects of sulfur and NO_x are not readily understood. In fact, it is not certain, but possibly, it is the mere presence of *other* particles whose effects are not accounted for [13]. But the correlations between these pollutants and certain health issues are established: the health impacts of ambient NO_2 include increased risk of respiratory tract infection through the pollutant's interaction with the immune system. Similarly, sulfur dioxide contributes to respiratory symptoms.
- Lead emissions: children are particularly vulnerable due to the disproportionate negative impact of a given Lead intake (when normalized by body weight). Lead impacts both mental capacity and functional abilities and academic progress [17–21]. Although lead as a fuel additive has been banned in the most western and developing countries, it is still used in many part of the world. Serious health issues appear to persist even with lower—otherwise thought safe—levels of Lead [17].

But how much of the above health concerns are traffic-generated? Most ground level pollutants that have recognized health effects are linked to motor traffic. There are, however, substantial differences. For example, all CO in

Northern Europe is from motor traffic. For the same region, it is the source of 75% of NO_x, 40% of PM10, and 25% of CO_2. Some of those numbers may decline or rise with time based on all contributing factors. Land development along with travel choices and technological changes to vehicles—both the mix and improvement to fuel efficiency of internal combustion engines—account for some of the large differences between the European and USA numbers. In some area, the technological improvements are offset by increase in number of vehicles and/or length of trips. Similar statistics are not available for many parts of the world, especially underdeveloped counties. In many parts of the world, such statistics and numbers are either not available or not accurate or complete. The cost of air pollution (due to all sources) to the world most advanced economies plus China and India is $3.5 trillion per year [22]. According to the same source, in OECD countries, the monetary impact of death and illness due to outdoor air pollution in 2010 is $1.7 trillion [22].

8.4.3 Psychological health impacts due to noise and other stress sources

The main source of these health issues are noise, stress from long commutes and difficult traffic conditions as chronic congestion, and post-trauma mental health impact both to the victims and to those around them.

- Noise: noise is a stressor, and transportation is a major noise source. Traffic noise in urban areas is a common and serious problem. Its symptoms include: irritability, stress, impaired social interaction. Noise impacts communication, school and work performance, sleep and temper, and cardiovascular effects and hearing impairments to name a few [13,23].

  High noise levels interfere with concentration and functioning of meetings and classrooms. For the elderly and hearing impaired individuals, this impact can be particularly pronounced. Noise interferes and reduces sleeping quality and sleep patterns and leads to fatigue and suboptimal performance. Children are particular vulnerable: noise can impair their reading acquisition, attention, and problem-solving ability [23]. Noise interferes and with and can compromise our ability to think and solve complex problems that require memory and concentration. Adaption strategies were found to lead to high blood pressure, increased stress and increased stress hormones [23]. Recent studies connected noise—through disturbing sleep patterns thus weakening the immune system—to Hodgkin Lymphoma among adults [24]. Orban *et al.* [25] reported on connections between residential traffic noise and depression.
- Long commutes: long commutes lead to long commutes stress, stress from road rage, reduced quality time for social interaction and connectivity. Long commutes, though, have some positive aspects: it can enhance well-being such as privacy, protected time, and the symbolic values of persona vehicle and freedom [8].
- Congestion: congestion induces tress, anxiety, reduced social interaction, and impaired communication at work and with family.

8.4.4 *Active transport*

Walking and cycling fall in this category (Chapter 9). Transit, though not an active mode of transport per se, was found to correlate with increased walking activities. For the purpose of this presentation, we label transit a semi-active transport mode. Beside its positive effects on health, by way of reducing volume of auto-trips thus reduced pollution levels and reduced commute stress, increased walking is a definite positive health benefit.

There are two key benefits to walking and cycling as modes of transport. The first includes all those associated with reduced auto trips; all those negative health impacts will fall. The second is the health benefits that come from the regular physical exercise that is a natural part of walking and cycling. Those health benefits have largely been overlooked and/or underrated—but not anymore. There is indisputable evidence to the positive impacts of physical activity on health [26–28]. The part that is still subject to research and discussion is how and what form of the physical exercise is best for the human body and how that is related to age, gender, and existing health conditions. But how much health benefits can we get from walking and cycling?

It's been known for long that walking and cycling to work have been shown to meet metabolic criteria for achieving health benefits from exercise [29]. Among the many specific health benefits are: 50% reduction in the risks of: coronary heart disease, adult diabetes, obesity; 30% risk reduction in hypertension; 10/8-mmHg (millimeter of mercury) decline in blood pressure; reduced risk of osteoporosis; relief of symptoms of depression and anxiety; and prevention of falls in the elderly [13]. But there are health risks with active and semi-active transport modes including crashes (with vehicles) and increased exposure to pollution. The benefits though outweigh the risks. For cycling, [30] quantified the balance between physical activity and air pollution and accident risks. They noted that that the "benefits outweigh the risks by an order of magnitude." Earlier, the British Medical Association observed that the benefits to life expectancy of cycling are 20 times the associate injury risks [31]. As to the increased risks due to air pollution, it may be true that some of the life expectancy gains from cycling would be lost due to increased exposure. However, the net benefit is positive. The key point here is that cycling increases physical activity. Without that there are only risks, and that, health-wise, is not positive. But this reinforces the point that reducing the risks is critical. Virtually, all other LCMOs help accomplish that, and as such they are doubly beneficial to health. The same is true for walking. The increased risks from crashes (with vehicles and cyclists) and exposure to pollution remain, on balance, insignificant. The USA Centers for Disease Control and Prevention contends that the benefits outweigh the risks especially when reasonable precautions are taken. As with cycling, reducing the risks to pedestrians is necessary to realize the full benefits of walking.

And it is not an inordinate amount of walking and cycle that is necessary to realize significant health gains. Old and new studies show that as little as 30 min of regular brisk walking (or 10,000 steps a day) or cycling reduces the risk of developing cardiovascular diseases, diabetes and hypertension and helps control body weight [32], or are associated with significantly higher health scores [33]. Better yet, the evidence now is that vigorous sustained exercise is not necessary to realize significant health benefits. That is a reinforcement of the notion that walking or cycling, simply as means/modes of transport to work, shop, socialize or visit parks is also a sure mean to attain public health goals. Where quantifications of the health benefits were attempted, the results are reassuring. For example, a Finish study estimated a 3%–7% reduction in deaths from coronary heart disease if 8% of the working population walked or cycled to work [34]. A 40% reduction in road causalities was attained with policies favoring walking and cycling in New York and London [35]. Earlier studies did not use similar procedure and as such the estimated benefits varied. A more systematic procedure that addressed many of the challenges professionals encountered earlier to quantifying the benefits of walking and cycling has been proposed [36] and will be briefly discussed in Section 8.5.

8.4.5 Community health, mental health, and social well-being

The following health issues fall in this category: post-traumatic stress from accidents, aggression and nervousness, reduced social life, constrains on physical space, and social interactions. These are not new subjects entirely; aspects of them are noted in other sections of this chapter. As these impacts span multiples of the earlier factors, they are presented here in the context of community health and social well-being.

Besides the significant physical toll of motor crashes that is monitored and documented in most countries, the long-term psychological effects are not. Studies have shown that sizable portions of motor vehicle crash survivors experience post-traumatic stress disorders and/or psychiatric problems for a long time after the crash [37–40]. Post-traumatic disorders can be serious and may involve re-experiencing the trauma through nightmares and flashback, uncontrollable, intrusive recollection; adopting avoidance tactics as keeping away from potential triggers, blocking feelings, and becoming detached or estranged; and excessive arousal with sleep difficulties and poor concentration and memory [40,41].

Besides aggression, nervousness, depression and increased irritability from noise, other aspects of traffic are equally troublesome. Repeated exposure to congestion impairs work performance, psychological adjustment and overall satisfaction with life [42,43]. The sense of confinement that comes with traffic congestion was found to increase blood pressure and frustration tolerance, something that reduces the well-being of those exposed. It can also lead to increased aggressive behavior, road rage, and the risk of being involved in a crash [44]. Up to one-third of study participants reported to have been participants in road rage [45].

Diminished social life is another price of transportation. The increased reliance on private automobile and suburbanization that comes with it precipitated increased isolation and reduced sense of community and level of interaction. In addition, it affected the social contact through what is called community severance effect where roads effectively divided those in the same locality. One of the most noted studies on the subject showed that after controlling for all external variables, the number of friends and acquaintances living on the same street diminished precipitously with increased traffic volume [46]. The relevance of this seminal study is even profound when we note that good social support groups and networks are critical for vulnerable groups as children and elderly people. A lack of social support was found to increase the mortality rates for coronary heart disease by up to four times [47].

Children development can be seriously restricted by traffic. Increased traffic volumes were also found to restrict the space within which children can move freely. Increased traffic was also found to restrict opportunities to walk or cycle. It also limited their experiences and hindered their personal development and their abilities to link with peers. The same study by Greenwood noted above found a clear connection between the time children spent outdoors and their sense of dangerousness and perceived attractiveness of their living environment [47]. The notion that modern communication tools and social media may have tempered the negative effects of traffic on social interactions and community health, and well-being is an interesting question that has not been fully answered. It is doubtful that the negative impacts on the more vulnerable groups, as children and the elderly, will be any different.

Modernity has its transport-health-related issues as well. Many wealthy and/or developed societies suffer from modern day non-communicable and serious diseases as obesity, diabetes, high blood pressure, strokes, cardiovascular problems, etc. In modern day, largely sedentary life style illnesses are associated with both a more sedentary life style, of which transport is a key enabler, and high caloric intake. The prevalence of these modern day diseases varies widely among countries and segments of population within a country. They are nonetheless a serious health threat [48].

8.4.6 Soil and water pollution and transport

Transport systems and activities can contaminate ground water and wetlands and soil due to accidents that involves hazardous material and goods and contamination from transportation specific infrastructure components. Examples of the later include fuel, salt, and other deicing agents, and roadway run-offs and storage facilities. The damage can reach humans and other living species through public water supplies, vegetation, and trees. The contribution of the transportation systems and the exact mechanism of occurrence of the contamination are not well understood due to the fact that it occurs over long time, and that it may originate from both point- and non-point sources. The presence of contamination in the food chain is often discovered long after the pollution itself has occurred. In so far as the health impacts, they are often better understood than the process and compounds that trigger them.

Significant progress has been made but considerable challenges remain and new ones appeared. LCMOs, by their own nature and the infrastructure they require, can only help and in big ways, in curbing issues related to soil and water pollution, and by extension, all of the health issues that stem from them. Reduced need for parking spaces and reduced roadway surface (thus reduced levels of polluted water runoff, lesser of salt and other deicing agents, less fuel spillage, and improved groundwater replenishment due to the reduction in impervious surfaces), reduced road construction and maintenance and accompanied reduction in chemical pollutant (from paint to bridges, road striping, and pesticide applications to road-side vegetation), and reduced chemical and biological contaminants present in storm water runoff, and reduced vehicles miles will mean less maintenance and fuel storage tanks and auto fueling stations are all good of LCMOs outcomes.

8.5 Assessing transport health impacts

All transport health impacts must be understood and accounted for. It is highly desirable that their economic value be estimated or measured. The necessity of the task stems from the fact that governmental units and agencies must incorporate health impacts of their projects and policies into the decision-making process including budgeting. This is a necessary first step to devise and implement solutions and counter measures. It is understood, though, that converting health impacts of different nature and from different specific causes into one economic unit is not an easy task. Part of problem is that there are no comprehensive metrics for health impacts. Also measuring health impact of transport options/choices is not an easy task, at least not in all cases and situations. In the case of traffic crashes and resulting injuries, it is easy. In the case of pollution, we made some progress; we have established some correlations between certain transport-related pollutants and certain health issues. Although the correlations in many cases are strong, those correlations still do not address the causality question.

The question of what metrics to use is especially important when the health impacts of different nature and appearance are to be combined. The current metrics used to measure and quantify health impacts use specific standard metrics: (i) years of life lost (YLLs), (ii) years lived with disability (YLDs), and (iii) disability-adjusted life years. This is the sum of the YLLs and YLDs. These metrics are consistent but reflect the negative definition of health introduced earlier. In the context of transportation, the holistic definition of health introduced earlier is more appropriate as transportation plays an integral positive role in our lives, both individuals and society. Incorporating metrics to capture community health, livability, and social well-being is necessary to fully capture and evaluate all relevant transport health impacts. Some progress has been made both on tools and measurements but much more remains to be researched and answered. Health Impact Assessments (HIAs) and the Health Economic Assessment Tool (HEAT), or Health Economic Assessment Tool are two recent developments aimed at formalizing the process of evaluation transport health impacts.

The HIA is a tool that can be used to inform the transportation planners of the potential health consequences of their plans and decisions [49]. The tool is still relative new. The extent and effectiveness of the interdisciplinary cooperation necessary to complete such studies has not been documented. Its usability and success is yet to be assessed. Dannenberg *et al.* [49] reported that 73 HIA studies were conducted in variety of projects in 22 states between 2004 and 2013. Conceptually similar but more active transport-oriented tools are in use in Europe, New Zealand, and Australia [50].

HEAT [51] is a tool designed to help conduct economic evaluation of walking or cycling. Similarly, tools are not available to do similar assessment of the negative transport health impacts outlined in the earlier parts of the chapter. Binary and Likert scale can be used but no standard process is yet available.

8.6 Summary and conclusions

Besides all the positive contribution of transport systems to societies and economies, they directly and indirectly impact public health. Transport as such is both a mean to prosperity and progress and a public hazard. This chapter makes the point that "transport has taken a low profile when it comes to making substantial contribution to public health improvement targets." And that, besides the known benefits of LCMO, there are significant public health benefits can be realized, especially in the long term. Not all aspects/impacts on public health, though, are well understood. The understanding in some areas is still crude. Further work and research is needed.

The definition of smart city as practiced today needs to expand to also mean "health/healthier". Even though health and healthy is sometimes implied, that is not enough. Health needs to be addressed and explicitly. Incorporating consideration of health in transportation projects, planning design, construction, operation, and maintenance is necessary. Significant negative impacts can be avoided; others can be minimized. Measuring health impacts, however, is not always easy; that contributes to uncertainty as to how much transportation is a part of public health. Either way it is certain that shifts towards LCMOs is a shift away from the negative health impacts of HCMOs. In some cases, certain LCMOs may have negative health impacts but the positives outweigh the negative ones by orders of magnitude.

Old and still current transport planning paradigms grounded in the "predict and provide" model are not sustainable nor are they efficient when issues beyond economic efficiency, as public health, are incorporated and accounted for. Those old paradigms are the roots of the HCMOs. HCMOs are the result of the dominance of economic growth, global trade, and "efficient" transport thinking. Here, economic growth is critical to economic and political stability and strength. It is also an enabler and facilitator of economic and strategic strength, thus the go-for-smooth-flow of goods and people eliminate bottlenecks. To accept otherwise is a cost to the economy. Time is money and therefore faster is better. This rationale drove the "predict (demand) and provide (infrastructure)" approach to transport planning. The unintended but real induced demand was overlooked, thus the never ending—actually increasing of—congestion. An alternative approach was inevitable. Transport demand management and sustainable transport became part of the transport planning thinking. But the current transport system is still carbon-intensive: continued emphasis and/or valuing of private car, higher speed and frequency and volume of travel [7]. A paradigm shift to LCMO is necessary, and a systems approach is a must to avoid unintended consequences. All activities in an urban environment are components of one interconnected systems. For it to function properly and optimally, all subcomponents (transport, land use, economy, politics, health system, etc.) must be considered as one whole. Devising solutions to any of those components must account for potential unintended side effects on other components.

As with all studies and coverages, no one chapter is enough to cover all that needs to be covered about transport impacts on health. This chapter is no exception. One thing became clear, though: transportation impacts on health, positive and negative, are pervasive and many of those impacts are not well understood. More research and analysis is still necessary before a fuller picture emerge of how urban transport systems impact health.

References

[1] IHME (Institute of Health Metrics and Evaluation). *Transport for Health: The Global Burden of Disease from Motorized Road Transport*, Institute for Health Metrics and Evaluation, University of Washington, 2014.

[2] Reynolds, G. *How Exercise Changes Our DNA*. NY Times, December 17, 2014.

[3] Aggrawal, A., A.J. Cook, R.A. Sequin, A.V. Moudon, P.M. Hurvitz, A. Drewnowski. Access to supermarkets and fruit and vegetable consumption. *American Journal of Public Health*. 2014; 104(5):917–923. doi: 10.2105/AJPH.2013.301763.

[4] Jiao, J., A.V. Moudon, S.Y. Kim, P.M. Murvitz, A. Drewnouski. Health Implications of Adults' Eating at and Living near Fast Food or Quick Service Restaurants. *Nutrition and Diabetes*. 2015;5(7):e171.

[5] Frieden, T. A Framework for Public Health Action: The Health Impact Pyramid. *American Journal of Public Health*. 2010;100(4):590–595.

[6] Holden, E. *Book Review, Moving Towards Low Carbon Mobility*. Editors M. Givoni and D. Banister. Edward Elgar Publishing: Cheltenham, UK, 2013.

[7] Givoni, M., D. Banister. *Moving Towards Low Carbon Mobility*. Editors M. Givoni and D. Banister. Edward Elgar Publishing: Cheltenham, UK, 2013.

[8] Novaco, R.W., O.I. Gonzalez. Commuting and Well-Being. In *Technology and Psychological Well-Being*. Cambridge University Press: Cambridge, UK. Yair Amichai-Hamburger (Ed.) 2011.

[9] WHO. *Global Status Report on Road Safety*, 2015. World Health Organizations. Online document.

[10] Boothe, V.L., T.K. Boehmer, A.M. Wendel, F.Y. Yip. Residential Traffic Exposure and Childhood Leukemia: A Systematic Review and Meta-analysis. *American Journal of Preventive Medicine*. 2014;46(4):413–422.

[11] Cone, M. U.S. *Neighborhoods Struggle with Health Threats from Traffic Pollution, Scientific American*. October 11, 2011.

[12] Levy, J.I., I.L. Jonathan, J.B. Jonathan, von S. Katherine. Evaluation of the Public Health Impacts of Traffic Congestion: A Health Risk Assessment. *Environmental Health*. 2010;9:65.

[13] WHO. *Transport, Environment and Health*. C. Dora and M. Phllips (Editors). WHO Regional Publications, European Series, No. 89. 2000.

[14] Morris, R.D., E.N. Naumova, R.L. Munasinghe. Ambient Air Pollution and Hospitalization for Congestive Heart Failure Among Elderly People in Seven Large US Cities. *American Journal of Public Health*, 1995;85:1361–1365.

[15] EPA (2016). http://www3.epa.gov/pm/. Accessed March 17, 2016.

[16] VCO (Mobility of the Future) (2013). http://www.vcoe.at. http://www.cleanair-europe.org/fileadmin/user_upload/redaktion/aktivitaeten/2013-02_FS_Feinstaub_engl.PDF

[17] Koller, K., T. Brown, A. Spurgeon, L. Levy. Recent Developments in Low-Level Lead Exposure and Intellectual Impairment in Children. *Environmental Health Perspectives*. 2004;112(9 (Jun)):987–994.

[18] Dietrich, K.N., O.G. Berger, P.A. Succop. Lead exposure and the motor developmental status of urban 6-year-old children in the Cincinnati prospective study. *Pediatrics*. 1993;91:301–307.

[19] Damm, D., P. Grandjean, T. Lyngbye, A. Trillingsgaard, O.N. Hansen. Early Lead Exposure and Neonatal Jaundice: Relation to Neurobehavioral Performance at 15 Years of Age. *Neurotoxicology and Teratology*. 1993;15: 173–181.

[20] White, R.F., R. Diamond, S. Proctor, C. Morey, H. Hu. Residual Cognitive Deficits 50 Years after Lead Poisoning During Childhood. *British Journal of Industrial Medicine*. 1993;50:613–622.

[21] Needleman, H.L., C. Gatsonis. Low Level Lead Exposure and the IQ of Children. *Journal of the American Medical Association*. 1990;263:673–678.

[22] UNEP (2014). http://www.unep.org/yearbook/2014/PDF/chapt7.pdf.

[23] WHO. Burden of Disease from Environmental Noise. *Quantification of Healthy Life Years Lost in Europe*. F. Theakston (Editor). World Health Organization. 2011.

[24] Sorensen, M., P.A. Harbo, M. Ketzel, D.S. Oksbjerg, S. Friis, O. Raaschou-Nielsen. *Road Traffic Noise and Risk for Non-Hodgkin Lymphoma Among Adults*. EuroNoise, Tenth European Conference on Noise Control, Maastricht, June, 2015.

[25] Orban, E., K. McDonald, R. Sutcliffe, B. Hoffmann, K.B. Fuks, N. Dragano, A. Viehmann. *et al*. Residential Road Traffic Noise and High Depressive Symptoms after Five Years of Follow-up: Results from the Heinz Nixdorf Recall Study. *Environmental Health Perspectives*, 2016;124(5):578–585.

[26] Buchard, C., S.N. Blair, W.L. Haskell. Why Study Physical Activity and Health? *Physical Activity and Health*. Bouchard *et al*. (Editors), Human Kinetics, Inc. 2012.

[27] McCartney, N. Physical Activity, Muscular Fitness, and Health. *Physical Activity and Health*. Bouchard *et al*. (Editors). Human Kinetics. 2012.

[28] CDC. *The Benefits of Physical Activity. US Center for Disease Control and Prevention*. 2015. http://www.cdc.gov/physicalactivity/basics/pa-health/. Accessed April 15, 2015.

[29] Oja, P., I. Vouri, O. Paronen. Daily Walking and cycling to work: their utility as health enhancing physical activity. *Patient Education and Counseling*. 1998;22(Supplement 1):87–94.

[30] de Hartog, J., H. Boogaard, H. Nijland, G. Hoek. Do the Health Benefits of Cycling Outweigh the Risks? *Environmental Health Perspectives*. 2010;118(8):1109–1116.

[31] BMA (British Medical Association). *Cycling: Towards Health & Safety*. Oxford University Press: Oxford, UK, 1992.

[32] Mourphy, M.H., A.E. Hardman. Training Effects of Short and Long Bouts of Briks Waling in Sedentary Women. *Medicine and Science in Sports and Exercise*. 1998;30(1):152–157.

[33] Raymond, J., S.T. Johnson, W. Diehl-Jones, J.K. Vallance. Walking, Sedentary Time and Health-Related Quality Life Among Kidney Transplant Recipients: An Exploratory Study. *Transplantation Proceedings*. 2016;48(1): 59–64.

[34] Vouri, I.M., P. Oja, O. Paronen. Physical Active Commuting to Work—Testing its Potential for Exercise Promotion. *Medicine and Science in Sports and Exercise*. 1994;26(7):844–850.

[35] BMA. *Road Transport and Health*. British Medical Association, London. 1997.

[36] Kahlmeier, F.R., F. Racioppi., N. Cavil, H. Rutter, P. Oja. Health in All Policies In Practice: Guidance and Tools to Quantifying the Health Benefits of Cycling and Walking. *Journal of Physical Activity and Health*. (Supplement 1)2010, 7(Suppl 1), S120–S125:12–125.

[37] Goldberg, L., M.A. Gara. *A Typology of Psychiatric Reactions to Motor Vehicle Accidents*. *Psychopathology*. 1990;23:15–20.

[38] Green, M.M., A.C. McFarlane, C.E. Hunter, W.M. Griggs. Undiagnosed Post-Traumatic Stress Disorder following Motor Vehicle Accidents. *Medical Journal of Australia*. 1993;159:529–544.

[39] Bush, S.S., T.E. Myers. Neuropsychological Services Following Motor Vehicle Collisions. *Psychological Injury and Law*. 2013;6(1 (March)):3–20.

[40] Wu, K.K., W.L. Patrick, L. Leung, W. Valda, SC Lawrence, SC Law. Posttraumatic Growth After Motor Vehicle Crashes. *Journal of Clinical Psychology in Medical Settings*. 2016;23(2 (June)):181–191.

[41] De Horne, L.D.J. Detection and Assessment of PTSD by Health and Legal Professions. *Psychiatry, Psychology and Law*. 1993;2(1):65–73.

[42] Novaco, R.W., M.M. Green, A.C. McFarlane, C.E. Hunter, W.M. Griggs. Objective and Subjective Dimension of Travel Impedance as Determinants of Commuting Stress. *American Journal of Community Psychology*. 1990; 18:231–257.

[43] Wener, R.E., G.W. Evans. *Comparing Stress of Car and Train Commuters. Transportation Research Part F: Traffic Psychology and Behaviour*. 2011; 14(2 (March)):111–116.

[44] Harris, P.B. Recklessness in Context Individual and Situational Correlates to Aggressive Driving. *Environment and Behavior*. 2010;42(1):44–60.

[45] Sansone, R.A., L.A. Sansone. Road Rage: What's Driving It? *Psychiatry*. 2010;7(7):14–18.

[46] Cassidy, T. *Environmental Psychology: Behavior and Experience in Context*. Psychology Press. 1997

[47] Greenwood, D.C., K.R. Muir, C.J. Packham, R.J. Madeley. Coronary Heart Disease: A Review of the Role of Psychological Stress and Social Support. *Journal of Public Health Medicine*. 1996;18:221–231.

[48] Bell, J. *The State of the UAE's Health*. 2016. http://gulfnews.com/gn-focus/special-reports/health/the-state-of-the-uae-s-health-2016-1.1658937.

[49] Dannenberg, A.L., A. Ricklin, C. Ross, M.A. Schwartz, J. West, S. White *et al. Use of Health Impact Assessment for Transportation Planning: Importance of Involving Transportation Agencies in Process*. Proceedings of the 2014 Transportation Research Board Meeting, Washington, D.C., 2014

[50] Mueller, N., D. Rojas-Ruedaa, T. Cole-Huntera, de Nazelled Audrey, E. Donse, R. Gerikeg. *et al.* Health Impact Assessment of Active Transportation: A Systematic Review. *Preventive Medicine*, 2015 July; 76:103–114.

[51] WHO (World Health Organization). 2014. http://www.heatwalkingcycling.org.

Glossary

BMA	British Medical Association
CDC	Centers for Disease Control and Preventions (USA)
CO	carbon monoxide
CO_2	carbon dioxide
CRHI	crash-related health impacts
HCMOs	high carbon mobility options
HEAT	health economic assessment tool
HIA	health impact assessment
GDP	gross domestic product
LCMOs	low carbon mobility options
NCRHI	non-crash-related health impacts
NO_x	nitrogen oxides
OECD	organization of economic cooperation and development
PMx	particulate matter with less than x nanometer (μm) in diameter
SO_2	sulfur dioxide
SWB	social well-being
TRB	Transportation Research Board

WHO	World Health Organization
VOCs	volatile organic compounds
YLD	years lived with disability
YLL	years of life lost

Further Reading

[1] Ryanolt, E., E. Christopher. How Does Transportation Affect Public Health? *Public Road*, FHWA-HRT-13-004, 2013; 76(6 (May/June)).

[2] Frank, L., S.Kavage, A.Devlin. *Health and the Built Environment, A Review*. The Canadian Medical Association, June 2012.

[3] Langlois, M., R.A.Wasfi, N.A.Ross, A.M.El-Geneidy. Can Transit-Oriented Developments Help Achieve the Recommended Weekly Level of Physical Activity. *Journal of Transport & Health*. 2012;3(2):181–190.

[4] Dora, C., J.Hosking, P.Mudu. *Urban Transport and Health*. World Health Organization, 2011.

[5] Frank, L., P.Engelke. *How Land Use and Transportation Systems Impact Public Health: A Literature Review of the Relationship Between Physical Activity and Built Form*. US CDC. http://www.cdc.gov/nccdphp/dnpa/pdf/aces-workingpaper1.pdf.

[6] Transport and Public Health. Europe Environment Agency, 2016.

[7] The Hidden Health Costs of Transportation. Report by Urban Design 4 Health and the American Planning Association (APA), 2010.

[8] Litman, T. Transportation and Public Health. *The Annual Review of Public Health*. 2013;34:22.1–22.17.

Chapter 9

Active transport: policy directions for creating built environments that support health and well-being

Susan Thompson[1] *and Michael A.P. Taylor*[2]

Abstract

The health and well-being of individuals and the communities to which they belong is strongly affected by travel behaviour, especially in car-dependent cities where sedentary lifestyles and social isolation are direct outcomes. Our chapter focuses on these issues, examining how planning policy, practice and design can facilitate health promoting, low carbon active transport such as public transport use, walking and cycling. We discuss the potential co-benefits emerging from the adoption of such mobility policies. The co-benefits are largely found in the areas of environmental impact, public health, social cohesion and economic productivity. A framework for identifying and assessing potential co-benefits is provided, as are case studies of planning and policy interventions to encourage low carbon active transport forms.

Keywords

Active transport; public health; co-benefits; walking; cycling; assessment; urban development

9.1 Introduction

Movement is at the heart of human health and well-being. When we move, we exercise the physical body and inevitably, meet with others as we travel from one place to another. The health and well-being of individuals and their communities is now known to be strongly affected by travel behaviour. This is especially the case in contemporary car-dependent cities where sedentary lifestyles and social isolation

[1]City Futures Research Centre, Faculty of Built Environment, The University of New South Wales, Sydney, Australia
[2]School of Natural and Built Environments, University of South Australia, Adelaide, Australia

are direct outcomes of a road-based transport system. Being inactive and lonely are significant risk factors for the major chronic diseases in epidemic proportions across the globe [1]. These are non-communicable diseases ranging from cancer, heart disease and diabetes, to asthma and depression. It is widely acknowledged that these conditions affect '... people of all ages, nationalities and classes' [2].

This chapter shows how health and well-being are affected by the way people travel, which is also related to environmental impact. Car-based journeys are at the core of this discussion. Not only do they contribute to risk factors for chronic disease, they are fuelling climate change and planetary degradation (Chapter 8). Health professionals, designers and urban planners are increasingly working together to address these inter-related issues. The built environment is being designed in a way that supports both human and planetary health. Referred to as co-benefits, they are emerging from the adoption of low carbon mobility (LCM) policies and the promotion of active transport modes – walking and cycling – in our cities. The co-benefits are largely found in the areas of positive environmental impact, public health, social cohesion and economic productivity. Our chapter provides a framework for identifying and assessing potential co-benefits. The chapter also includes case studies of planning practice and policy interventions to encourage low carbon active transport forms such as walking, cycling and public transport usage.

9.2 Linking the built environment, travel modes and health

We now turn to consider how the built environment influences health and well-being in terms of travel modes. First, it is necessary to set the broader context for this discussion. There is a large body of research evidence demonstrating a direct relationship between the shape of the built environments and people's health [3,4]. In previous work, Kent and Thompson [4] devised a useful way of conceptualising the relationship between the built environment and health. This is via three domains which are where urban planning can most effectively and efficiently focus support for people's health and well-being. These three domains relate to specific behaviours needed to achieve good health:

- *Getting people active* (physical activity) – to reduce obesity and risk of heart disease, diabetes, some cancers and depression.
- *Connecting and strengthening communities* (social interaction) – to reduce risk of mental illness, particularly depression.
- *Providing healthy food options* (nutrition) – to reduce obesity and risk of heart disease and other chronic conditions.

Table 9.1 lays out the basis for a healthy built environment based on these three domains. Table 9.2 expands the specific actions for the 'Getting people active' domain, using two sets of questions about how the built environment can contribute to healthy behaviours in terms of utilitarian physical activity and recreational physical activity. Table 9.3 indicates how activity can also be encouraged and supported, in less direct ways, through the domains of 'Connecting and strengthening communities' and 'Providing healthy food options'.

Table 9.1 The basis of a healthy built environment – key domains and actions to support healthy behaviour (adapted from [5])

Key domain	Specific actions for healthy supportive behaviour
Getting people active The built environment can get people active by facilitating both utilitarian physical activity ('active transport') and recreational physical activity	*Facilitate utilitarian physical activity* Increasing the accessibility of destinations via active transport modes (i.e. by walking, cycling, and using public transport), and ensuring the experience of walking, cycling and public transport is of high quality, will assist the use of these modes relative to more sedentary car travel *Facilitate recreational physical activity* Providing facilities for formal and informal, and individual and group physical recreation in public spaces and via commercial and non-commercial organisations will assist in increasing overall levels of physical activity
Connecting and strengthening communities The built environment can connect people and strengthen communities – and thus support mental health by fostering feelings of belonging and restoration – by facilitating incidental neighbourhood social interactions, by making community spaces, and by building for crime prevention	*Facilitate incidental neighbourhood interaction* Ensuring public spaces are 'friendly' (busy, comfortable, safe and open to all) and with clear expectations as to appropriate behaviour included in the design of public spaces (e.g. via the provision of facilities and signage) will assist in encouraging positive incidental interactions between individuals and groups *Make community spaces* Providing explicit and well-designed (accessible, comfortable, safe) spaces open to all will allow for gatherings and other activities by the community as a whole and for particular interest groups; similarly, access to natural green environments will extend the notion of community to include the psychological restorative effect of nature *Build for crime prevention* Designing the built environment to discourage crime and help people feel safe (while still facilitating social interactions) will assist an overall sense of belonging, caring and community commitment
Providing healthy food options The built environment can provide healthy food options by facilitating access to healthy food, and by responsible food advertising	*Facilitate access to healthy food* Ensuring supermarkets, green grocers and farmers' markets are accessible relative to fast food outlets, pubs and convenience stores (e.g. through zoning and land-use regulation, and subsidised spaces) will promote the consumption of healthy foods and discourage purchase of unhealthy alternatives *Promote responsible food advertising* Marketing, advertising and promoting the visibility of healthy foods (e.g. near schools and other community locations, and relative to unhealthy foods) will have positive influences on consumption habits

Table 9.2 Specific planning-related questions and measures for the key 'Getting people active' domain (adapted from [5])

Specific health supportive action	Built environment contribution to healthy behaviour	Rationale for the built environment measure
Facilitate utilitarian physical activity	Do people engage in active transport modes?	Relative number of trips by active transport (defined as walking, cycling and public transport use) indicates levels of non-sedentary means of transport. It is also important to distinguish between types of active transport, as shown below
	Do people use public transport?	Relative number of public transport trips is an indicator of an aspect of active transport
	Is public transport viable (convenient, comfortable, safe and affordable)?	Easy-to-access public transport can encourage active travel modes. Facilities, such as bus shelters and bicycle racks, can also encourage public transport use. Access to destinations further afield may be facilitated by public transport
	Do people walk or cycle as a means of transport?	Relative number of trips by walking or cycling indicates two forms of active transport. Similar health benefits arise if people cycle for (otherwise walkable) trips
	Is walking a viable transport form (convenient, comfortable and safe)?	The grouping, layout and travel distance between land uses, particularly destinations accessed daily, influence walking, and therefore propensity to walk regularly for transport
	Do people cycle as a means of transport?	Relative number of trips by cycling indicates levels of active, rather than sedentary, means of transport
	Is cycling viable for active transport (convenient, comfortable and safe)?	Travel distance between land uses, particularly destinations accessed daily, influences cycling, and therefore the propensity to cycle
	Do people use stairs?	Using stairs, rather than a lift or escalator, provides an opportunity for incidental physical activity
	Is use of stairs viable?	Stair visibility and ease of accessibility and convenience increase the propensity to use stairs

Facilitate recreational physical activity	Do residents or workers walk for recreational physical activity?	Recreational walking is a good way to achieve the required minimum hours of physical activity to keep healthy
	Is walking viable for recreational physical activity: convenient, comfortable, safe and enjoyable?	The provision of comfortable, safe, convenient and attractive routes encourages walking for recreational physical activity. Green parklands and access to natural areas such as waterways, are particularly important
	Do residents cycle for recreational physical activity?	Recreational cycling is a good way to achieve the required minimum hours of physical activity to keep healthy
	Is cycling viable for recreational physical activity: convenient, comfortable, safe and enjoyable?	The provision of comfortable, safe, convenient and attractive routes encourages cycling for recreational physical activity
	Does public open space provide for recreational physical activity?	An important function of public open space is to provide facilities for a variety of recreational physical activities that need a lot of space, particular equipment, and use by teams and groups
	Are other facilities available (by either public or private providers) for recreational physical activity?	Some recreational physical activities will not be provided for within public spaces at the neighbourhood scale (e.g. sporting ovals, large parklands, extensive walking trails) – but still need to be accessible to invite use. Private spaces (e.g. indoor gymnasiums, yoga studios) if accessible (considering distance, transport options for access, operating hours, provision of child care, etc.) allow additional opportunities for recreational physical activity

Table 9.3 How 'Community connection' and 'Access to healthy food' can contribute to physical activity as part of a healthy built environment (adapted from [5])

Domain	Facet	Questions	Actions
Community connection	Nature	Does the design and governance of public and private space allow contact with nature?	Providing opportunities for contact with nature (green features, water amenities, and opportunities to garden in common areas) can support mental health by fostering feelings of restoration and relaxation
	Accessibility	Is formal public and semi-public space accessible to the community at large?	Real and perceived accessibility of neighbourhood destinations will promote use for physical activity and social interactions, as well as reduce vehicular trips and increase neighbourhood cohesion and safety
	Safety	Is the use of public space for active transport, incidental and organised physical exercise, and social interactions hindered by actual or perceived threats to personal security?	People will not interact within, or feel part of, a community they perceive to be unsafe. Actual and perceived levels of security can inhibit or promote choices to actively travel, participate in recreational physical activity and/or engage in social interaction. Fostering a sense of belonging, caring and commitment to public space also requires a good level of perceived safety
Access to healthy food	Garden	Can residents grow healthy food?	The provision of space and resources can encourage people to grow their own food. This facilitates access to fresh, affordable and nutritious food. It also raises interest and awareness about healthy eating generally. Working in a garden growing food together involves physical activity, as well as social interaction

These three domains of healthy built environments are closely connected and 'interlock' with each other. These interlockings also comprise their own important, active entities of spatial (built environment) and behavioural interactions. For instance:

- it is easier to be active when you are doing things with supportive companions (socialising)
- nutrition is important in maximising the benefits from keeping active (eating well), and
- food is a good catalyst for maintaining social connections (eating together) and for activity (community gardening, getting to local markets by active transport).

When these six entities (the three domains and the three connecting interactions) are combined, they create a 'synergistic' whole that is more powerful in generating healthy outcomes than any of the individual components in isolation [5] (Figure 9.1).

Built environments that support human health need to respond to specific locational and community characteristics. There are nevertheless, urban forms that, although not specifically developed to support health, result in environments that are generally conducive to well-being. These urban forms or models include walkable communities, transit-orientated development, urban villages, liveable neighbourhoods, smart-growth and new urbanism. Various handbooks and check-lists exist for these approaches, some of which appear in the case study section of this chapter. Features of such urban forms include:

- a close connection between land-use planning and transport planning
- population densities that are able to support effective public transport
- mixed residential, non-residential and community land uses clustered around, or within, walking distance of public transport stops
- public spaces and streets that are pleasant, inviting and well maintained – and as such invite use
- streets designed with not just the motor car in mind, making it pleasant and safe for other users ('complete streets')
- daily neighbourhood activities clustered to make them convenient to access, and to promote a 'bustle' that encourages socialisation, and
- in new developments, up-front provision of retailing, parks and other community facilities to ensure patterns of behaviour conducive to good health are established when the first residents arrive [5].

9.3 Defining co-benefits

In the context of planning and policy interventions that encourage LCM and active transport forms such as walking, cycling and public transport use, it is necessary to identify the associated co-benefits. Put simply, it means that multiple benefits are possible from one policy initiative designed to do something else [6]. The notion of co-benefits is being developed significantly in relation to action on climate change where it is acknowledged that low carbon ways of living are also healthy ways of

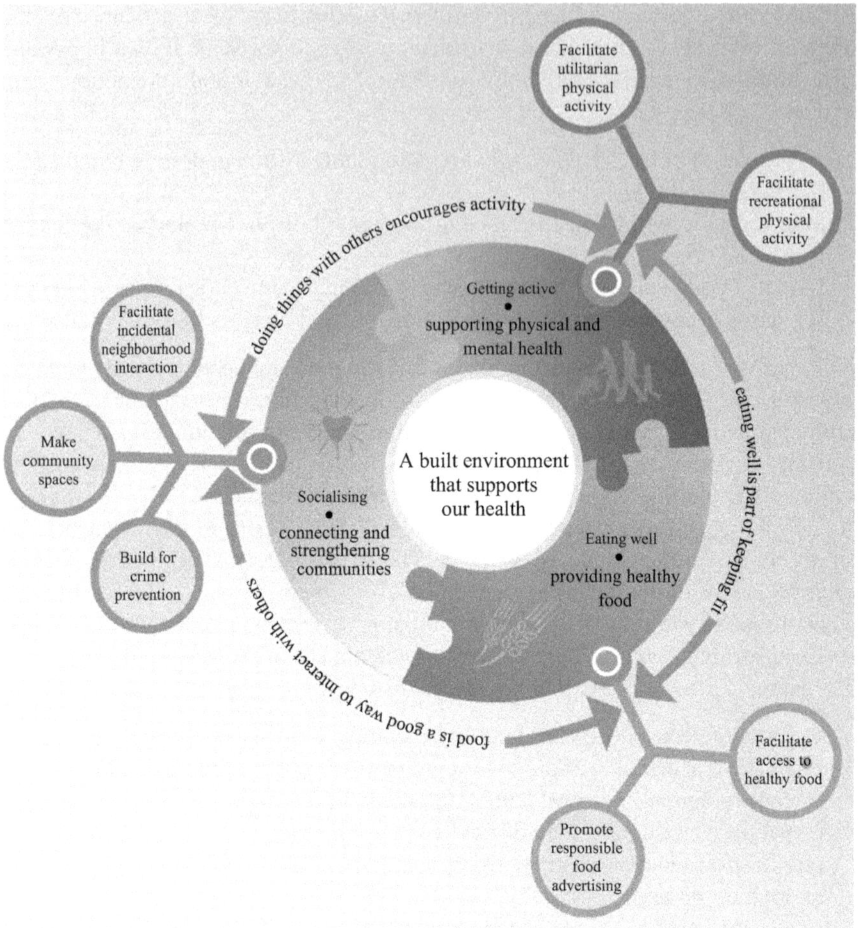

Figure 9.1 Supporting health through the built environment. Source: [5]. Greg Paine and Susan Thompson own copyright

living [7–9]. While co-benefits are defined in different ways, they are 'derived from the intentional decision to address air pollution, energy demand and climate change in an integrated manner, but also consider the other unspecified benefits that may arise such as improved transport and urban planning, reduced health and agricultural impacts, improved economy or reduced overall policy implementation cost' [10]. Thompson and Capon [11] used the diagram in Figure 9.2 (adapted from [12]) to describe the co-benefit concept by considering human health impacts:

> *All human activities have potential direct positive and negative human health impacts. This is through pathways such as nutrition and levels of physical activity, and indirect human health impacts through the health of*

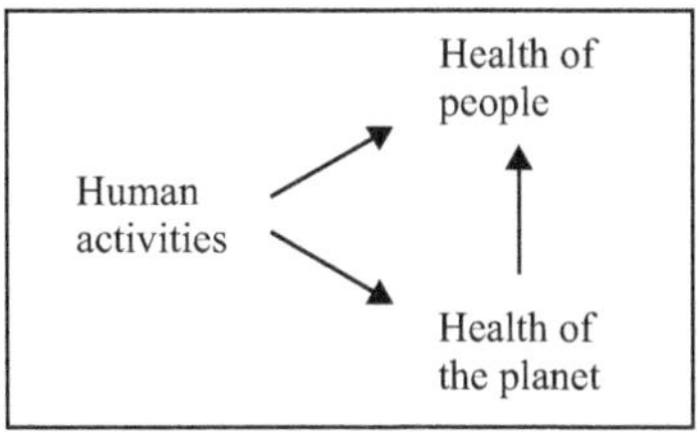

Figure 9.2 Concept of co-benefits for health. Source: [11,12].

the planet (for example, the climate system). It follows that there will be co-benefits for health from actions to tackle climate change. For clarity, the arrows are presented as uni-directional, however there are relationships in both directions.

Our focus in this chapter is on the co-benefits that can be derived from facilitating mobility in ways that are health supportive, economically efficient and environment friendly. Increasingly this involves the provision of a broad range of travel modes from the private car, to public transport, cycling and walking. Incorporating physical activity as part of transportation results in greater opportunities to redress sedentary behaviour, which as discussed earlier, is a significant risk factor for chronic disease. Active transport encompasses walking and cycling, and the use of public transport. Several studies – both in Australia and elsewhere – have shown that walking to and from public transport can achieve close to one-third of the daily recommended physical activity for an adult [13,14]. As well, a decline in private-car use has environmental benefits. This occurs by lowering carbon dioxide emissions which positively impacts on health by reducing associated risks of respiratory and cardiovascular disease [15]. Economic gains result from less expenditure on medical interventions. These financial benefits are also associated with bicycling infrastructure which can reduce traffic congestion and parking costs [16,17]. Fewer cars will improve safety in relation to traffic accidents and as there are more people on the street, walking and cycling in their neighbourhoods, community safety can also be anticipated, benefiting social and cultural capital. These achievements are a clear illustration of co-benefits from active transport policies.

We now turn to explore in greater detail the expected benefits from adopting active transport options. These have been classified under four descriptive categories which are addressed below:

- public health
- environmental
- quality of life/social, and
- economic/productivity.

To put this in context, we first consider the broader co-benefits context for active transport planning and policy.

9.4 Planning and policy directions for active transport

North American and Australasian cities are typically composed of low density, dispersed suburbs, which are highly car dependent. This is resource and carbon emission intensive and therefore unsustainable in the long term [18]. Planners and policy makers are seeking the means for the transition of these cities to low carbon alternatives with more efficient forms and functions with respect to carbon emissions. A low carbon city must improve the energy efficiency of its buildings and transport systems [19]. A key aspect of this transition is transportation; the level and intensity of the demand for which depends greatly on the land-use system and the land-use planning instruments that have been applied. These all, in turn, affect the lifestyles and health outcomes of the population, and the physical, social and economic environments pervading the city.

Transport systems dictate our mobility, that is, our ability to move from one place to another in order to achieve our needs for participation in activities and utilisation of facilities, the locations of which depend on the distribution and intensity of land uses across a region. Transportation provides access to jobs, education and social interactions, for example, all of which are fundamental to human development [20]. Public awareness and perception are two major barriers for the transition to low carbon transport options. Banister [21] emphasises the need to understand the importance of time and how people want to use time in travel compared to other activities. Aditjandra *et al.* [22] suggest that residents have to be well informed about the available opportunities for them to pursue sustainable travel choices.

The shape and form of our built environment, along with lifestyle choices, strongly influence the use of the private car in urban areas. Newton and Newman [23] addressed this issue in their consideration of the carbon benefits that can accrue from more compact urban forms where public and active transport modes are generally of most benefit. They considered the need for urban design innovation in Australia, largely focusing on energy demands in housing and transport, as well as available alternative energy technologies and fuel types. The 'post war' suburbs, that is those areas of our cities first developed in the latter half of the twentieth century which are heavily car dependent, provide the most challenges in transitioning to a lower carbon future. The Newton–Newman framework for low carbon technology interventions is based around the consideration of appropriate low carbon technologies applied to housing and transport in suburban and inner urban areas. The framework identifies key differences in consideration between higher density urban regions and suburban developments. An extended version of the framework, explicitly identifying the active transport modes as key factors for both urban and suburban transport, is presented in Figure 9.3.

Given the known trade-off between housing type and location, with resulting impacts on demand for transport which are largely met at present by the private car in suburban locations, strategies to reduce urban carbon emissions need to be directed at providing substantial alternatives to that mode, based around greater use of the active modes and public transport. To make this direction feasible requires in-depth consideration of built form, land-use-transport interaction, land-use mixes,

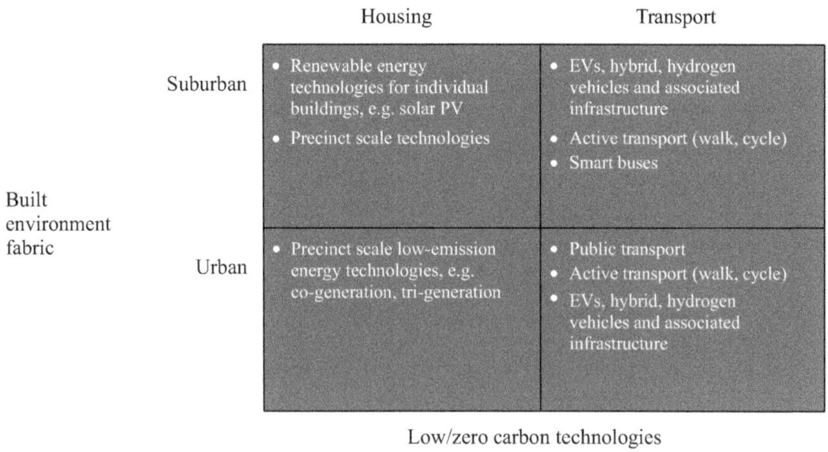

Figure 9.3 *Modified Newton–Newman model framework for low carbon technology interventions in urban and suburban forms, including active transport as a key factor in both urban and suburban transport (adapted from [23]). The original of this figure is also included in Chapter 6 and permission to re-produce has already been obtained. The journal is open source*

the location and intensity for services and facilities, as well as the supporting infrastructure, and precinct planning and design. The latter includes the relationships between neighbouring precincts, and between precincts and major activity centres such as the central business district.

9.4.1 Planning interventions overview

Urban form can have a significant effect on the carbon intensity of travel, with some urban forms showing a greater capacity than others to reduce the rate of carbon emissions per capita. Newton *et al.* [24] underscore the importance of integrated land-use and transport planning for growing cities. They provide examples where transport planning and development planning have happened independently resulting in reliance on private vehicles or excessive pressure on public transport services. A major transformation in the way transport planning is carried out is required. A new approach towards environmental and liveability aspects, together with a focus on achieving carbon efficiency in transport and urban precincts is essential [25]. Beyond reducing carbon emissions, changing the way we undertake urban and transport planning offers the potential to generate other important long-term benefits. These include open space preservation, improved air quality and public health, and reduced infrastructure investments, leading to improved quality-of-life in urban areas [20].

The travel decisions that people make are strongly associated with built environment factors such as density, location, mix of land uses and precinct design [26].

Changes are starting to occur in the structure of cities as they reach a limit to car use growth, and there are shifts away from traditional urban culture and economic paradigms [27]. As such there exists the potential for urban design and planning to play a major part in establishing a low carbon transport future.

Many theories exist regarding the ideal layout for low carbon cities; however, there is a huge gap between theory and practice [26]. The current understanding is that urban form and layout set the 'envelope of possibilities' for travel. Further, residential layout and supporting facilities have a complex relationship with resident attitudes, preferences and perceptions [22,26]. The 'compact city' approach attempts to bring activities closer to residents so that they can fulfil their needs by using low carbon options such as active travel or public transport. Problems exist in maintaining quality of life and space in high population density environments, as well as available options for built environment interventions where high density living is not feasible.

9.4.2 Policy interventions overview

Santos *et al.* [28] consider a wide range of LCM policy options, which they group into three categories:

- physical policies, concerning infrastructure and service provision,
- soft policies, aimed at bringing about behavioural change, and
- knowledge policies, emphasising the role of investment in research and development for future sustainable mobility.

They argue that policy integration is the key and that optimal blending of mutually reinforcing policies is essential. Banister [29] discusses the concept of the sustainable mobility paradigm, which again requires integrated and mutually supporting sets of policies. Banister *et al.* [30] suggest a possible policy framework for LCM systems. However, this still requires determination of relevant policy sets. A major question is: just what policies should be considered, and how can policy settings be optimised? Nakamura and Hayashi [31] review international developments on LCM policies and strategies. They conclude that the specific development processes of individual cities significantly affect the feasibility and effectiveness of different policies. As an example, Whitehead [32] provides the case of Stockholm, Sweden where a congestion charging scheme was introduced with an exemption for low-emission vehicles, in conjunction with the introduction of financial incentives for the purchase of low-emission vehicles. As the number of low-emission vehicles increased, the effectiveness of the congestion reduction charging was severely diminished. As such, the exemption policy was phased out less than 18 months after being introduced.

9.5 Co-benefits generated from low carbon mobility plans

We now turn to discuss the expected benefits from adopting active transport options.

9.5.1 Public health benefits

Walking and cycling provide zero carbon impact and numerous public health co-benefits. Many studies have investigated the impact of active transport on public health [26,33–36]. Providing opportunities for regular physical activity, such as walking and cycling, can make a big impact on improving public health and life expectancy [37–42]. In fact, the quantified health benefits of active transportation can outweigh any risks associated with these activities by as much as 77 to one, and add more years to our lives than are lost from inhaled air pollution and traffic injuries [43,44]. Cities with higher rates of cycling and walking to work also appear to have a greater percentage of the population meeting recommended levels of physical activity and have lower rates of obesity, high blood pressure and diabetes [35]. Active transport not only improves our physical health, but also our mental well-being and ability to focus [45–48]. Specifically in relation to children's health, Mackett [49] reviewed the impact of active travel on school children. The study showed that children who walk more than they travel by car have greater energy in all activities over a larger proportion of the day, on average, than those who use the car more than they walk [49,50].

The task remains to find an effective way to evaluate these benefits. Shaw *et al.* [51] conducted a systematic review of the public health and carbon benefits from various international transport policies. They concluded that the quality of studies was generally poor, raising validity issues with respect to reported results, due to issues including:

- limited methodological detail
- poor or no justification for selection of controls
- no information about the validity of measurement instruments
- inadequate measurement of transport use and health outcomes
- poor statistical analysis
- short follow-up period, and
- failure to consider and manage confounding.

Nevertheless, there is a strong body of evidence that clearly shows health benefits from active transport.

9.5.2 Environmental benefits

The most harmful pollutants are emitted within minutes of starting a car (the 'cold start' phenomenon), meaning that short trips may pollute more by distance and have a bigger impact on our overall health than longer trips [52]. These short trips are often ones that can be undertaken using a zero carbon option such as walking or cycling.

The cost of the health impact of outdoor air pollution in OECD countries, both deaths and illness, was about USD 1.7 trillion in 2010, with road transport accounting for about 50% of this cost, almost USD 1 trillion [53].

Reduction in noise pollution would occur with less car use due to decreased car numbers and reduced congestion.

9.5.3 Quality of life and social benefits

Social inclusion is also recognised as a major benefit from low carbon planning and encouraging the use of active transport modes. Taylor and Ampt [54] found that where voluntary travel behaviour change interventions were undertaken, aimed at reducing car usage and increasing levels of active transport, targeted areas experienced:

- increased activity and interaction in streets
- reduced reports of local crime
- increased social interaction and trust at a community level
- increased recognition of local heritage and culture
- increased numbers of cultural products in local shops
- decreased numbers of 'complaints-without-solutions' to local councils, and
- increased community initiated projects.

9.5.4 Economic and productivity benefits

Years of planning and building streets for cars have left many communities severely lacking in bicycle and pedestrian infrastructure. Building new facilities for cycling and walking can be a boost for the economy via factors such as:

- new jobs
- rising property values
- increased business at local establishments, and
- reduced traffic congestion [55].

There is also a direct economic incentive to individuals. The costs of running a car far outweigh active transport options and can result in significant household savings [56].

Reduced income from carbon-based fuel taxes may be recouped through increased public transport patronage [54]. Improvements in productivity for urban freight operations are also possible under policies and plans for better integration of land-use and transport [57,58].

The reduction of direct and indirect health benefits discussed above also has an economic impact through reduced burden on the health care system.

9.6 Case studies of current policy and practice

There are many examples of policy and practice initiatives that can result in improved community and environmental well-being. It is interesting to note that many of the initiatives originated in public health agencies and did not have carbon reduction as a primary aim. Indeed, in many cases, there is no mention of environmental sustainability related outcomes. Yet, LCM features strongly in them, from the perspectives of encouraging physical activity through active transport provision. Urban planning and design are used to make walking and cycling more attractive and readily accessible transport options. An understanding of the co-benefits framework can initiate an awareness of the multiple benefits possible from one policy initiative designed to do something else. Another important awareness raising outcome is interdisciplinary appreciation. This emerges as different professional groups are exposed to very different policy environments which can have benefits for them. This is manifest for those working, researching and studying in the low carbon area who are not aware of what is being done in public health. The co-benefits framework can facilitate cross-disciplinary exchange and knowledge. We now consider several of these initiatives and contributions in the Australian context.

9.6.1 National Heart Foundation

The Heart Foundation of Australia is a major not-for-profit health stakeholder contributing to the evidence about the role that the built environment plays in supporting good health – particularly active living. In 2014, the Foundation published its 'Blueprint for an active Australia' [59]. This aims to support the broad range of public and private sector professionals who have a role in the planning, design, development and maintenance of the public realm, making it easier to incorporate design considerations that will have a positive impact on health and well-being. The Blueprint provides a set of design considerations set within the context of the important role that local government and planning professionals play in creating health supportive environments. The design considerations relate to the following:

- walking and cycling routes
- streets
- local destinations
- open space
- public transport
- seating, signage, lighting, fencing and walls, and
- fostering community spirit.

The Heart Foundation has also undertaken work on residential and commercial density, examining how different density levels create varying types of neighbourhoods with a range of characteristics that influence whether an environment is supportive of, or hinders, an active lifestyle. Research is presented in three documents which examine the health benefits of denser neighbourhoods:

- *Does Density Matter?* The role of density in creating walkable neighbourhoods (2014).
- *Low-Density Development:* Impacts on physical activity and associated health outcomes (2014).
- *Increasing Density in Australia:* Maximising the health benefits and minimising harm (2012).

Not surprisingly, these reports present research evidence which suggests that higher densities encourage greater physical activity. Nevertheless, the reports caution that density has to be implemented carefully so that neighbourhood amenity is enhanced – the provision of green open space, quality public realm, good design and proximity of destinations and public transport are all critical. Busy, heavily trafficked and highly polluted roads are not conducive to health and should not be the location for high density development, a message that will be difficult for decision makers currently putting medium and high rise residential buildings along major transport corridors.

9.6.2 Walk21

Walk21 is an international movement which 'exists to champion the development of healthy, sustainable and efficient communities where people choose to walk' [60]. Annual conferences are held and, in 2014, the NSW State Government and City of Sydney Council hosted Walk21. This put the spotlight on what is happening across Australia to support walking for health and environmental sustainability. The conference presented research, policy and practice initiatives to help Australian planning and public health advocates better understand what motivates and encourages walking. The decision to host Walk21 was made shortly after the release of 'Sydney's Walking Future' [61]. This report recognises the broader social and economic benefits of walking, aiming to make it 'the transport choice for quick trips under two kilometres'. Walk21 showcased a wealth of research evidence, practice initiatives and inspiration to achieve this outcome.

9.6.3 'Health in all policies' South Australia

South Australia's 'Health in All Policies' (HiAP) is an approach which emphasises the fact that health and well-being are largely influenced by measures that are often managed by government sectors other than health [62]. HiAP seeks to highlight the connections and interactions between health and policies from other sectors. HiAP explores policy options that contribute to the goals of non-health sectors which will improve health outcomes. By considering health impacts across all policy domains such as agriculture, education, the environment, finances, housing and transport, population health can be improved, and the growing economic burden of the health care system can be reduced. The health sector's role is to support other sectors to achieve their goals in a way which also improves health and well-being.

The link between transport and health is listed as a justification for this approach and used to highlight the need for policy makers in all sectors to be aware of the impact of their decisions on population health and to incorporate considerations of health into their policies. HiAP seeks agreement on the policy focus and utilises robust methods of assessment and analysis to explore the links between the policy area and health and well-being of the population.

9.6.4 NSW 'Healthy Urban Development Checklist'

The NSW Ministry of Health has recognised the need to develop its capacity to influence healthy urban design and development and to more actively participate in, and influence, urban planning and development processes. The 'Healthy Urban Development Checklist' is one of the measures now being used to do this [63]. The purpose of the checklist is to assist health professionals to provide advice on urban development policies, plans and proposals. It is intended to ensure that the advice provided is both comprehensive and consistent. The checklist focuses on helping health professionals to answer the questions: (1) what are the health effects of the urban development policy, plan or proposal, and (2) how can it be improved to provide better health outcomes?

There are ten characteristics of healthy urban development in the checklist. Each one has detailed information about its importance in supporting human health and how it can be incorporated into the built environment (both in terms of policies and land-use decisions). The topics are as follows:

- healthy food
- physical activity
- housing
- transport and physical connectivity
- quality employment
- community safety and security
- public open space
- social infrastructure
- social cohesion and social connectivity, and
- environment and health.

The checklist is a tool for reviewing and commenting on policy proposals and development application plans, and providing advice on health-related issues at an early stage of the urban planning and policy process. Although not exclusively, the primary users of the checklist are locally based health professionals who use the document to:

- provide a standardised tool to guide and inform feedback and advice to, for instance, local government and developers on urban development policies and plans in NSW
- evaluate the health aspects of urban developments
- support engagement between urban planners, developers and health professionals, and
- inform others (planners, developers and policy makers) about the range of factors that need to be considered in creating healthy urban developments.

More recently an additional component, 'Environmental Sustainability and Climate Change', has been proposed. Although this has not been formally incorporated into the original checklist, it is available for use [64]. With this update, the 'Healthy Urban Development Checklist' embraces the notion of co-benefits and puts an understanding of the inter-relationships between action on climate change and improvements in human health firmly on the agenda.

9.6.5 Sydney Metropolitan Strategy

The previous case studies are largely predicated at the state/province level. One example of a metropolitan-scale initiative is the Sydney Metropolitan Strategy [65], which embraces health as an important outcome of the state's planning decisions. The Strategy's Goal 3: 'A great place to live with communities that are strong, healthy and well connected' includes an entire section on health. Direction 3.3 'Create healthy built environments' presents key ways that the built environment can support healthy communities. In addition, the Sydney Plan includes reference to co-benefits (Direction 4.3) in relation to the development of 'Green Cover Design Principles' which are to 'incorporate vegetated, permeable and reflective surfaces into urban settings, to address thermal loading in the built environment and provide co-benefits such as reduced energy costs for cooling, stormwater management, cleaner air and biodiversity habitat' (p. 105). There are likely to be co-benefits for human health, although they are not specifically referenced in the plan.

9.7 Estimating co-benefits

This policy and practice review has outlined the possible co-benefits associated with a shift to low carbon active transportation modes and the potential for LCM options to improve community and individual well-being. However, the review has also revealed that there is little direct policy action being undertaken to reduce our carbon outputs. Understanding the value of carbon reduction actions and the associated co-benefits would provide impetus for policy makers and practitioners to further the carbon reduction agenda. A streamlined method is required that will enable this important information to be quantified, so that it can be fed into low carbon living decision-making.

The World Health Organization (WHO) recognises that the promotion of active transport is an important approach to address the challenge of high levels of physical inactivity around the world. Further, achieving this requires support from the transport and urban planning sectors. Economic appraisal is an established practice in planning; however, health effects of transport interventions are often excluded from these analyses due to their complex nature. To address this issue, WHO [66] developed the health economic assessment tool (HEAT) for walking and cycling. This instrument provides guidance and a practical, web-based tool for economic assessment of the health effects of walking and cycling. The tool estimates the value of reduced mortality resulting from specified amounts of walking or cycling.

Crawford and Whyte [67] provide a case study application of HEAT to increased levels of cycling in Glasgow. They found significant economic benefits from reduced mortality as a result of greater uptake of cycling. They concluded that the tool could be used to provide more comprehensive and meaningful cost–benefit analysis of plans and proposals for new infrastructure and transport service provision.

As indicated earlier in this chapter, co-benefits from active transport implementations can be found in a number of areas besides public health. A unified framework is required to capture and relate the co-benefits and to avoid issues of 'double counting' in an economic evaluation. The United Nations University (UNU) [68] has proposed such a framework. It combines an economic cost–benefit analysis model of the transport sector with an institutional evaluation. This assesses not just impacts of transport policies and plans but also determines barriers to implementation of policies and plans. The UNU tool has three features which make it a flexible instrument for use in urban planning:

- it is based on a scalable approach, so that the tool can be used to investigate the effects of different types of changes
- it can be used in both ex-post (project data assessment) or ex-ante (scenario-based assessment) with explicit consideration of environmental co-benefits as the intended benefit. As such decision makers can use the tool to decide on suitable options before selecting a specific policy or project, and
- it may be used to assist with policy implementation, so that policy makers can identify suitable alternative development scenarios for a given city.

Although the UNU tool itself is not appropriate for direct application to LCM, its conceptual model provides a basic framework that can be adapted for that purpose. A LCM version of this framework was developed by Philp *et al.* [69], and is shown in Figure 9.4. The model has six parts:

1. *intended benefits*, which cover the broad aims of the proposed plan or implementation
2. *policy instruments*, which describe the tools and policy measures available (or considered) for use
3. *targets*, which identify the specific objectives to be met
4. *indicators*, which provide metrics and benchmarks for the realisation of the targets
5. *modal focus*, which identifies and collects the individual contributions of different travel modes, vehicle types and technologies to the costs and benefits of the alternative scenario, and
6. *co-benefits*, which provide quantitative estimations of the performance of each tested scenario or alternative design in terms of the nominated benefits (taken as greenhouse gas emission reductions as the primary benefit), with health improvements and economic productivity improvements as the co-benefits.

The model presented in Figure 9.4 is designed for the assessment of public health and economic productivity co-benefits from the development and implementation of LCM projects and design measures as an integral part of urban development or redevelopment. The conceptual model is complex – because of the inherent complexity of the interactions between the various factors to be considered – so the diagrammatic representation of the model is also complex. Nevertheless, this model provides a useful framework for the development of a co-benefits calculator.

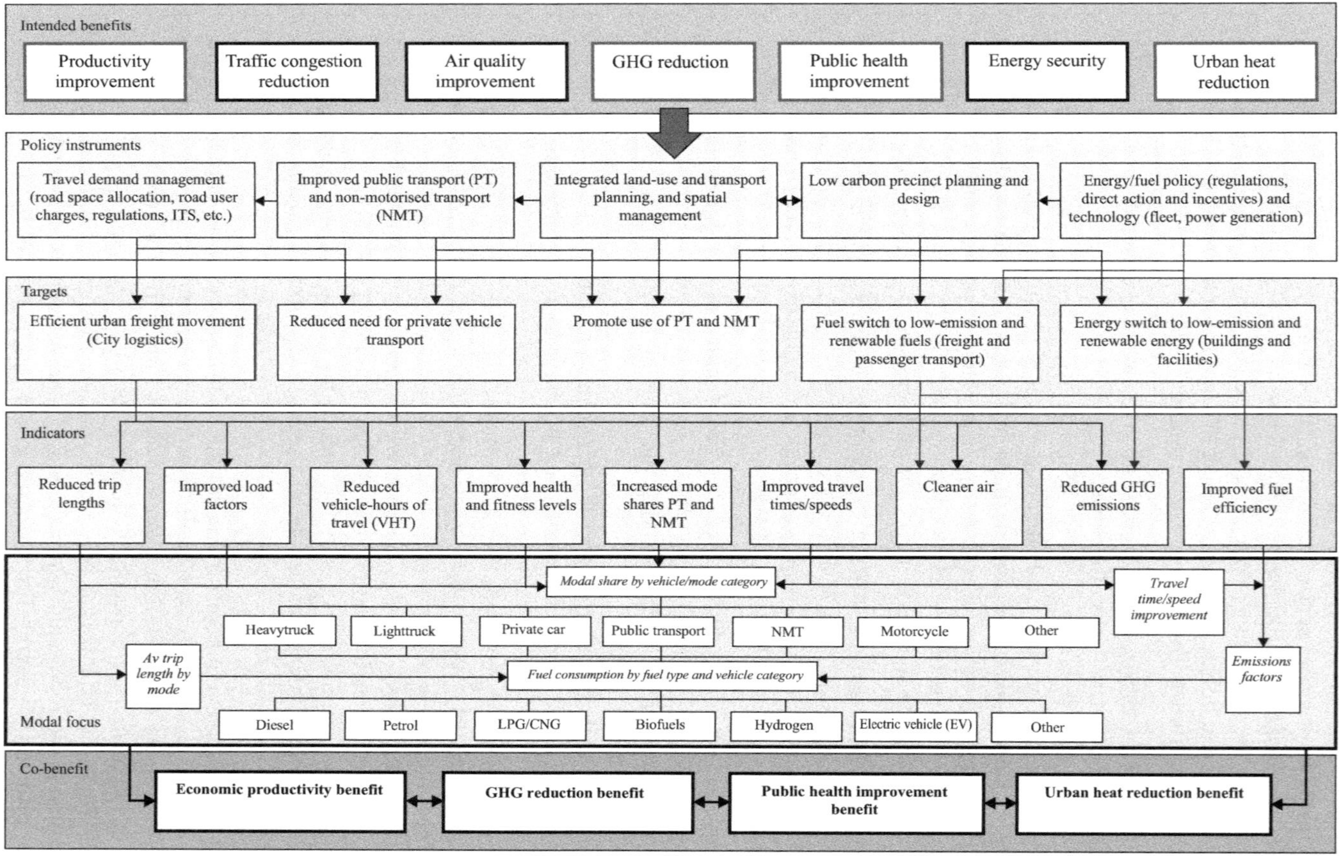

Figure 9.4 Framework for measuring co-benefits for low carbon mobility projects, modified and adapted from [68]. Unpublished and produced by Michael A.P. Taylor – Michael A.P. Taylor holds copyright

9.8 Conclusions

Health and well-being are affected by the way people travel, which itself is significantly influenced by the shape and form of the built environment. In turn, this largely determines the accessibility and ease-of-access of different travel modes. Our chapter has identified a set of three domains relating to specific behaviours that support the achievement of good health. The 'getting people active' domain has direct connections with travel and transport systems in urban areas. Walking and cycling, as active transport modes, make major contributions to individuals' health outcomes, both through the utilitarian physical activity involved in daily life and recreational physical activity undertaken for fitness and enjoyment. The chapter includes sets of basic questions and corresponding actions which planners can use to help create built environments that encourage physical activity to bring about positive community health outcomes, which also have environmental and productivity benefits, particularly around the reduction of car-based congestion.

The research evidence continues to demonstrate a strong relationship between people's health, climate change and the way the built environment is designed, planned and managed. Adopting a co-benefits approach to professional policy, practice and community action provides a strong platform to consolidate and extend effective and efficient approaches to addressing these issues. The need for integrated urban development and transport policies was explored in the chapter, with optimum blending of mutually reinforcing policies identified as a key requirement. Recognition of the roles of the active transport modes in meeting people's accessibility needs is a vital component of urban policy formulation and implementation.

Environmental sustainability is an established and fundamental principle of planning, underpinning contemporary practice. Contributing to a healthy built environment is still evolving but with the use of the 'co-benefits framework', policy makers in urban planning can see that multiple dividends will result from a single spend in one area.

Adopting a co-benefits framework can facilitate cross-disciplinary exchange, mutual understanding and a better appreciation of the synergies between the disciplines of health, transport and environmental sustainability. Further, employing co-benefits can help professionals to link different policy environments, facilitating an ability to see what they can contribute to their own policy and research endeavours. This is critical to support policy and practice initiatives for low carbon travel which supports health for humans and the planet upon which they reside. The chapter provides a number of case studies that are focussed on improving community well-being and health outcomes, along with reductions in carbon emissions. The chapter further suggests an analytical framework for estimating the co-benefits that can result from policy initiatives. This framework is explicitly designed for the assessment of public health and economic productivity co-benefits from the development of policies and projects aimed at encouraging, enhancing and reinforcing the use of active transport modes in our cities.

References

[1] Y. Rydin, A. Bleahu, M. Davies, *et al.*, "Shaping cities for health: complexity and the planning of urban environments in the 21st century", *Lancet* 379 (June), pp. 2079–2108, 2012.

[2] A.S. Daar, P.A. Singer, D.L. Persad, *et al.*, "Grand challenges in chronic non-communicable diseases", *Nature* 450 (7169), pp. 494–496, 2007.

[3] H. Barton, S. Thompson, S. Burgess, and M. Grant (eds.). *The Routledge Handbook of Planning for Health and Well-being: Shaping a Sustainable and Healthy Future*. London: Routledge, 2015.

[4] J.L. Kent and S.M. Thompson, "The three domains of urban planning for health and well-being", *Journal of Planning Literature* 29 (3), pp. 239–256, 2014.

[5] G. Paine and S.M. Thompson. *Healthy Built Environment Indicators*. Sydney: The University of NSW, City Wellbeing Program, CFRC, 2016.

[6] A. Smith. *The Climate Bonus: Co-Benefits of Climate Policy*. Milton Park: Earthscan from Routledge, 2013.

[7] N. Watts, W.N. Adger, P. Agnolucci, J. Blackstock, P. Byass, "Health and climate change: policy responses to protect public health", *The Lancet Commissions*, 386 (10006): 1861–1914. Published Online June 23, 2015. Available at: http://dx.doi.org/10.1016/S0140-6736(15)60854-6

[8] S. Whitmee, A. Haines, C. Beyrer, *et al.*, "Safeguarding human health in the Anthropocene epoch: Report of The Rockefeller Foundation/Lancet Commission on Planetary Health", *Lancet* 386 (10007), pp. 1973–2028, 2015.

[9] D. Ganten, A. Haines and R. Souhami, "Health co-benefits of policies to tackle climate change", *Lancet* 376 (9755), pp. 1802–1804, 2010.

[10] C.K. Castillo, D.C. Sanqui, M. Ajero and C. Huizenga, *The Co-benefits of Responding to Climate Change: Status in Asia.* Mandaluyong City, Philippines: Clean Air Initiative for Asian Cities, 2007.

[11] S. Thompson and A. Capon, "Healthy people, places and planetary systems: the co-benefits framework for understanding and action on climate change", in H. Barton, S. Thompson, M. Grant and S. Burgess (eds.), *The Routledge Handbook of Planning for Health and Well-being: Shaping a Sustainable and Healthy Future.* London: Routledge, 2015.

[12] S. Boyden, "Biohistory," in S.J. Singh, H. Haberl, M. Chertow, M. Mirtl, M. Schmid (eds.), *Long Term Socio-Ecological Research: Studies in Society-Nature Interactions Across Spatial and Temporal Scales*, New York: Springer, 2013, 386 (10006): 1861–914. Available at: http://www.biosensitive futures.org.au/content-images/bio-triangle-2.jpg/view

[13] M.A.P. Taylor and S.M. Thompson, *An analysis of current levels of active transport usage in Australia – towards a measure of baseline activity*. Research Project 2015 – Carbon Reductions and Co-benefits: Final Report – Part II, Co-operative Research Centre for Low Carbon Living, 2015. Available at: http://www.lowcarbonlivingcrc.com.au/resources/crc-publications/crclcl-project-reports/rp2015-carbon-reductions-and-co-benefits-final.

[14] C. Morency, M. Trepanier and M. Demers, "Walking to transit: an unexpected source of physical activity", *Transport Policy*, 18, pp. 800–806, 2011.

[15] IPCC, *Human health: impacts, adaptation, and co-benefits, in climate change 2014: impacts, adaptation, and vulnerability*. IPCC Working Group II Contribution to AR5. Inter Governmental Panel on Climate Change, 2014. Available at: http://ipcc-wg2.gov/AR5/images/uploads/WGIIAR5-Chap11_FGDall.pdf, accessed 1 May 2014.

[16] RTA and DECC, *Evaluation of the costs and benefits to the community of financial investment in cycling programs and projects in New South Wales.* Roads and Traffic Authority of NSW and the Department of Environment and Climate Change, 2009. Available at: http://www.pcal.nsw.gov.au/__data/assets/pdf_file/0008/90899/Evaluation_of_NSW_cycling_study.pdf, accessed 28 September 2015.

[17] A. Davis and J. Parkin, "Active travel: its fall and rise", in H. Barton, S. Thompson, S. Burgess and M. Grant (eds.), *The Routledge Handbook of Planning for Health and Well-Being: Shaping a Sustainable and Healthy Future*. London: Routledge. 2015.

[18] V. Rauland and P. Newman, "Decarbonising Australian cities: a new model for creating low carbon, resilient cities", in F. Chan, D. Marinova and R.S. Anderssen (eds.), *MODSIM2011: 19th International Congress on Modelling and Simulation*, pp. 3073–3079. Perth, WA: Modelling and Simulation Society of Australia and New Zealand; Dec 12–16, 2011.

[19] A. Chavez and A. Ramaswami, "Progress toward low carbon cities: approaches for transboundary GHG emissions' footprinting", *Carbon Management* 2 (4), pp. 471–482, 2011.

[20] P. Donoso, F. Martinez and C. Zegras, "The Kyoto Protocol and sustainable cities-potential use of clean-development mechanism in structuring cities for carbon-efficient transportation", *Transportation Research Record 1983*, pp. 158–166, 2006.

[21] D. Banister, "City transport in a post carbon society", in M. Givoni and D. Banister (eds.), *Moving Towards Low Carbon Mobility*, Cheltenham: Edward Elgar Publishing, pp. 255–266, 2013.

[22] P. Aditjandra, C. Mulley and J. Nelson, "The influence of neighbourhood design on travel behaviour: empirical evidence from North East England", *Transport Policy* 26, pp. 54–65, 2013.

[23] P.W. Newton and P. Newman, "The geography of solar photovoltaics (PV) and a new low carbon urban transition theory", *Sustainability* 5, pp. 2537–2556, 2013.

[24] P.W. Newton, A. Pears, J. Whiteman and R. Astle, "The energy and carbon footprints of urban housing and transport: current trends and future prospects", in R. Tomlinson (ed.), *Australia's Unintended Cities: The Impact of Housing on Urban Development*, Melbourne: CSIRO Publishing, pp. 153–189, 2012.

[25] R. Hickman and D. Banister, "Looking over the horizon: transport and reduced CO_2 emissions in the UK by 2030", *Transport Policy* 14, pp. 377–387, 2007.

[26] R. Hickman, "Urbanization and future mobility", in M. Givoni and D. Banister (eds.), *Moving Towards Low Carbon Mobility*. Cheltenham: Edward Elgar Publishing, pp. 60–74, 2013.

[27] P. Newman, J., Kenworthy and G. Glazebrook, "Peak car use and the rise of global rail: why this is happening and what it means for large and small cities", *Journal of Transportation Technologies* 3, pp. 272–287, 2013.
[28] G. Santos, H. Behrendt and A. Teytelboym, "Part II: policy instruments for sustainable road transport", *Research in Transportation Economics* 28 (1), pp. 46–91, 2010.
[29] D. Banister, "Cities, mobility and climate change", *Journal of Transport Geography* 19 (6), pp. 1538–1546, 2011.
[30] D. Banister, K. Anderton, D. Bonilla, M. Givoni and T. Schwanen, "Transportation and the environment", *Annual Review of Environment and Resources* 36, pp. 247–270, 2011.
[31] K. Nakamura and Y. Hayashi, "Strategies and instruments for low-carbon urban transport: an international review on trends and effects", *Transport Policy* 29, pp. 264–274, 2013.
[32] J. Whitehead, "Making the transition to a 'green' vehicle fleet – an analysis of the choice and usage effects of incentivizing the adoption of low emission vehicles", unpublished Licentiate thesis in Transport Science, Royal Institute of Technology, Stockholm, 2013.
[33] B. Giles-Corti and R. Donovan, "Socioeconomic status differences in recreational physical activity levels and real and perceived access to a supportive physical environment", *Preventive Medicine* 35 (6), pp. 601–611, 2002.
[34] J. Sallis, L. Frank, B. Saelens and M. Kraft, "Active transportation and physical activity: opportunities for collaboration on transportation and public health research", *Transportation Research Part A: Policy and Practice* 38 (4), pp. 249–268, 2004.
[35] L. Wen and C. Rissel, "Inverse associations between cycling to work, public transport, and overweight and obesity: findings from a population based study in Australia", *Preventive Medicine* 46 (1), pp. 29–32, 2008.
[36] J. Woodcock, P. Edwards, C. Tonne, *et al.*, "Public health benefits of strategies to reduce greenhouse-gas emissions: urban land transport", *Lancet* 374 (9705), pp. 1930–1943, 2009.
[37] R. Buehler, J. Pucher, D. Merom and A. Bauman, "Active travel in Germany and the USA: contributions of daily walking and cycling to physical activity", *American Journal of Preventive Medicine* 40 (9), pp. 241–250, 2011.
[38] P. Gordon-Larsen, J. Boone-Heinonen, S. Sidney, B. Sternfeld, D.R. Jacobs Jr. and C.E. Lewis, "Active commuting and cardiovascular disease risk: the CARDIA study", *Archives of Internal Medicine* 169 (13), pp. 1216–1223, 2009.
[39] M. Hamer and Y. Chida, "Active commuting and cardiovascular risk: a meta-analytic review", *Preventive Medicine* 46 (1), pp. 9–13, 2008.
[40] P. Oja, S. Titze, A. Bauman, *et al.*, "Health benefits of cycling: a systematic review", *Scandinavian Journal of Medicine and Science in Sports* 21, pp. 496–509, 2011.
[41] J. Pucher, R. Buehler, D. Bassett and A. Dannenberg, "Walking and cycling to health: recent evidence from city, state, and international comparisons", *American Journal of Public Health* 100 (10), pp. 1986–1992, 2010.

[42] R. Shephard, "Is active commuting the answer to population health?", *Sports Medicine* 39 (9), pp. 751–758, 2008.

[43] D. Rojas-Rueda, A. Nazelle, M. Tainio and M. Nieuwenhuijsen, "The health risks and benefits of cycling in urban environments compared with car use: health impact assessment study", *British Medical Journal* 343, 2011, doi: 10.1136/bmj.d4521.

[44] P. Jacobsen and H. Rutter, "Cycling safety", in J. Pucher and R. Buehler (eds.), *City Cycling*. Cambridge MA: MIT Press, pp. 141–156, 2012.

[45] J. Garrard, C. Rissel and A. Bauman, "Health benefits of cycling", in J. Pucher and R. Buehler (eds.), *City Cycling*. Cambridge, MA: MIT Press, pp. 31–56, 2012.

[46] A. Singh, L. Uijtdewilligen and J.W.R. Twisk, "Physical activity and performance at school", *Archives of Pediatrics and Adolescent Medicine*, 166 (1), pp. 49–55, 2012.

[47] L. Chaddock, K.I. Erickson and R.S. Prakash, "Basal ganglia volume is associated with aerobic fitness in preadolescent children", *Developmental Neuroscience* 32, pp. 249–256, 2010.

[48] C.H. Hillman, D.M. Castelli and S.M. Buck, "Aerobic fitness and neurocognitive function in healthy preadolescent children", *Medicine and Science in Sports and Exercise* 37, pp. 1967–1974, 2005.

[49] R. Mackett, "Children's travel behaviour and its health implications", *Transport Policy* 26, pp. 66–72, 2013.

[50] A. Goodman, R. Mackett and J. Paskins, " Activity compensation and activity synergy in British 8–13 year olds", *Preventive Medicine* 53, pp. 293–298, 2011.

[51] C. Shaw, S. Hales, P. Howden-Chapman and R. Edwards, "Health co-benefits of climate change mitigation policies in the transport sector", *Nature Climate Change* 4, pp. 427–433, 2014.

[52] FHWA, *Report to the US Congress on the Outcomes of the Non-motorized Transportation Pilot Program: SAFETEA-LU Section 1807*, 2012. Available at: www.fhwa.dot.gov/environment/bicycle_pedestrian/ntpp/2012_report/page00.cfm.

[53] OECD, *The Cost of Air Pollution: Health Impacts of Road Transport*. OECD Publishing, Paris, 2014. dx.doi.org/10.1787/9789264210448-en.

[54] M.A.P. Taylor and E. Ampt, "Travelling smarter down under: policies for voluntary travel behaviour change in Australia", *Transport Policy* 10 (3), pp. 165–177, 2003.

[55] F. Creutzig and D. He, "Climate change mitigation and co-benefits of feasible transport demand policies in Beijing", *Transportation Research Part D* 14, pp. 120–131, 2009.

[56] T. Litman, *Quantifying the benefits of non-motorized transport for achieving TDM objectives*. Victoria Transport Policy Institute. Available at: www.vtpi.org, 1999.

[57] M.A.P. Taylor, "The City Logistics paradigm for urban freight transport", Proceedings of the 2nd State of Australian Cities Conference. December, Brisbane, Urban Research Program, Griffith University, 2005.

[58] TLIAP, *Collaborative urban logistics: the synchronized last mile*, TE White Paper, The Logistics Institute – Asia Pacific, National University of Singapore. Available at: www.tliap.nus.edu.sg, 2013.
[59] Heart Foundation, *Blueprint for an Active Australia* (2nd ed.). Heart Foundation of Australia, 2014. Available at: http://heartfoundation.org.au/for-professionals/physical-activity.
[60] Walk21, 2016. Available at: http://www.walk21.com/.
[61] Transport for NSW, 2013. Available at: http://www.transport.nsw.gov.au/sites/default/files/b2b/publications/sydneys-walking-future-web.pdf.
[62] SA Health, *Health in All Policies*, SA Health, Adelaide, 2014. Available at: www.sahealth.sa.gov.au/wps/wcm/connect/public+content/sa+health+internet/health+reform/health+in+all+policies.
[63] NSW Department of Health, *Healthy urban development checklist – a guide for health services when commenting on development policies, plans and proposals*, NSW Department of Health, Sydney. 2009. Available at: www.health.nsw.gov.au.
[64] Sydney West Local Health District, 2015. Available at: http://www.sswahs.nsw.gov.au/populationhealth/hud/.
[65] NSW Planning and Environment, 2014. Available at: http://www.planning.nsw.gov.au/Plans-for-Your-Area/Sydney/A-Plan-for-Growing-Sydney.
[66] WHO, "Health economic assessment tools (HEAT) for walking and for cycling: economic assessment of transport infrastructure and policies. Methodology and user guide, 2014 update", World Health Organization Regional Office for Europe, Copenhagen, 2014.
[67] F. Crawford and B. Whyte, "Assessing the health benefits of cycling in Glasgow", 2014 STAR Conference. Scottish Transport Applications and Research, 2014. Available at: www.starconference.org.uk/star2014.html.
[68] UNU, *Guidebook: The Co-benefits Evaluation Tool for the Urban Transport Sector*. Yokohama: United Nations University Institute of Advanced Studies, 2014. Available at: www.ias.unu.edu.
[69] M. Philp, M.A.P. Taylor and S.M. Thompson, *RP-2015 Carbon Reductions and Co-benefits: Final Report – Part I, Literature and practice review of Australian policies relating urban planning and public health*. Cooperative Research Centre for Low Carbon Living, Sydney, 2015. Available at: www.lowcarbonlivingcrc.com.au.

Glossary

CBD	central business district
CNG	compressed natural gas
EV	electric vehicle
GHG	greenhouse gas
HEAT	health economic assessment tool

HiAP	health in all policies
LCM	low carbon mobility
LPG	liquefied petroleum gas
NMT	non-motorised transport
NSW	New South Wales
OECD	Organisation for Economic Cooperation and Development
PT	public transport
TOD	transit-oriented development
UNU	United Nations University
USD	United States dollar
VHT	vehicle-hours of travel
Walk21	International organisation supporting and promoting walking
WHO	World Health Organization

Further Reading

[1] Arup, *Cities Alive: Towards a Walking World*, Arup, London, 2016. Available at: http://publications.arup.com/publications/c/cities_alive_towards_a_walking_world.

[2] Barton, H., Thompson, S., Burgess, S. and Grant, M. (eds.). *The Routledge Handbook of Planning for Health and Well-Being: Shaping a Sustainable and Healthy Future*. London: Routledge, 2015.

[3] Kent, J.L. and Thompson, S.M. "The three domains of urban planning for health and well-being", *Journal of Planning Literature* 29 (3), pp. 239–256, 2014.

[4] Philp, M., Taylor, M.A.P. and Thompson, S. RP-2015 *Carbon Reductions and Co-benefits: Final Report – Part I, Literature and practice review of Australian policies relating urban planning and public health*. Cooperative Research Centre for Low Carbon Living, Sydney. Available at: www.lowcarbonlivingcrc.com.au, 2015.

[5] Smith, A. *The Climate Bonus: Co-Benefits of Climate Policy*. Milton Park: Earthscan from Routledge, 2013.

[6] Thompson, S. and Capon, A. Healthy people, places and planetary systems: the co-benefits framework for understanding and action on climate change. In Barton, H., Thompson, S., Grant, M. and Burgess, S. (eds.), *The Routledge Handbook of Planning for Health and Well-being: Shaping a Sustainable and Healthy Future*. London: Routledge, 2015.

[7] Thompson, S. and Kent, J.L. "Healthy Planning: The Australian Landscape", *Built Environment*, 42 (1), pp. 90–106, 2016.

[8] Watts, N., Adger, W.N., Agnolucci, P., Blackstock, J., Byass, P. "Health and climate change: policy responses to protect public health", *The Lancet*

Commissions, 386 (10006): 1861–914. Published Online June 23, 2015. Available at: http://dx.doi.org/10.1016/S0140-6736(15)60854-6.
[9] Wedderburn, M. *Improving the cost-benefit analysis of integrated PT, walking and cycling*. Research Report 537, NZ Transport Agency, Wellington. Available at: www.nzta.govt.nz, 2013.
[10] WHO. *Health economic assessment tools (HEAT) for walking and for cycling: economic assessment of transport infrastructure and policies*. Methodology and user guide, 2014 update. World Health Organization Regional Office for Europe, Copenhagen, 2014.
[11] Woodcock, J., Edwards, P., Tonne, C., *et al*., Public health benefits of strategies to reduce greenhouse-gas emissions: urban land transport. *Lancet* 374 (9705), pp. 1930–1943, 2009.

Chapter 10

Mobility and the sharing economy: industry developments and early understanding of impacts

Susan A. Shaheen[1], Apaar Bansal[1], Nelson Chan[1] and Adam Cohen[1]

Abstract

Shared mobility—the shared use of a vehicle, bicycle, or other mode—is an innovative transportation strategy that enables users to gain short-term access to transportation modes on an "as-needed" basis. Shared mobility includes various forms of carsharing, bikesharing, ridesharing, for-hire driver services, and microtransit. In addition, smartphone and mobile "apps" aggregate and optimize these mobility services and are critical to many shared mobility modes. Courier network services connect couriers using their personal vehicles or bicycles with freight and seek to disrupt the existing package and food delivery industry. The emergence of automated vehicles into shared mobility could further transform the passenger and freight transportation system, with greater emphasis on shared mobility. This chapter describes the different models that have emerged in the shared mobility space and reviews research that has quantified the environmental, social, and transportation-related impacts of these services. The authors also project future trends as automated vehicles begin to emerge.

Keywords

Shared mobility; sustainable transportation; sharing economy; business models; automated vehicles

10.1 Introduction

Shared mobility—the shared use of a vehicle, bicycle, or other mode—is a growing sector of the sharing economy. Although the concept of sharing is not new, economic

[1]Transportation Sustainability Research Center, University of California, Berkeley, Berkeley, CA, USA

models have emerged, particularly since the 1990s, that are based on peer-to-peer (P2P) sharing or the collaborative consumption of resources. Other names for the sharing economy include the "collaborative economy" and "P2P economy." Sharing has been facilitated through online social networking platforms and global positioning system (GPS[1])-enabled mobile technology. Sharing economy business models include the sharing of lodging, labor, equipment, food, and transportation. Consumers are opting for accessing goods and services without the cost, burden, and environmental impacts commonly associated with personal ownership [1]. PricewaterhouseCoopers estimated that five sectors of the sharing economy (e.g., equipment, housing, books, DVDs, and cars) generated US$15 billion in global revenue in 2014, and it is poised to grow to US$335 billion in 2025 [2].

10.1.1 Shared mobility and the sharing economy

Although the latter half of the twentieth century in North America and western Europe emphasized personal vehicle ownership, information and communication technologies (ICT) and new business models have spawned innovative modes that fill the gap between private autos and public transit. Travelers can now hail a driver and vehicle (e.g., Lyft, Uber); rent a car or bicycle for a short trip (e.g., car2go, Vélib', Zipcar); ride a private shuttle on-demand or on a crowdsourced route (e.g., Bridj, Chariot, Via); and have groceries or takeout food delivered in someone's personal vehicle (e.g., Postmates and Instacart)—all using Internet-enabled smartphones, tablets, and other mobile devices. Such innovative mobility services fall under the umbrella term of shared mobility.

Shared mobility is the shared use of a vehicle, bicycle, or other mode that enables users to have short-term access to transportation modes on an "as-needed" basis. It includes carsharing; personal vehicle sharing (PVS); scooter sharing; bikesharing; for-hire driver services (including taxi e-Hail and ridesourcing or transportation network companies (TNCs)); ridesharing (i.e., carpooling, vanpooling); microtransit; and courier network services (CNS).

Figure 10.1 categorizes these key areas of shared mobility depending on the service. Carsharing, scooter sharing, and bikesharing (Figure 10.1, left) are services that enable the sharing of vehicles or equipment. Ridesharing, for-hire driver services, and microtransit (Figure 10.1, middle) facilitate passenger rides. Finally, CNS, Figure 10.1, right, enable the use of private vehicles for delivery trips.

The impacts of shared mobility have been documented in cities worldwide—cost savings and convenience, reduced vehicle miles traveled (VMT)/vehicle kilometers traveled (VKT) and personal vehicle ownership, as well as decreased greenhouse gas (GHG) emissions. Moreover, there are often economic benefits to shared mobility, such as increased economic activity near multimodal hubs and commercial areas [3,4]. However, some forms of shared mobility, such as for-hire driver services, microtransit, and CNS, have not been extensively studied.

[1] Commonly referred to as Global Positioning System (GPS) in North America, Galileo in the European Union, GLONASS in Russia, BeiDou in China, and IRNSS in India

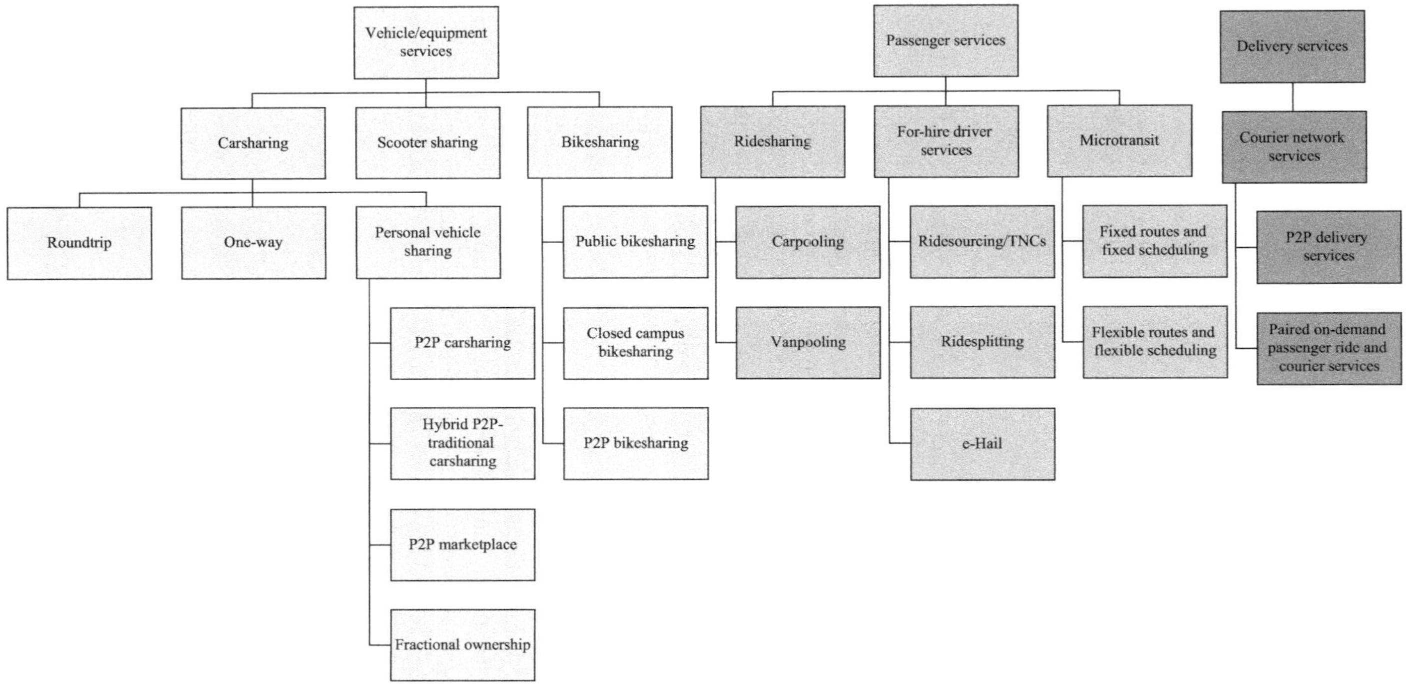

Figure 10.1 Key areas of shared mobility

Many issues regarding equity, accessibility, disabled access, and the digital divide remain unaddressed. As the spectrum of shared mobility options expands in the future, public policy must evolve by enacting new or revised legislation to protect public safety, ensure equity, and provide guiding policies to harness the positive impacts of shared mobility.

This chapter presents an overview of the key areas of shared mobility. Within each section, we present the history and current understanding of each shared mobility mode, with a focus on the energy and environmental impacts. Finally, the chapter discusses the future of urban transportation as automation and shared mobility become more integrated into a sustainable, multi-modal transportation system.

10.2 Carsharing

Carsharing members gain the benefits of a private vehicle user without the cost and responsibilities of ownership (e.g., fuel, maintenance, insurance). Instead, members are able to access a fleet of shared vehicles on an as-needed basis and pay a usage- and/or membership-based fee. The first carsharing program can be traced back to Switzerland in 1948. More popularized European programs launched in the 1980s, and since then, the industry has grown worldwide. In October 2014, there were 4.8 million carsharing members worldwide, 2.2 million of which were in European programs and 1.6 million in North American programs [5]. There are various carsharing business models in existence, the earliest of which is the roundtrip model. More recently, one-way carsharing and PVS have emerged. The following sections discuss these business models.

10.2.1 Roundtrip carsharing

Roundtrip carsharing provides its members access to a fleet of shared vehicles on an hourly basis, and it requires vehicles be returned to the same location from where they were accessed. Because roundtrip carsharing has the longest history, there have been several studies on its impact on VMT/VKT, vehicle ownership, GHG emissions, and individual transportation costs. From a 2004 study of City CarShare members in the San Francisco Bay Area, 30% of members shed one or more of their personal vehicles, and two-thirds postponed purchasing a vehicle after using the service for two years [6]. A 2008 survey by researchers at the University of California (UC), Berkeley of over 6,000 members of North American carsharing organizations documented the impact of roundtrip carsharing on modal shift [7]. Although there was an overall decline in public transit use among carsharing members, the study noted substantial increases in non-motorized and sustainable travel—walking, bicycling, and traditional carpooling. Similarly, another later study found that carsharing members in Montreal, Canada had a modal split with private car usage significantly lower than that of non-carsharing members [8].

It is important to note that the aggregate-level data presented are not necessarily generalizable on a municipal or regional basis (e.g., reflecting impacts of the built environment), as the analysis reflects the combined impacts across the U.S. and Canadian study populations. This was due to data agreements with participating companies that allowed for limited data disaggregation. The member survey, although imperfect in measuring changes in large numbers, is an important instrument for obtaining a before-and-after measure of carsharing impacts ($n =$ 6,281) [9,10]. For this reason, despite advances in technology that improve approaches to travel behavior measurement (i.e., activity data and revealed preference), surveys will likely continue to play a fundamental role in assessing causes of change and in providing critical inputs to its measurement. A similar discussion is relevant to impact analyses of the other shared modes discussed in this chapter.

10.2.2 One-way carsharing

One-way carsharing, also known as point-to-point or free-floating carsharing, allows members to pick up a vehicle at one location and drop it off at another. In 2012, one-way carsharing began expanding rapidly to seven countries worldwide [11]. As of October 2014, there were 851,988 one-way carsharing members globally, 372,466 of which were in Europe, 445,722 were in North America, 29,600 in Asia, 3,500 in South America, and 700 in Oceania [5]. As of January 2015, 35.7% of North American fleets allowed one-way trips, and 30.8% of members had access to such fleets [12]. An advantage of one-way carsharing is its increased flexibility and potential to further enhance first- and last-mile connectivity to public transit and other modes. Moreover, as one-way trips are short city trips, electric vehicles (EVs) can be used to reduce GHG emissions. To serve these mobility opportunities, companies and cities have launched one-way carsharing programs, beginning in Europe in the 1970s, East Asia in the 1990s, and continuing in North America from the 1990s until today.

Procotip was the first recorded one-way carsharing experiment, launched in August 1971 in Montpellier, France with 35 cars and 19 stations. However, it closed in May 1973 due to technological and financial issues [13]. Liselec launched in 1993 in La Rochelle, France with 50 EVs at seven stations. Liselec was successful and still exists today as Yelómobile, and it is sponsored by Peugeot-Citroen and receives financial support from the government. Praxitèle launched in 1997 as a one-way carsharing system with connections to public transit. Fifty EVs were placed in Saint-Quentin-en-Yvelines, France, allowing its 500 members to make one-way trips between 14 stations located in neighborhoods, near offices, and at public transit stations. By March 1999, 90% of total trips were one way [14]. Although Praxitèle succeeded in its technical implementation, the program struggled with costs and ended in July 1999.

In Japan, automakers have experimented with mobility services and one-way carsharing. In 1998, Honda Motor Company deployed the Intelligent Community Vehicle System, which included both roundtrip and one-way carsharing with connections to public transit [15]. Similarly, in 1999, Toyota Motor Company launched the Crayon System in Toyota City, Japan. Fifty EVs were placed at public transit stations and other locations for its 700 members. Both Honda and Toyota employed emerging technologies including: smartcards, automatic vehicle location, vehicle information and communication systems, and a management system for reservations and recharging [16].

Early experiments with one-way carsharing in the United States began with station car programs, serving primarily rail and public transit commuters. In 1999, CarLink I launched at the Dublin/Pleasanton Bay Area Rapid Transit (BART) station in the East Bay of the San Francisco Bay Area [17,18]. Twelve compressed natural gas Honda Civics were available to drive between the BART station and the Lawrence Livermore National Laboratory. Similarly, CarLink II was based at the Caltrain station in Palo Alto, California with 27 Honda Civics shared among 100 users [19]. The CarLink programs facilitated one-way trips between rail stations and home- and work-based trips. Some who lived near the transit station took BART into the central business district (CBD). Others took public transit to their suburban workplace, using CarLink as the last-mile link from the transit station to work. These commuters then returned their vehicles to the station in the evening, and home-based users picked up these vehicles after taking public transit from the CBD. At the end of the CarLink II pilot, Flexcar took over the service in 2002. However, it ceased operations in 2003 due to concerns with cost recovery and its limited scale.

The UC piloted several one-way carsharing projects as well. UC Irvine operates the Zero Emission Vehicle Network Enabled Transport (ZEV · NET) system. Launched in 2002, ZEV · NET facilitates trips between the Irvine Transportation Center commuter rail terminal, four employers, and the UC Irvine campus. In addition, its fleet entirely comprises EVs. More recently, 30 Toyota iQ EVs were added in March 2013, and the system still operates today [20]. UCR Intellishare operated around the UC Riverside campus and the Downtown Riverside Metrolink station from 1999 to 2010 [21].

Since the 1970s, despite many successful one-way carsharing projects that were well received, many ceased operations. The main reasons include economic viability (e.g., CarLink); underuse (e.g., Praxitèle); and insufficient technology (e.g., Procotip). Yet, others have succeeded and still operate today (e.g., Yélomobile). Early one-way carsharing attempts established the foundation for existing carsharing services today.

At present, one-way carsharing reflects two main models: (1) free-floating carsharing and (2) station-based carsharing. Free-floating carsharing enables shared vehicles to be picked up and dropped off anywhere within a designated geo-fenced operating area. In contrast, station-based models require users to return the vehicle to any designated station. Although this model may be perceived as less flexible, station-based carsharing limits vehicle search to select locations.

Experts in carsharing forecast one-way systems to continue to grow. Interviews with leaders in the carsharing industry have indicated that future innovations in one-way carsharing will involve its integration with public transit and its use of low- or zero-emission vehicles [22]. Much of the recent carsharing literature has focused on vehicle rebalancing, personnel management, and the broader logistics of one-way operations (e.g. [23,24]). A study on car2go users in five North American cities found that each carsharing vehicle removed between 7 and 11 vehicles from the road. Two to 5% of members sold a vehicle and 7%–10% postponed a vehicle purchase (see Figure 10.2). Moreover, estimated VMT impacts due to carsharing ranged from −6% to −16% per car2go household, and GHG emissions changed by −4% to −18% [25]. For disaggregated results by city, please refer to Martin and Shaheen [25]. Nevertheless, more research is needed to better understand one-way carsharing's potential impact on VMT/VKT, GHG emissions, and automobile ownership [22].

10.2.3 Personal vehicle sharing

PVS is another carsharing service model characterized by short-term access to privately owned vehicles (as opposed to an operator-owned vehicle fleet). PVS companies facilitate transactions between car owners and renters by providing the infrastructure needed to make the exchange possible and seamless, such as an online platform, customer support, auto insurance, and vehicle technology. Members access vehicles either through a direct key transfer from the owner to the renter or through operator-installed in-vehicle technology that enables unattended access. There are four distinct models of PVS: (1) P2P carsharing, (2) hybrid P2P-traditional carsharing, (3) P2P marketplace, and (4) fractional ownership [26].

10.2.3.1 Peer-to-peer carsharing

P2P carsharing employs privately owned vehicles made temporarily available for sharing by an individual or members of a P2P company. Although still heavily focused in urban areas and cities, P2P carsharing operations are not as geographically confined as other types of carsharing because the users provide the floating vehicle fleet. In addition, P2P carsharing may serve a more diverse population than traditional station-based carsharing services. In a study of P2P carsharing use in Portland, Oregon, 37% of families in poverty lived in a census block group that contained at least one P2P vehicle, whereas only 13% lived in a

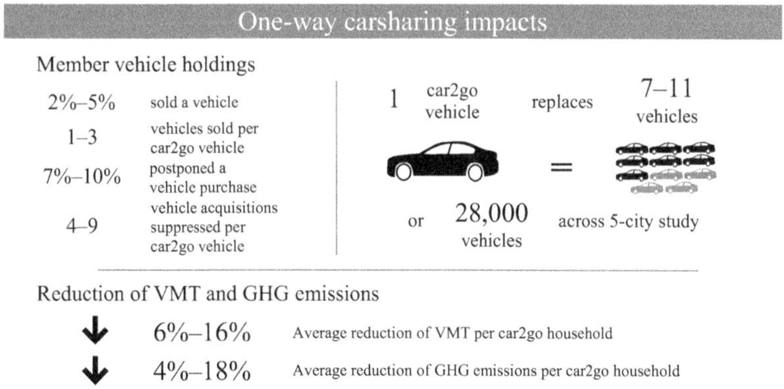

Figure 10.2 Impacts of one-way carsharing [25]

census block that had a station-based carsharing vehicle [27]. Furthermore, another recent study projected that P2P carsharing will have more pronounced impacts on below-median income consumers than above-median income users [28]. Examples of P2P carsharing operators in the United States include, Getaround and Turo (formerly RelayRides). Pricing and rental terms for P2P carsharing services vary, typically determined by owners listing their vehicles for rent. P2P operators generally take a portion of the rental amount in return for facilitating the exchange and providing third-party insurance. As of May 2015, there were eight active P2P operators in North America and 15 active operators in Europe.

10.2.3.2 Hybrid P2P-roundtrip carsharing

In the hybrid P2P-roundtrip carsharing model, individuals access vehicles by joining an organization that maintains its own fleet, but it also includes private vehicles throughout a network of locations.

10.2.3.3 P2P marketplace

P2P marketplace enables direct exchanges among individuals via the Internet, including pricing agreements. Terms are generally decided among parties of a transaction, and disputes are subject to private resolution.

10.2.3.4 Fractional ownership

In the fractional ownership model, individuals sublease or subscribe to a vehicle owned by a third party. These individuals have "rights" to the shared vehicle service in exchange for taking on a portion of the operating and maintenance expenses. This enables access to vehicles that individuals might otherwise be unable to afford, and it results in income sharing when the vehicle is rented to non-owners. Fractional ownership is often facilitated through a dealership or a partnership with a carsharing operator. Fractional ownership is typically used with luxury cars and recreational vehicles. At present, this segment of the industry is small, and it remains to be seen whether fractional ownership can compete with existing carsharing models and personal vehicle ownership overall. Fractional ownership companies in the United States currently include Curvy Road, Gotham Dream Cars, and CoachShare. In December 2014, Audi launched its "Audi Unite" fractional ownership model in Stockholm, Sweden. Audi Unite offers multiparty leases with pricing based on model, yearly mileage, and the number of drivers ranges from two to five. In February 2016, Ford launched a fractional ownership program in Austin, Texas that leases vehicles to self-organized groups of up to six people. Co-owners can reserve a vehicle, check on its status, exchange messages, make vehicle payments, and monitor their accounts via an app and in-vehicle device.

10.3 Scooter sharing

Scooter sharing is a recent variation on the vehicle-sharing business model. As of September 2015, there were two scooter sharing systems in Europe: Motit in

Barcelona and Enjoy in Milan. Moreover, there were two systems in North America: Scoot Networks in San Francisco, California, and Scootaway in Columbia, South Carolina. All these systems offer both one-way and roundtrip short-term scooter sharing and include insurance and helmets. There are several other systems in Europe being tested, including CityScoot in Paris, eMio in Berlin, and Scoome in Munich and Cologne. In addition to scooters, Scoot Networks also offers electric motorcycle sharing and Scoot Quads (Renault's Twizy EV). As of October 2015, Scoot had over 400 scooters and 10 Quads in its network, and Scootaway's fleet comprised 350 scooters. At that time, Scoot's vehicles were being driven over 70,000 mi each month [29].

Due to the lower speed of scooters, which must stay on city streets versus highways, scooter sharing systems have remained in urban areas. Although scooter sharing represents a smaller market relative to other shared modes, it provides urban mobility with fewer carbon emissions than automobiles. Similar to public bikesharing, scooters require less parking space at public transit stations, which could potentially increase ridership. Future research could test multimodal linkages, similar to station car and one-way carsharing pilots of the past.

10.4 Bikesharing

Bikesharing systems allow users to access bicycles on an as-needed basis from either a network of bikesharing stations (station-based) or free-floating within a geo-fenced area. Most operators are responsible for bicycle maintenance, storage, and parking costs. Bikesharing can also include P2P systems, where private owners lease their bicycles for short-term use through third-party hardware and applications. There are three main types of bikesharing systems: (1) public bikesharing, (2) closed campus bikesharing, and (3) P2P bikesharing [30]. The majority of bikesharing systems in the world are public, where anyone is able to access a bicycle for a nominal fee (and typically with a credit/debit card on file). To better serve low-income communities, some public bikesharing systems offer subsidies and cash memberships [31]. Closed-campus bikesharing systems are increasingly being deployed at university and office campuses; they are only available to the particular campus community they serve. P2P bikesharing services are available in urban areas for bike owners to rent out their idle bikes for others to use and are also growing due to companies, such as Spinlister and Bitlock.

Bikesharing has emerged as one of the fastest growing transportation innovations in many global cities. As of April 2016, there were 1,019 cities with information technology-based public bikesharing systems in the world comprising 1,324,530 bikes, the majority of which were in China. The United States had 75 programs spread over 99 cities, with approximately 32,000 bikes and 3,400 stations at that time [32].

Bikesharing has the potential to impact public transit and other modes, serving as an effective and efficient first- and last-mile connection with no energy use. Researchers at UC Berkeley conducted a two-part study of public bikesharing

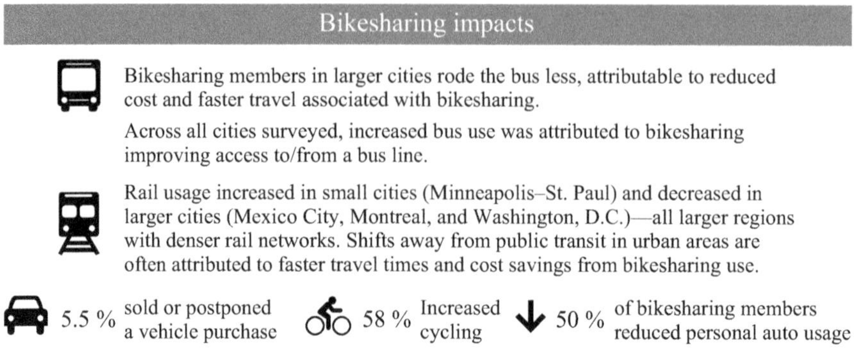

Figure 10.3 Aggregate impacts of North American public bikesharing [10]

programs in North America to determine the program impacts on modal split [31,33]. The results suggest that public bikesharing in larger cities (e.g., Mexico City, Montreal, and Washington, D.C.) takes riders off of crowded buses, whereas bikesharing in smaller cities (e.g., Minneapolis-St. Paul, Minnesota) improves access to and egress from bus lines. Moreover, respondents reported that rail usage decreased in larger cities due to faster travel speeds and cost savings from bikesharing. Half of all bikesharing members reported reducing their personal automobile use [31]. Aggregate-level impacts of bikesharing are summarized in Figure 10.3 based on a number of cities analyzed in North America ($n = 6{,}168$).

10.5 Ridesharing

Ridesharing facilitates shared rides among drivers and passengers with a common origin and/or destination. Traditionally defined, ridesharing includes vanpooling (the grouping of seven to 15 persons commuting together in one van) and carpooling (groups of seven or less traveling together in one car), which have been alternative transportation strategies for decades. Because innovations in the area of sharing a passenger ride have burgeoned only recently, the terminology remains unclear. To differentiate carpooling and vanpooling (ridesharing) from the emerging services, the terms "ridesharing" and "for-hire driver services" are used.

Ridesharing has been classified into the following categories: (1) acquaintance-based, (2) organization-based, and (3) ad hoc [34]. Acquaintance-based ridesharing consists of carpools that are formed by people who are already acquaintances (i.e., carpools among family ("fampools") and coworkers). Organization-based carpools require participants to join the service either through membership or by visiting a website. Ad-hoc ridesharing involves more unique forms of ridesharing, including casual carpooling that is also known as "slugging." Although traditional modal share in the United States declined from 20.4% in 1970 to 9.3% in 2013, it remains the second largest travel mode, after driving alone [35].

Since the late 1960s, public agencies particularly in the United States and Canada have tried to increase ridesharing's modal share through policies (e.g., transportation demand management, trip reduction ordinances); high-occupancy vehicle (HOV) infrastructure (e.g., HOV lanes, park-and-ride facilities); and technology (computerized ridematching). The combination of those strategies proved successful—as of July 2011, there were 638 ridematching services in North America [34]. However, as mobile technology has proliferated and improved in this decade, companies have increasingly targeted ridesharing modal share. Examples include Carma Carpooling, Carzac, and Scoop. Moreover, ridesourcing companies (discussed in more detail below) have launched services to target commuters to use their apps to facilitate ridesharing. Lyft began Driver Destination in 2014, and Uber with uberCOMMUTE in China in 2015, which targets drivers with longer commutes to pick up passengers along their commute. Lyft also partnered with the Metropolitan Transportation Commission in the San Francisco Bay Area to pilot Lyft Carpool in March 2016. Lyft Carpool ceased operations in August 2016 due to insufficient use.

10.6 For-hire driver services

For-hire driver services differ from traditional ridesharing, in that these services involve the passenger requesting a ride through a mobile device and a mobile app, and often its drivers do not have an incidental origin or destination as the passengers. For-hire driver services have experienced notable growth in the last few years in the midst of an evolving regulatory and policy climate. They include ridesourcing (also known as TNCs), ridesplitting or "pooling" within ridesourcing services, and e-Hail services for taxis with medallions.

10.6.1 Ridesourcing

Ridesourcing (or TNC) services use mobile apps to connect community drivers with passengers. There are various terms used for this emerging transportation option—ridesourcing among transportation academics, TNCs among practitioners, and ride-hailing and ride-booking among the popular press. Examples of these services include: Lyft and Uber (specifically, uberX, uberXL, and UberSELECT), as well as specialized services for children and the older population. These services can provide various vehicle types (e.g., sedans, sports utility vehicles, vehicles with car seats, wheelchair-accessible vehicles, and vehicles where the driver can assist older or disabled passengers). Although taxis are often regulated to charge static fares, ridesourcing often uses market-rate pricing, popularly known as "surge pricing" when prices usually go up during periods of high demand to incentivize more drivers to take ride requests.

Lift Hero is a specialized ridesourcing service that caters toward older adults and those with disabilities in the San Francisco Bay Area. The drivers are specially trained in caring for such riders. Similarly, Lyft has partnered with National MedTrans Network to operate Concierge, a platform to provide non-emergency medical transportation in New York City. Another specialized for-hire vehicle service is HopSkipDrive, which provides rides for children either to or from school or afterschool activities. The drivers are either mothers or those with a background in childcare. At present, HopSkipDrive is available in the Los Angeles and San Francisco Bay Areas. Shuddle similarly catered to children in the San Francisco Bay Area, but it ceased operations in April 2016 due to financial difficulties.

There are mixed responses from cities and public agencies regarding the popularity of ridesourcing in urban areas. Although some have attempted to prohibit ridesourcing operations, others are interested in its potential to complement the existing public transit system and reduce parking demand. In March 2016, a suburb of Orlando, Florida began a pilot to subsidize Uber rides within its city and to and from its commuter rail station.

Since ridesourcing services began in 2012 in San Francisco, California, relatively little research, other than internal studies of companies' proprietary data, has documented their impacts. An exploratory study was conducted in 2014 of 380 ridesourcing users in San Francisco [36]. Ridesourcing users were generally younger and more highly educated than the city average (84% had a bachelor's degree or higher). UberX provided the majority of trips (53%), whereas other Uber services (black car, SUV, etc.) represented another 8%. Lyft provided 30% of trips, Sidecar (now defunct) provided 7%, and the remainder of trips were provided by other services. Of respondents who owned a car, 40% stated that they had reduced their driving due to the service. A summary of the study findings is provided in Figure 10.4.

Ridesourcing may also complement public transit, serving specific times when, and areas where, public transit is sparser. A 2016 study of the American Public Transportation Association found that ridesourcing is most often used between the hours of 10 PM and 4 AM, when public transit is infrequent or even unavailable. Moreover, the study found that ridesourcing appeared to substitute for private auto trips rather than trips via public transit among shared mobility users [37].

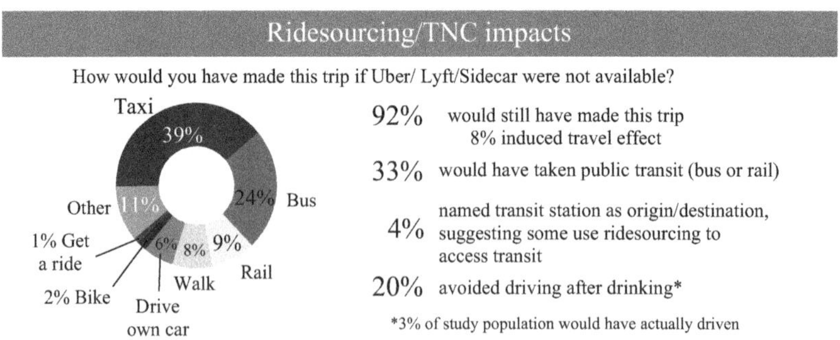

Figure 10.4 Impacts of ridesourcing/TNCs in San Francisco, California [10]

10.6.2 Ridesplitting or pooling

A more recent development in ridesourcing is the introduction of ridesplitting (or pooling), which is the splitting of a ride and its fare with another party taking a similar route. Ridesourcing companies operate ridesplitting services, such as Lyft Line and UberPOOL, which match riders with similar origins and destinations together. These shared services enable routes to be changed dynamically as passengers request pickups in real time. Lyft Line has experimented with "Hot Spots" in the San Francisco Bay Area that encourage passengers to congregate at select intersections in exchange for discounted fares as a means of consolidating operations and making them more efficient. Similarly, UberPOOL has tested "Smart Routes," where users can get a discounted fare starting at US$1 off the normal UberPOOL price in return for walking to a major arterial street. This allows drivers to make fewer turns and complete ride requests faster [38].

Uber also launched uberHOP in Seattle, Washington, and Toronto, Canada—an on-demand pooling service for peak-period travel. Travelers request a ride through the app and walk to a designated pickup location and share the ride with other commuters. The uberHOP service spans ridesplitting and on-demand microtransit (discussed below). Like microtransit, it pools more riders together and uses predetermined pickup and dropoff locations, but the service employs community vehicles and drivers, which is closer to ridesplitting.

10.6.3 e-Hail services

The taxi industry has responded to the rising popularity of ridesourcing with its own mobile device apps known as "e-Hail." Travelers can use e-Hail apps, which are maintained either by the taxi company or a third-party provider, to electronically hail a taxi. There has been a dramatic increase in use of e-Hail services, such as Arro, Bandwagon, Curb, Flywheel, Gett, Hailo, and iTaxi, in the United States. For example, as of October 2014, Flywheel was used among 80% of San Francisco taxis (1,450 taxis). This has brought taxi wait times closer in line with those of ridesourcing [39]. Increasingly, taxi and limousine regulatory agencies are developing e-Hail pilot programs and mandating e-Hail app compatibility. In February 2015, Flywheel was operating in six cities with over 5,000 drivers, and Curb (acquired by Verifone) served approximately 60 U.S. cities with 35,000 cabs. The Bandwagon app combines ridesplitting with e-Hail to facilitate taxi splitting. It matches riders going in a similar direction in taxis and provides a platform for splitting the fare. As of Spring 2016, Bandwagon operates at LaGuardia Airport and John F. Kennedy International Airport in New York City. In April 2016, the Gett app introduced a fare splitting feature for use in over 15,000 black cabs in London. Because regulated taxis charge static fares, e-Hail services also employ locally regulated taxi rates and do not use "surge pricing" during periods of high demand, as ridesourcing services often do.

Partnerships are starting to form among public agencies, taxi companies, and app developers to create a common platform for travelers to use e-Hail apps in various countries. UpTop is a global taxi network developed by a partnership of IRU and the Taxicab, Limousine and Paratransit Association. UpTop has been partnering with app companies. More recently, it added Curb, The Ride, and zTrip apps to its network, and it covers 500,000 taxis, or 10%, of all taxis worldwide.

10.7 Microtransit

Other transportation options have existed in addition to traditional, fixed-route public transit: paratransit, jitneys, dollar vans, and feeder shuttles operated by employers or public transit agencies. Recently, a form of technology-enabled transit has emerged, called microtransit, which incorporates flexible routing, flexible scheduling, or both. Microtransit services are typically operated by private companies, but public transit agencies are gaining interest in operating their own microtransit services. These services operate similarly to jitneys of the past but are enhanced with ICT [40]. Microtransit operators primarily target commuters, connecting residential areas with urban and suburban job centers. Microtransit's use of mobile technology avoids traditional and costly methods of reserving rides, such as call centers or even booking websites, thus potentially lowering operating costs for services that target special populations, such as the disabled, older adults, and low-income groups. Microtransit services can be categorized into the following two models: (1) fixed route, fixed schedule and (2) flexible route with on-demand scheduling.

10.7.1 Fixed routes and fixed scheduling

Fixed-route and fixed-schedule microtransit operates similarly to fixed-route public transit or even vanpooling. An example of such a service is Chariot (acquired by Ford in September 2016), which runs 14-seat vans along predefined routes in San Francisco, California. Chariot allows its customers to request new routes through its mobile app as a way to determine new crowdsourced routes. Although these services may appear to be like vanpooling, microtransit drivers are professional drivers rather than vanpool participants who share driving responsibilities. To some public transit agencies, microtransit seems to directly compete with bus routes. However, in congested cities, microtransit may alleviate overcrowded buses or provide feeder services to transit trunk lines. Research is needed to better understand this dynamic.

10.7.2 Flexible routes and on-demand scheduling

Flexible-route, on-demand microtransit operates more similarly to ridesplitting and paratransit services. An example includes Bridj in Boston, Massachusetts; Kansas City (both Missouri and Kansas); and Washington, D.C. Its app enables users to request rides in select neighborhoods. Based on the requests received, Bridj selects a central meeting spot, which passengers walk to and share a ride with other passengers traveling along a similar route. According to Bridj, the service is able to transport 22 passengers per vehicle per hour. Notably, *Ride KC: Bridj* in Kansas City is the first public–private partnership between a shared mobility company

(Bridj), an automaker (Ford Motor Company), and a public transit agency (Kansas City Area Transportation Authority). Another on-demand microtransit service is Via in New York City. Via users request rides in real time and share a ride with passengers going a similar direction. Due to the one-way avenue design of New York City, passengers are often asked to walk to an adjacent one-way avenue to minimize route deviation and improve operational efficiency. Via has served 1.5 million rides since its launch in late-2013 and expanded to Chicago, Illinois.

The public sector has also been experimenting with this model. In Helsinki, Finland, publicly operated microtransit service Kutsuplus began in 2012 as a pilot service similar to Bridj, except that it had passengers gather at bus stops for their pick up. However, due to growing public subsidy costs and low ridership, the service shut down in late-2015. Another example was the Santa Clara Valley Transportation Authority's VTA FLEX pilot launched in early-2016 in San Jose, California, which was an on-demand bus service that routed buses to pick up passengers at existing public bus stops. Similar to fixed-route microtransit, on-demand microtransit could provide first- and last-mile service for public transit, but research is needed to understand impacts.

10.8 Courier network services

CNS (also referred to as flexible goods delivery) provide for-hire delivery services using an online platform (such as a website or mobile app) to connect couriers using their personal vehicles, bicycles, or scooters with freight (e.g., packages, food). Although the business models in this realm are evolving, two general approaches appear to have emerged: (1) P2P delivery services and (2) paired on-demand passenger ride and courier services.

10.8.1 P2P delivery services

In P2P delivery services, couriers use their own vehicle, bicycle, or scooter to conduct a delivery. Within P2P delivery services, there are a variety of business models. For example, Postmates has its couriers deliver groceries, takeout food, or goods from almost any restaurant or store in a city. They charge a delivery fee in addition to a 9% service fee based on the cost of the goods being delivered. Instacart is similar to Postmates, but it is limited to grocery delivery and charges a delivery fee between US$4 and $10 depending on the time given to complete the delivery. DoorDash is a takeout food delivery service, charging a flat delivery fee of US$7. Roadie is another courier service, primarily for inter-city goods movement rather than same-day intra-city deliveries. Finally, Shipbird is a shipping service that connects everyday commuters with individuals seeking couriers. Couriers provide the Shipbird app with their availability, commuting route, and the distance they are willing to deviate from their commute route to complete a delivery. Shipbird then matches these couriers with the requested delivery jobs. The proliferation of P2P delivery services, where couriers use their personal vehicles or travel modes, could reduce the need for delivery companies to maintain their own fleet for operations.

10.8.2 Paired on-demand passenger ride and courier services

The second CNS model that has emerged is one in which for-hire driver services (e.g., ridesourcing) also conduct package deliveries. Deliveries via these modes can either be made in separate trips or in mixed-purpose trips (e.g., for-hire drivers can transport packages and passengers in the same trip). Sidecar was the only ridesourcing company that conducted mixed-purpose trips in addition to dedicated goods delivery trips, through a service called Sidecar Deliveries. In February 2015, the company claimed that 10% of its passenger rides in San Francisco included package deliveries [41]. Sidecar Deliveries also allowed couriers on foot, bicycles, and scooters into its network. The service partnered with Yelp Eat24, a food ordering service, to assist with deliveries. The company claimed to have cut estimated delivery times in half. Sidecar ceased operations in December 2015.

Uber has also entered the delivery services market with UberEATS (food) and UberRUSH (bike, foot, and vehicle messenger goods delivery service). UberEATS charges a US$3 flat delivery fee (US$4 in New York City) in addition to the food cost, whereas UberRUSH charges a 20% delivery fee in New York City and 25% in San Francisco and Chicago. Uber piloted UberRUSH in New York City in 2014, first as a bike messenger service where couriers pick up an item from the requester and deliver it somewhere within a coverage area within the same day. This is now being expanded to merchant delivery, where items are picked up from stores and delivered either to the requester or to a third party by foot or vehicle. It has recently expanded to San Francisco and Chicago [42]. Uber is also experimenting with UberCARGO in Hong Kong for moving and delivery needs (e.g., mattress delivery to a new house). For one day in June 2015, Lyft ran a promotion with Starbucks where they delivered free iced coffee. Finally, both Lyft and Uber are now partnering with Walmart on same-day grocery delivery. Thus, three major ridesourcing operators have tried to expand their ride services to include package/item delivery, food delivery, or both. As the model is still new, this paired passenger and courier model will likely continue to evolve over the coming years.

10.9 Future of shared mobility

Smartphones are redefining transportation accessibility and mobility in the twenty-first century. Demographic shifts, advancements in geo-spatial routing and computing power, the use of cloud technologies, faster wireless networks capable of carrying greater bandwidth, concerns about congestion, and heightened awareness about the environment and climate change are changing the way people think about mobility. Mobility consumers are increasingly using smartphone applications dubbed "apps" and mobile-capable websites for an array of transportation functions. More people are starting their trips with smartphones, engaging in activities such as route planning, finding departure information for the next bus or railcar, or requesting a driver (e.g., Lyft, Uber, or taxi). The increasing capability, availability, and affordability of intelligent transportation systems, GPS, wireless, and cloud technologies—coupled with growing data availability, sharing, and interoperability—are causing more people to use smartphone apps to assist with their transportation needs. Time savings (e.g., HOV lanes available to users of dynamic ridesharing); financial savings (e.g., dynamic pricing providing discounts for peak and off-peak travel and for choosing low-volume routes); incentives (e.g., offering points, discounts, or lotteries); and gamification (e.g., use of game design elements in a non-game context) are among the key factors driving the growth of transportation apps. This section focuses on shared mobility's future including use of trip planning mobile apps, potential opportunities with automated vehicles, and the concept of mobility as a service (MaaS) or mobility on demand (MOD).

10.9.1 Mobility-related smartphone apps

Many of the shared mobility services mentioned in this chapter are only accessible by a smartphone or mobile app. Smartphone apps have helped give rise to innovative business models, services, and ways to access destinations. These are broadly summarized in the five categories below:

1. Apps that have enabled on-demand for-hire driver services (e.g., ridesourcing and e-Hail taxis);
2. Public transit-related apps that give real-time arrival information and can help plan routes and transfers;
3. Apps that enable business-to-consumer transactions (e.g., roundtrip or one-way carsharing);
4. Apps that facilitate P2P transactions, such as those with carsharing or bike-sharing modes; and
5. Apps that aggregate different modes, routes, and real-time travel conditions to assist with trip planning.

The fifth category listed above, or trip-planning apps, can aid the use of the other four app services. Trip planning apps help travelers identify their preferred travel route and mode based on cost, convenience, time considerations, and environmental impact. They can also provide step-by-step assistance as users navigate their chosen route. In this way, they can act as an enabling technology for shared mobility use. Initial research indicates that 80% of such app users took modes other than their personal cars, mostly opting for public transit [43].

Trip planning apps can be grouped into two general categories: (1) single-mode trip planning and (2) multimodal trip aggregators. A survey of 82 common U.S. transportation apps, which include the ones discussed below, found that 93% were offered free of charge [44].

10.9.1.1 Single-mode trip planning

Trip planning apps can be designed for a single mode, usually with public transit and driving route assistance. Increasingly, mobility apps are using real-time information. Transit trip planning apps augment static maps and timetables with real-time information about delays. For example, the Washington, D.C. Metro has its own app that provides real-time delay information about its trains. Driving-related single mode apps include Waze and Metropia. These apps use real-time traffic congestion and incident data to generate optimal travel routes, and they also give turn-by-turn assistance during the journey. The turn-by-turn guidance is similar to Google Maps and Apple Maps, although those applications also show modes other than driving when generating route options for a particular trip.

10.9.1.2 Multimodal trip aggregators

Multimodal trip aggregators offer a single platform for planning trips involving different modes including: public transit, taxi services, carsharing, ridesharing, for-hire driver services, bicycling, walking, and personal vehicles. Travelers can quickly view the time, cost, and even calories burned while using different modes and routes. These apps also use real-time information to provide accurate departure and arrival times. For bicyclists, some apps also show the flattest route to take. These apps can also quantify carbon dioxide (CO_2) emission savings and provide a platform for integrated payments across modes. A few additional examples of trip aggregation apps include Citymapper, Moovit, Nimbler, Swiftly (formerly Swyft), Transit App, and TripGo.

Companies are also partnering with cities to create apps specific to modes and services available in that particular city or region. One example is Go LA, a multimodal transit app developed in collaboration between Xerox and the City of Los Angeles. This app was launched in January 2016. Xerox launched a similar app, Go Denver, for the City of Denver in February 2016 [45].

10.9.1.3 Gamification

Some mobile apps use gamification or game theory and game mechanics to incentivize positive travel behaviors. These apps achieve this by way of leaderboards, badges, levels, points, and progress bars, in addition to other tools [46]. For example, Waze gives its users points for providing traffic data and warnings of road hazards (construction, police, cameras) for other drivers. Metropia provides commuting routes, but it also offers incentives for people to take alternative routes and departure times to reduce traffic on certain routes. Incentives, provided through community sponsors, include online music, gift cards to local and online shops, etc. The app also tracks how many pounds of CO_2 the user saves. A Metropia pilot study on its users in Los Angeles found that after six weeks of use, 86% of commuters reported time savings, and over 60% of users changed their regular departure time. Users who changed their departure time and route experienced between a 20% and 30% reduction in commute times [47].

10.9.1.4 Impact on travel behavior

The use of smartphone apps for transportation can have economic, social, and psychological impacts on users. First, they reduce the cognitive burden of a user trying to plan a trip after considering public transit options and delays, as well as route preference and current road traffic conditions. Another often overlooked benefit of trip planning apps is giving perceived control to the user, which may make him or her more satisfied with their trip regardless of whether there was an objective improvement in comfort [44]. For example, several studies have shown that bus riders without real-time arrival data perceived their wait time to be longer than was experienced by riders with real-time data, suggesting that the presence of real-time information can increase the perceived trip satisfaction [48]. Further, smartphone apps, such as Swiftly, help users consider the menu of options available and can facilitate public transportation use. Smartphones apps that match riders to each other, to drivers, or to P2P shared vehicles can also help foster a sense of trust in the network. This is done by employing a system of ratings as well as by linking social media profiles. Taking the presence of the embedded social network a step further, travel apps also leverage social dynamics to influence behavioral adjustments. This is done by apps, such as Waze, Moovit, and GasBuddy, which use points, competition, or ratings to shape behavior. The behavioral mechanisms employed by transportation applications are worth greater study as an increasing number of users consult travel applications before starting a trip. Findings from future research could build upon anecdotal evidence that suggests mobility-related apps are successful in affecting travel behavior [44].

10.9.2 Impact of automated vehicles on shared mobility

Automated vehicles (AVs) move passengers or freight without human intervention. Research and development activities on AVs are increasing (Chapter 11), with many high-profile companies and research institutions developing AVs or partnering to develop them. Some have predicted that AV technology will be emergent by 2020, accepted by the 2030s, and predominant by 2050. As shared mobility use has grown exponentially in recent years, so has the prospect of shared automated vehicles (SAVs). By 2020, several automakers are planning to release limited self-driving AVs. In this scenario, AVs could drive up to carsharing users, park, and refuel/recharge themselves, saving time and increasing user convenience. Moreover, with enhanced safety features, AVs in a shared vehicle setting may reduce an operator's insurance costs. These savings could be passed onto shared mobility consumers. Higher levels of automation will likely enhance the capabilities and convenience of SAV savings [49].

At present, research on SAVs is limited. Initial results of an agent-based model of SAVs estimate that one SAV could replace 11 conventional vehicles (Chapters 11, 13). Such a system could also increase VMT by approximately 11% due to travel between user trip requests [50]. Another study has postulated that small electric SAVs, when combined with a low-carbon electric grid could reduce per-mile (or km) GHG emissions by 90% compared to today's vehicles. In addition, right-sized SAVs can be deployed according to passenger occupancy needs, further reducing excess GHG emissions [51]. Recent research has argued that AVs have the potential for significant emission and energy reductions, but those reductions are not guaranteed. Various scenarios were created to determine the net impacts of AVs on energy use, with wide variation—from halving emissions to doubling them—due to different assumptions on the social and behavioral response to this new technology [52]. Research, evaluation, and adaptive policies are needed to achieve maximum AV benefits.

10.9.3 Mobility as a service

With the proliferation of different mobility services accessible through the smartphone, it can be challenging to implement an integrated approach to managing transportation services that are simple for end users to understand. This has led to the mobility paradigm referred to as "mobility as a service," where travelers can gain access to an increasingly diverse ecosystem of mobility options through a single platform or a bundle of individual services repackaged and sold through a single service provider (Chapter 13). This new way of envisioning urban transportation reflects changing attitudes toward mobility—from one that emphasizes ownership to one of access. Instead of the various mobility choices mentioned above operating disparately, they can be presented and seamlessly tailored to specific users and trips. This concept is similar to the emergence of multimodal trip aggregators, but it aims to go further with the level of integration on the city or regional scale, where payment and operation can be done through the single portal. The MaaS or MOD provider can be the public transportation agency or specialized MaaS providers [53]. MaaS is emerging as a concept that is only beginning to be deployed [54].

10.10 Conclusion

Shared mobility, a subset of the larger sharing economy, enables users to gain short-term access to transportation modes on an as-needed basis. Carsharing, bikesharing, for-hire driver services, microtransit, and CNS are changing how urban travelers access transportation, make connections to other modes, and send and receive goods. Studies have found a number of environmental, social, and transportation-related impacts of shared mobility services, such as a reduction of vehicle miles/kilometers traveled, vehicle use, and vehicle ownership. Users have often cited cost savings and convenience over private vehicle ownership as reasons for using a shared mode. This has been demonstrated for roundtrip and one-way carsharing and public bikesharing; however, more research is needed to better understand the impacts of for-hire driver services, microtransit, and CNS.

Despite shared mobility's impacts and opportunities, challenges remain particularly in the area of public policy. Cities and states have confronted difficulties in enacting and/or revising policies to keep pace with shared mobility innovations while addressing issues of public safety, insurance and liability, and fair labor practices. All too often, equity considerations remain unanswered—early anecdotal evidence is emerging that suggests that lower-income populations may lack sufficient access to certain shared mobility services. In addition, labor issues concerning for-hire vehicle drivers raise questions regarding whether drivers should be considered independent contractors or employees. Public–private partnerships, such as the Ride KC: Bridj pilot, which employs drivers from the Kansas City Area

Transportation Authority, may offer guidance on how public agencies can address labor issues. As automated vehicles are poised to dramatically transform urban transportation, SAVs may too become a major public transportation mode, filling gaps in the existing fixed-route network and allowing first- and last-mile connectivity. Although research and development into AVs has increased, its impacts on travel and emissions remain speculative. Future research should focus efforts on better understanding the longer-term impacts of this burgeoning industry on future transportation networks.

Acknowledgment

The authors of this chapter would like to thank the California Department of Transportation, the Federal Highway Administration, and the Transportation Sustainability Research Center at the UC Berkeley for funding the research, which led to this chapter. The contents of the chapter reflect the views of the authors and do not necessarily indicate acceptance by the sponsors.

References

[1] R. Botsman and R. Rogers, *What's Mine Is Yours: The Rise of Collaborative Consumption*. New York: Harper Collins, 2010.

[2] PwC, "The sharing economy—sizing the revenue opportunity," Price waterhouseCoopers, 2014. Available at: http://www.pwc.co.uk/issues/megatrends/collisions/sharingeconomy/the-sharing-economy-sizing-the-revenue-opportunity.html, accessed 15 September 2015.

[3] X. Wang, G. Lindsey, J.E. Schoner and A. Harrison, "Modeling bike share station activity: effects of nearby businesses and jobs on trips to and from stations," *Journal of Urban Planning and Development*, vol. 142, no. 1, 2016

[4] R. Buehler and A. Hamre, "Business and bikeshare user perceptions of the economic benefits of Capital Bikeshare," TRB 94th Annual Meeting Compendium of Papers, Transportation Research Board, 2015.

[5] S. Shaheen and A. Cohen, "Innovative mobility carsharing outlook," Transportation Sustainability Research Center, University of California, Berkeley, Winter 2016.

[6] R. Cervero and Y. Tsai, "City CarShare in San Francisco, California: second-year travel demand and car ownership impacts," *Transportation Research Record: Journal of the Transportation Research Board*, vol. 1887, pp. 117–127, 2004.

[7] E. Martin and S. Shaheen, "The impact of carsharing on public transit and non-motorized travel: an exploration of North American carsharing survey data," *Energies*, vol. 4, no. 11, pp. 2094–2114, 2011.

[8] L. Sioui, C. Morency and M. Trépanier, "How carsharing affects the travel behavior of households: a case study of Montréal, Canada," *International Journal of Sustainable Transportation*, vol. 7, no. 1, pp. 52–69, 2013.

[9] E. Martin and S. Shaheen, "Greenhouse gas impacts of carsharing in North America," Mineta Transportation Institute, Report 09–11, 2010.
[10] S. Shaheen and N. Chan, "Mobility and the sharing economy: impacts synopsis, shared-use mobility definitions and impacts, special edition," Transportation Sustainability Research Center, University of California, Berkeley, Spring 2015.
[11] S. Shaheen and A. Cohen, "Innovative mobility carsharing outlook: carsharing market overview, analysis, and trends," Transportation Sustainability Research Center, University of California, Berkeley, Fall 2012.
[12] S. Shaheen and A. Cohen, "Innovative mobility carsharing outlook," Transportation Sustainability Research Center, University of California, Berkeley, Winter 2015.
[13] V. Biau, "Montpellier 1971–1974: une expérience de 'transport individuel public'," *Transports Urbains*, vol. 38, no. 72, pp. 21–25, 1991.
[14] M.-H. Massot, "Praxitèle, un concept, un service, une expérimentation, bilan d'un prototype," *TEC, Transport, Environment, Circulation*, no. 2000, pp. 25–32, 2011.
[15] Honda Motor Co., Ltd. "Honda announces the development of Intelligent Community Vehicle System (ICVS) management technology and vehicles," Honda Motor Co., Ltd., Tokyo, 10 September 1998. Available at: http://world.honda.com/news/1998/c980910.html, accessed 1 March 2016.
[16] M. Barth, S. Shaheen, T. Fukuda and A. Fukuda, "Carsharing and station cars in Asia: an overview of Japan and Singapore," *Transportation Research Record: Journal of the Transportation Research Board*, vol. 1986, pp. 106–115, 2007.
[17] S. Shaheen, "Dynamics in behavioral adaptation to a transportation innovation: a case study of CarLink - a smart carsharing system," Institute of Transportation Studies, University of California, Davis, UCD-ITS-RR-99-16:232, 1999.
[18] S. Shaheen, J. Wright, D. Dick and L. Novick, "CarLink—a smart carsharing system field test report," Institute of Transportation Studies, University of California, Davis Publication UCD-ITS-RR-00-4:182, 2000.
[19] S. Shaheen and L. Novick, "Framework for testing innovative transportation solutions: case study of CarLink, a commuter carsharing program," *Transportation Research Record: Journal of the Transportation Research Board*, vol. 1927, pp. 149–157, 2005.
[20] UC Irvine News, "UC Irvine's car-sharing program charges ahead," University of California, Irvine, 2013. Available at: http://news.uci.edu/press-releases/uc-irvines-car-sharing-program-charges-ahead/, accessed 21 March 2013.
[21] M. Barth, M. Todd and H. Murakami, "Intelligent transportation system technology in a shared electric vehicle program," *Transportation Research Record: Journal of the Transportation Research Board*, vol. 1731, pp. 240–245, 2000.

[22] S. Shaheen, N. Chan and H. Micheaux, "One-way carsharing's evolution and operator perspectives from the Americas," *Transportation*, vol. 42, no. 3, pp. 519–536, 2015.

[23] M. Nourinejad, S. Zhu, S. Bahrami and M.J. Roorda, "Vehicle relocation and staff rebalancing in one-way carsharing systems," *Transportation Research Part E: Logistics and Transportation Review*, vol. 81, pp. 98–113, 2005.

[24] S. Weikl and K. Bogenberger, "A practice-ready relocation model for free-floating carsharing systems with electric vehicles—mesoscopic approach and field trial results," *Transportation Research Part C: Emerging Technologies*, vol. 57, pp. 206–223, 2015.

[25] E. Martin and S. Shaheen, "Impacts of car2go on vehicle ownership, modal shift, vehicle miles traveled, and greenhouse gas emissions," Working Paper, UC Berkeley, Berkeley, CA, July 2016. Available at: http://innovativemobility.org/wp-content/uploads/2016/07/Impactsofcar2go_FiveCities_2016.pdf?platform=hootsuite, accessed 3 August 2016.

[26] S. Shaheen, M. Mallery and K. Kingsley, "Personal vehicle sharing services in North America," *Research in Transportation Business & Management*, vol. 2416, no. 3, pp. 71–81, 2012.

[27] J. Dill, "Early insights into peer-to-peer carsharing," Transportation Insight for Vibrant Communities, Portland State University, 31 October 2014. Available at: http://trec.pdx.edu/blog/early-insights-peer-peer-carsharing, accessed 2 March 2015.

[28] S. Fraiberger and A. Sundararajan, "Peer-to-peer rental markets in the sharing economy," NYU Stern School of Business Research Paper, 2005. Available at: http://ssrn.com/abstract=2574337, accessed 2 March 2015.

[29] Scoot Networks, Personal communication, 2015.

[30] S. Shaheen and M. Christensen, "Shared-use mobility summit: retrospective of North America's first gathering on shared-use mobility," Transportation Sustainability Research Center, University of California, Berkeley, June 2014.

[31] S. Shaheen, E. Martin, N. Chan, A. Cohen and M. Pogodzinski, "Public bikesharing in North American during a period of rapid expansion: understanding business models, industry trends and user impacts," Mineta Transportation Institute, Report 12–29, 2014.

[32] R. Meddin, Personal communication, 2016.

[33] S. Shaheen, E. Martin, and A. Cohen, "Public bikesharing in North America: early operator and user understanding," Mineta Transportation Institute, Report 11–26, 2012.

[34] N. Chan and S. Shaheen, "Ridesharing in North America: past, present, and future," *Transport Reviews*, vol. 32, no. 1, pp. 93–112, 2012.

[35] U.S. Census Bureau, "Sex of workers by means of transportation to work," American Community Survey 1-Year Estimates, Table B08006, American FactFinder, 2013. Available at: http://factfinder2.census.gov, accessed 5 April 2015.

[36] L. Rayle, D. Dai, N. Chan, R. Cervero and S. Shaheen, "Just a better taxi? A survey-based comparison of taxis, transit, and ridesourcing services in San Francisco," *Transport Policy*, vol. 45, pp. 168–178, 2016.

[37] APTA, "Shared mobility and the transformation of public transit," American Public Transportation Association, March 2016. Available at: http://www.apta.com/resources/hottopics/Pages/Shared-Use-Mobility.aspx, accessed 17 March 2016.

[38] C. de Looper, "Uber testing bus-like 'Smart Routes'," *Tech Times*, 2015. Available at: http://www.techtimes.com/articles/79084/20150824/uber-testing-bus-smart-routes.htm, accessed 14 March 2016.

[39] K. Steinmetz, "Taxi drivers are using apps to disrupt the disruptors," *TIME*, 2014. Available at: http://time.com/3119161/uber-lyft-taxis/, accessed 15 March 2016.

[40] R. Cervero, *Paratransit in America: redefining mass transportation*, Westport, CT: Greenwood Publishing Group, 1997.

[41] T. Lien, "Sidecar to expand package delivery service," *Los Angeles Times*, 9 February 2015. Available at: http://www.latimes.com/business/la-fi-0210-sidecar-delivery-service-20150210-5-story.html, accessed 11 February 2015.

[42] A. Cutherbertson, "Uber planning same-day merchant delivery service through UberRush," *International Business Times*, 29 April 2015. Available at: http://www.ibtimes.co.uk/uber-planning-same-day-merchant-delivery-service-through-uberrush-1498887, accessed 3 May 2015.

[43] J. Gossart and A. Whitney, "RideScout T 76 IDEA grant," IDEA Program Final Report: Transportation Research Board, 2014. Available at: http://onlinepubs.trb.org/onlinepubs/IDEA/FinalReports/Transit/Transit76.pdf, accessed 28 March 2016.

[44] S. Shaheen, A. Cohen, I. Zohdy and B. Kock, "Smartphone applications to influence travel choices: practices and policies," Federal Highway Administration, Report FHWA-HOP-16-023, July 2016. Available at: http://www.ops.fhwa.dot.gov/publications/fhwahop16023/index.htm, accessed 3 August 2016.

[45] B. McKee and J. Palmeroni, "Los Angeles commuters to usher in new era of daily trip planning," Press Release, Norwalk, CT: Xerox Corporation, 27 January 2016. Available at: http://news.xerox.com/news/City-of-LA-introduces-new-Xerox-Go-LA-app, accessed 20 March 2016.

[46] A. Marczewski, *Gamification: a simple introduction and a bit more*, Seattle, WA: Amazon Digital Services, 2012.

[47] X. Hu, Y. Chiu, S. Delgado, L. Zhu, R. Luo, P. Hoffer, *et al.*, "Behavior insights for an incentive-based active demand management platform," TRB 93rd Annual Meeting Compendium of Papers, Transportation Research Board, 2014.

[48] K.E. Watkins, B. Ferris, A. Borning, G.S. Rutherford and D. Layton, "Where is my bus? Impact of real-time information on the perceived and actual wait time of transit riders," *Transportation Research Part A: Policy and Practice*, vol. 45, no. 8, pp. 839–848, 2011.

[49] J.B. Greenblatt and S. Shaheen, "Automated vehicles, on-demand mobility, and environmental impacts," *Current Sustainable Renewable Energy Reports*, vol. 2, no. 3, pp. 74–81, 2015.
[50] D.J. Fagnant and K.M. Kockelman, "The travel and environmental implications of shared autonomous vehicles, using agent-based model scenarios," *Transportation Research Part C*, vol. 40, pp. 1–13, 2014.
[51] J.B. Greenblatt and S. Saxena, "Autonomous taxis could greatly reduce greenhouse-gas emissions of US light-duty vehicles," *Nature Climate Change*, vol. 5, pp. 74–81, September 2015.
[52] Z. Wadud, D. MacKenzie and P. Leiby, "Help or hindrance? The travel, energy and carbon impacts of highly automated vehicles," *Transportation Research Part A*,vol. 86, pp. 1–18, April 2016.
[53] S. Hietanen, "'Mobility as a service'—the new transport model?" *Eurotransport*, vol. 12, no. 2, pp. 2–4, 2014.
[54] M. Kamargianni, M. Matyas, W. Li and A. Schafer, "Feasibility study for 'mobility as a service' concept in London." University College London Energy Institute, May 2015. Available at: https://www.bartlett.ucl.ac.uk/energy/docs/fs-maas-compress-final, accessed 27 July 2016.

Glossary

Automated vehicles	Vehicles that move passengers or freight without human intervention
Bikesharing	A system where users access bicycles on an as-needed basis for one-way (point-to-point) or roundtrip travel
Carpooling	A formal or informal arrangement through which commuters share a vehicle for trips from a common origin, destination, or both
Car rental	A non-membership-based service or company that rents cars or light trucks, typically by the day or week. Traditional rental car services include storefronts that require an in-person transaction through a rental car attendant. Rental car services may also use "virtual storefronts," allowing unattended vehicle access similar to carsharing
Carsharing	Program where individuals have short-term access to a vehicle without the costs and responsibilities of ownership. Individuals typically access vehicles by joining an organization that maintains a fleet of cars and light trucks
Closed-campus bikesharing	A bikesharing system deployed on university and office campuses. Bicycles are available only to members of the particular campus community being served

Courier network services	For-hire delivery services that use an online application or platform (such as a website or smartphone application) to connect delivery drivers using personal transportation modes with package/item or food delivery requests. These services can also be used to pair package delivery with existing passenger trips
e-Hail	Services operated by a taxi company or third-party provider that allow travelers to request a taxi via a smartphone app
Employer shuttles	Employer-sponsored shuttles that transport employees between workplaces and public transit stations
Fractional ownership	Model where individuals sublease or subscribe to access to a motor vehicle or low-speed mode owned by a third party
Hybrid peer-to-peer access model	Model where individuals access vehicles or low-speed modes by joining an organization that maintains its own fleet, but it also includes privately owned autos or low-speed modes
Microtransit	A privately owned and operated shared transportation system, typically comprised of vans and buses, which can have fixed routes and schedules or flexible routes and on-demand scheduling
Mobility as a service (MaaS)	A mobility paradigm whereby travelers gain access to several mobility options through a single platform
Modal split	The proportion of total person trips for each transportation mode (e.g., auto, public transit, bicycle, walking)
Multimodal	Use of more than one transportation mode to complete a trip
One-way carsharing	A form of carsharing that enables members to pick up a vehicle at one location and drop it off at another. This is also called a point-to-point carsharing service. One-way carsharing can be station based or free floating
Peer-to-peer (P2P) bikesharing	A system where users rent out their privately owned bicycles
Peer-to-peer (P2P) carsharing	A program that uses privately owned vehicles made temporarily available for shared use by an individual or members of a peer-to-peer carsharing company. Also called a peer-to-peer access model
Personal vehicle sharing	Sharing of privately owned vehicles where companies broker transactions between car owners and renters

Ridesharing	Shared rides among drivers and passengers with a common origin and/or destination
Ridesourcing/ Transportation Network Companies (TNCs)	Programs that provide prearranged and on-demand transportation services for compensation by connecting drivers of personal vehicles with passengers through mobile applications
Ridesplitting/pooling	A form of ridesourcing where riders with similar origins and destinations are matched to the same driver and vehicle in real time, and the ride and costs are split among users
Roundtrip carsharing	Program that allows members hourly access to shared vehicles that must be returned to the same location where the vehicle was picked up
Scooter sharing	Program that provides members access to private scooters for roundtrip or one-way trips
Shared mobility	Shared use of a motor vehicle, bicycle, or other low-speed mode
Slugging	Informal carpooling between strangers (a hybrid between commuter carpooling and hitchhiking). This is also known as casual carpooling
Taxis	Type of for-hire vehicle service (pre-arranged or on-demand) with a driver and used by one or more passengers
Vanpooling	Program consisting of vans, small buses, and other vehicles operating ridesharing services with capacity for a minimum of seven passengers. Typically, participants split the vehicle and operational costs and may share the responsibility of driving
Vehicle miles traveled (VMT)/vehicle kilometers traveled (VKT)	Number of miles/kilometers traveled annually by vehicles

Further Reading

Bikesharing

"Riding the Bike-Share Boom: Top Five Components of a Successful System Info Graphic". Institute of Transportation and Development Policy. ITDP Bike Share Planning Guide, 2013.

Nair R, Miller-Hooks E, Hampshire R, and Busic A. Large-Scale Vehicle Sharing Systems: An Analysis of Velib. *International Journal of Sustainable Transportation*. 2013; vol. 7, pp. 85–106.

Shaheen S, Guzman S, and Zhang H. "Chapter Nine: Bikesharing across the Globe". *City Cycling*. The MIT Press, pp. 183–210, 2012.

Carsharing

Elliot M, Shaheen S. "Impacts of car2go on Vehicle Ownership, Modal Shift, Vehicle Miles Traveled, and Greenhouse Gas Emissions: An Analysis of Five North American Cities. Transportation Sustainability Research Center", 2016. Available at: http://innovativemobility.org/wp-content/uploads/2016/07/Impactsof car2go_FiveCities_2016.pdf.

Ridesharing

Deakin E, Frick KT, Shivley K. "Dynamic Ridesharing". *Access Magazine*. 2012; vol. 40, pp. 23–28.

Ridesourcing, Taxis, and Transportation Network Companies

Daus MW, Klein S, Mallah S, Matera A, Mischel J, and Taye B. "Study for a Centralized Application for Taxis in Montréal". City of Montreal Taxi Bureau, 2016. Available at: www.windelsmarx.com/resources/documents/Study%20for% 20a%20Centralized%20Application%20for%20Taxis%20in%20Montreal%20-% 20April%202016.pdf.

Matthew DW, Mischel JR. "Accessible Transportation Reform: Transforming the Public Paratransit and Private For-Hire Ground Transportation Systems". *The Transportation Lawyer*. 2014; 17 (2): 37–41. Available at: http://www.beneschlaw. com/Files/Publication/0f84dd11-9c46-4ccc-b1e9-5fed7d22db70/Presentation/Pub lication Attachment/603fac98-36e9-4b0f-a446-6a63f40d74cf/TTL%20October% 202015_Web.pdf.

The Sharing Economy, Shared Mobility, Smartphone Applications, and Millennials

Shaheen, S, Cohen A, Zohdy I, Kock, B. "Smartphone Applications to Influence Travel Choices: Practices and Policies", 2016. Available at: www.ops.fhwa.dot. gov/publications/fhwahop16023/fhwahop 16023.pdf.

Sakaria, N, Stehfest, N. Millennials and Mobility: Understanding the Millennial Mindset and New Opportunities for Transit Providers. *Transit Cooperative Research Program*, 2013. Available at: http://onlinepubs.trb.org/onlinepubs/tcrp/tcrp_w61.pdf.

"Between Public and Private Mobility: Examining the Rise of Technology-Enabled Transportation Services". Transportation Research Board, 2016. Available at: http://onlinepubs.trb.org/onlinepubs/sr/sr319.pdf.

Chapter 11

Autonomous vehicles and shared mobility: shaping the future of urban transport

Hussein Dia[1]

Abstract

The fast pace of development of automated vehicles is being driven by a number of converging forces including self-driving technologies, mobile computing, and on-demand vehicle and ride sharing. The coming together of these powerful trends is shaping an urban mobility future inspired by a vision of zero road injuries and low carbon living. The introduction of self-driving vehicles will have major socio-economic impacts on our society – some good and some bad. Although self-driving vehicles are expected to pave the way to drastic improvements in road safety and urban mobility, they can potentially have adverse impacts on the liveability of our cities if they are not planned as part of a holistic approach to promote low carbon mobility. This chapter provides an environmental scan and an objective review and analysis of the technological, social and economic impacts and trends surrounding (ultimately) autonomous vehicles, and how they are likely to influence the transport and mobility industries and marketplace now and into the future. The chapter also outlines the technological, societal and regulatory challenges that must be overcome before widespread adoption and deployment. The chapter gathers and collates information from a wide body of literature to help city planners, infrastructure engineers and other stakeholders in responding to the likely needs and impacts of this disruptive technology and to inform their policy positions in shaping the future of urban mobility.

Keywords

Autonomous vehicles; shared on-demand mobility; disruptive technologies; smart mobility

[1]Department of Civil Engineering, Swinburne University of Technology, Melbourne, Australia

11.1 Introduction

According to the World Economic Forum [1], we stand today at the brink of a technological transformation (Fourth Industrial Revolution) which will fundamentally change the way we live, work and relate to one another. In its scale, scope and complexity, the transformation will be unlike anything humankind has experienced before [1]. This transformation is characterised by a number of factors which make it different from the previous three industrial revolutions: speed of changes, scope and systems impact [1]. The speed of current breakthroughs has no historical precedent and is evolving at an exponential rather than a linear pace. Moreover, it is disrupting almost every industry in every country. And the breadth and depth of these changes herald the transformation of entire systems of production, management and governance.

The possibilities of billions of people and objects connected by mobile devices, with unprecedented processing power, storage capacity and access to knowledge, are unlimited. And these possibilities will be multiplied by emerging technology breakthroughs in fields such as artificial intelligence (AI), robotics, the Internet of Things, autonomous vehicles, energy storage, and quantum computing to name a few. Already, AI is all around us. Impressive progress has been made in AI in recent years, driven by exponential increases in computing power and by the availability of vast amounts of data.

As part of these developments, it is anticipated that vehicles which drive themselves may very well be on our roads within 5–10 years [2]. Vehicles with varying levels of self-driving capabilities are already available to consumers today [3], and the transition to full autonomous operation is expected to be gradual taking up to 15–30 years [4]. The pace of change will depend in part on acceptance by consumers, regulators and the wider industries which may be disrupted by the changes [5].

Although autonomous vehicles are still a few years away, they have already started to shape some visions for a mobility transformation, beyond the technology itself, driven by a number of converging forces including automated self-driving, vehicle electrification, cloud computing, the Internet of Things, mobile computing and on-demand vehicle and ride sharing (Figure 11.1). The coming together of these powerful trends is shaping an urban mobility future inspired by a vision of zero road injuries and low carbon living.

The fast pace of breakthroughs in autonomous driving technologies and scientific advances is relentless and continues to unfold on many fronts. As the list of 'next big thing' in smart and disruptive mobility grows longer, organisations are increasingly finding it challenging to identify how these developments may unleash disruptive changes and impact the growth and performance of their businesses and industries. This chapter aims to address this by developing a better understanding of the likely impacts of autonomous vehicles, and how policy directions can be shaped to enable their adoption for a safer and more sustainable mobility future.

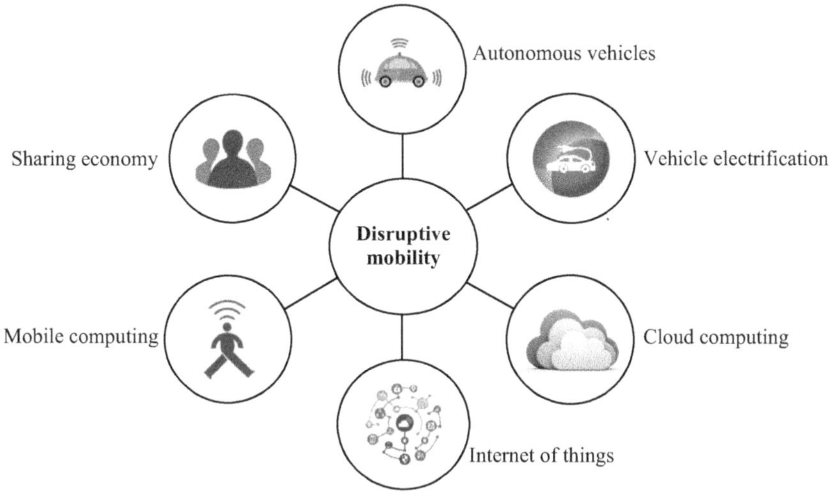

Figure 11.1 The key forces shaping the future of urban mobility.

Autonomous vehicles are a reality. When they arrive, they will have socio-economic impacts and other effects on our society. This chapter will provide a best practice review of their current state of development, timelines for technology maturity and deployment, and identify their role in delivering new mobility solutions as part of a holistic approach to improve road safety and promote low carbon mobility.

This chapter also includes an environmental scan of the current and anticipated external factors that are of direct relevance to the environment of low carbon mobility in urban areas. Specifically, the chapter covers the following topics: (1) It identifies and describes the autonomous vehicles components and the technologies needed to facilitate their introduction; (2) it identifies the potential wide-spread impacts of introducing autonomous vehicles and how they will affect traffic safety, vehicle ownership, congestion patterns, on-demand mobility, freight delivery, vehicle insurance industry, parking, emissions and air pollution; (3) it investigates how automation of the driving task is likely to disrupt the transport industry workforce, and also the new business models and opportunities that it may introduce; (4) it identifies the regulatory and legal enablers needed to facilitate the introduction of autonomous vehicles and describes how careful policymaking in this rapidly evolving and continuously changing field will be necessary to maximise the social benefits that this technology will enable, while minimising the disadvantages; (5) it describes lessons learnt from a number of jurisdictions overseas (e.g. California and Nevada in the United States) who have already introduced legislation and licensed the operation of autonomous vehicle trials on their roads; and (6) it makes recommendations for how stakeholders can continue to monitor and be kept abreast of developments, and the need for future research in relevant key areas.

11.2 Vehicle technologies and autonomous driving functions

Vehicle automation is part of a much larger disruption in automation and connectivity. Autonomous vehicles include a range of sensors, radars and cameras which work together to provide emergency braking, lane departure warning, cross-traffic warning, pedestrian detection, collision avoidance, blind spot detection, rear collision, surround views and traffic signal recognition in addition to other functions. Autonomous vehicles operate on a design concept known as 'Sense-Plan-Act' which is a predominant and well-established robot control methodology.

The technology for automated driving is quite mature. Currently available prototypes of automated vehicles apply state-of-the-art sensors to gather information about the world and surroundings. Combined with high accuracy maps, they allow on-board systems to identify appropriate navigation paths, as well as obstacles and relevant signage. The vehicles also include increasingly sophisticated algorithms (based on AI) to process sensor data and control the vehicle. The computational power of computing chips and processors has permitted this level of development and allows for the processing to be run and acted upon in real time.

Vehicle manufacturers and technology companies are working towards a vision of fully autonomous vehicles, and that vision includes taking the human driver out of the loop. They have already made huge advancements in this space. For example, the self-driving software that has been developed, based on 'deep neural networks', includes millions of virtual neurons that mimic the brain.

The neural nets do not include any explicit programming to detect objects in the world. Rather, they are trained to recognise and classify objects using millions of images and examples from data sets representing real-world driving situations. The computational powers of on-board supercomputers and processors have permitted this level of development and allow for the processing to be run and acted upon in real time.

Although robotic systems are very good at collecting data about the environment, making sense of that data remains probably the hardest part of developing an ultra-reliable autonomous vehicles. Although the past few years have seen some scientific advances and developments which have overcome many of the limitations of these systems (e.g. accuracy), it is still not entirely clear what types and combination of sensors is likely to emerge as the best option for functionality and price – particularly for vehicles that function in full autonomous modes. To allow full autonomous operation without an alert backup driver ready behind the steering wheel, the technology will need to degrade gracefully in such a way that crashes are avoided. For example, if an element of the system fails in the middle of a curve in busy traffic, there must be a sufficiently robust back-up system so that even with the failure, the vehicle can manoeuver to a safe stop. Developing this level of reliability is challenging but is something many vehicle manufacturers are currently working on and hoping to resolve over the next few years.

11.3 Levels of driving automation

The ability for vehicles to sense and act in response to their environment covers a range of autonomous capabilities according to the International Levels of Driving Automation by the Society of Automotive Engineers (SAE). An autonomous vehicle must collect information, make a decision based on that information, and execute that decision. The information comes from vehicle equipment, physical and digital infrastructure, any of which may be public or private. Many of these technologies exist today and are capable of guiding vehicles and in some cases drive vehicles with minimal or no driver input and across diverse driving environments.

The SAE International Levels of Driving Automation identify how the driving task is divided between human and machine [6]. For example, Level 0 (no automation – driving is performed entirely by a human driver); Level 5 (full automation – diving is entirely performed by an automated driving system). In the 'middle', this task is shared simultaneously or sequentially, raising questions and issues around human–machine interaction. Although Level 5 is expected within few years, most technology providers are currently focused on the middle of the spectrum.

Efforts towards full automation will probably follow one of two incremental paths [6]. The first involves gradually improving the automated driving systems available in conventional vehicles so that drivers can shift more of the dynamic driving task to these systems. The second involves deploying vehicles without a driver and gradually expanding this operation to more contexts. These two approaches have been described as 'something everywhere' and 'everything somewhere', respectively [6]. Many of today's production vehicles are capable of driver assistance (Level 1), typically through the use of adaptive cruise control to adjust speed based on following distance. Some luxury vehicles also incorporate an active lane-keeping assist feature in a way that makes them capable of partial automation (Level 2). Both of these levels assume that the human driver continues to actively monitor the driving environment.

The threshold between partial automation (Level 2) and conditional automation (Level 3) corresponds to what many have accepted as the line between non-automated and automated driving. Conditional automation in Level 3 raises particularly difficult issues of human–machine interaction that have not been satisfactorily solved. Levels of automation beyond conditional automation (Level 3) may operate on the basis of inputs solely from vehicle sensors (self-sensing) or via a combination of vehicle and infrastructure information that are communicated to the vehicle in near real time. The latter approach requires available data transmission frequencies, low-latency, trusted, secure and fail-safe data transmission protocols and a harmonised data syntax that ensures safe interoperability [6]. High automation (Level 4) describes a driving system that can always revert to a 'minimal risk condition' should a human driver not resume active driving [6]. This may be easier in some contexts (e.g. low-speed parking facilities) than in others (e.g. urban expressways). For this reason, a highly automated driving system is capable of operating in some, but not necessarily all, contexts or 'driving modes'. High automation (Level 4) is that from which the 'everything somewhere' strategy begins. At the moment, the custom vehicles that operate without any real-time input from human drivers are limited to highly specific contexts, including particular routes and low speeds.

A key challenge is introducing these vehicles to more geographic areas, roadway types, traffic conditions, weather conditions, and events/incidents. Once they eventually function in most traffic and weather conditions, they would have approached full automation (Level 5) and would be capable of handling roadway and environmental conditions that can be managed by a human driver [6].

11.4 Timelines of deployment

There is yet no consensus on the commercial maturity of highly automated and ultimately fully automated vehicles. Some manufacturers have announced the arrival of highly automated and possibly fully automated vehicles by as early as 2019. In December 2015, Tesla Motors announced that it is only 2 years away from a fully self-driving vehicle [7]. Others have proposed different timelines extending to 2030, with Japan's motor industry represented by Toyota and Nissan having pegged 2020 as the year they hope to release fully autonomous vehicles to the public. Audi has estimated that it will be ready to release fully autonomous vehicles by 2025. A number of studies [4,6,8] have identified different timelines for the introduction of autonomous vehicles that extend further into the future. It is worth noting that some of the technologies are here today, some may be distant, and some will depend on specific technical innovations or particular policy choices.

11.5 A cyber-physical system – how does the autonomous vehicle work?

The autonomous vehicles now in development use a variety of sensors to read the world around them. These include radar sensors, video cameras, global positioning system (GPS), ultrasonic sensors and LiDAR (light detection and ranging) laser sensors. The radar and LiDAR sensors do most of the work looking for other vehicles, pedestrians, and other obstacles, while cameras typically read street signs and lane markers. The ultrasonic sensors (position estimators) also work along with the GPS to help the vehicle map its position. The technology that serves as the eyes of a self-driving vehicle is typically a mix of high-end laser beams, radar technology and cameras. The most prominent and advanced sensor in this group of technologies is the LiDAR which was once used in the Apollo 15 lunar mission in the 1970s. This sensor helps the autonomous vehicle to see and be smart enough to navigate obstacles, other vehicles, people and objects in the real world. It captures 360 degree images and millions of data points every second.

The reason the Laser or LiDAR is important is that GPS is not quite reliable enough and cannot detect objects or draw a map around the vehicle as the LiDAR can when the vehicle is travelling around the road network. Developed initially as a remote sensing technology, LiDAR has been around since the 1960s. Back then, it required enormous scanners that filled planes and cost hundreds of thousands of dollars. Although it is still an expensive item of technology (it costs around $80,000 per unit in the Google vehicle), it has shrunk considerably in size and cost over the years making it easier for researchers and vehicle manufacturers to bring the 3D vision mapping technology to vehicles. It is increasingly becoming more affordable, and in some configurations can fit inside an object the size of a tennis ball. The LiDAR technology has evolved over the years and today police use LiDAR speed guns to measure vehicles that exceed the speed limit. It is also still being used for widespread surveying, mapping for geological research and archaeology.

The technology works on the following principles. A scanner transmits infrared lasers that reflect or bounce off solid matter on surfaces. The distance measured helps create a 3D image. In the case of robots such as the autonomous vehicle, that 3D image is interpreted by a sophisticated computer system, which is coded with instructions for how the robot should react. Together with other types of sensors, navigational systems and detailed maps, LiDAR helps direct self-driving vehicles by rendering 3D images of people and objects surrounding the vehicle as it travels.

Central to the operation of the autonomous vehicle is high-fidelity 3D maps of the roads where the vehicles drive. Those maps include details like the exact position of the curbs and lane lines, trees and signs, along with local speed limits and other relevant rules. The more an autonomous vehicle knows about an area, the more it can focus its sensors and computing power on detecting temporary obstacles, like people and other vehicles, in real time. These maps have another advantage: The vehicle can use them to determine with good accuracy where it is at any given moment. For example, consider that the vehicle cannot see the lane lines, but it can see a nearby stop sign which is on the map. Its LiDAR scanner can tell it exactly how far it is from the sign and then quickly determine how far it is from the lane lines.

The technology is not without limitations. A recent study from Oxford University has shown that LiDAR sensors can be ‘misguided’ from a short distance away by someone using a specially modified laser pointer [11]. Hackers could conceivably make a self-driving vehicle stop or swerve by ‘convincing’ the LiDAR that it is about to hit an imaginary obstacle like a vehicle, person or wall. This is because LiDAR uses light, an unregulated spectrum that is accessible to anyone. Fortunately, there are ways to mitigate these attacks, such as identifying and removing outliers from the raw LiDAR data. With its nearly instantaneous 360-degree sweeps, however, LiDAR is ideal for covering large areas quickly.

Given that LiDAR is still expensive, researchers are looking at alternatives. A group of researchers at the University of Cambridge recently announced they may have found an inexpensive fix for LiDAR [11]. The UK team has developed a new sensor system that relies on camera technology readily found in any smartphone. The technology can quickly and accurately ‘see’ what is happening outside any vehicle by scanning its surroundings and sorting features into recognisable objects, everything from telephone poles to other vehicles on the road and buildings. The new system relies on deep learning algorithms, which require teaching the system how to tell objects apart by feeding it tens of thousands of different images [12].

11.5.1 The autonomous vehicle brain – artificial intelligence algorithms

The previous sections described how the imagery and data are captured using the different radar and sensor technologies. These data, on their own, have no value without sophisticated software that can accurately and reliably analyse the data and act on it [13]. This is where consumers are going to start to see some differentiation in the vehicles which automobile and technology companies will manufacture. All the hardware described previously is relatively similar for the different prototypes, but it is the software that will ultimately make the real difference.

The software, typically based on artificial neural networks (ANNs) and deep learning algorithms, is what makes the vehicle adaptive when it sees new situations it has not encountered before. The more sophisticated the software, the more adaptive, fault-tolerant and self-healing the vehicle becomes. These algorithms help the vehicle in three ways: situational awareness, path finding and learning.

ANN algorithms essentially make up the brain of these smart vehicles. The initial step in setting up the ANN is to train the algorithm in the vehicle using examples of real-world driving situations and responses on the road network in which the vehicle is expected to operate. For situational awareness and path finding, for example, the vehicle digitises the video images collected from the training and records the appropriate steering and driver responses to different scenarios. By training the algorithm on a very large number of scenarios and under different environmental conditions, the algorithm would then be ready to produce the appropriate steering and response without human intervention. Some of the more sophisticated algorithms can also potentially allow the vehicle to learn (on-line) by having a human driver correct its responses in real time.

As described before, self-driving vehicles use a broad spectrum of sensors to understand their surroundings. An example of a technology that is applied in autonomous driving is the NVIDIA DRIVE product which can process the inputs of video cameras, radar and ultrasonic sensors at a rate of 24 Trillion deep learning operations per second, which is equivalent to the computing power of 150 MacBook Pros. It fuses them to accurately detect objects, identify them, determine where the vehicle is relative to the world around it, and then calculate its optimal path for safe travel. This complex work is facilitated by a suite of software tools, libraries and modules that accelerates development and testing. Some tools enable sensor calibration, acquisition of surround data, synchronisation, recording and then processing streams of sensor data through a complex pipeline of algorithms running on specialised and general-purpose processors. Software modules are included for every aspect of the autonomous driving pipeline, from object detection, classification and segmentation to map localisation and path planning. These essentially will become the brain of future autonomous vehicles that will be continuously alert, and eventually achieve high levels of situational awareness.

Most of the key automakers have ramped-up their investment when it comes to machine-learning technologies and expertise which could help them in deploying self-driving vehicles. Google, Tesla, and Ford have all made notable moves in recent months including setting-up research centres in Silicon Valley. Toyota announced in November 2015 that it will also be joining its competitors in California pledging $1 billion for its project over the next 5 years [14]. This is an area where major developments and breakthroughs are expected over the coming few years as we see more vehicle manufacturers align themselves with technology companies who specialise in machine learning, with the aim of producing smart and sophisticated AI algorithms that assist driving, connectivity and interactivity and help vehicles avoid crashes.

11.6 Impact on road safety – the moral imperative

Nearly 1.2 million people die in road traffic crashes worldwide annually [15] (Chapter 2). Staggeringly, some 70%–90% of motor vehicle crashes are caused at least in part by human error [16]. A large proportion of these crashes could be avoided by using autonomous vehicles, and there is considerable logic in removing humans – the key source of distraction and collisions – from the driving equation. Driven by AI algorithms, these vehicles will not make errors of judgement the way a human driver does, and they will not take as long to make crucial split-second decisions. They will not drive drunk. They will not get distracted by SMS and smart phones. And they will not fall asleep while driving. In addition, vehicles' sensors will be able to see objects from very far, in every direction, in the most unforeseeable scenarios even using radars, and always calculating the best course of action in milliseconds. That would make them very safe chauffeurs. Take the Google self-driving vehicles as an example: A recent report [17] shows these vehicles have travelled over 1.7 million miles over 6 years and experienced only 11 minor accidents. In December 2015, the company averaged around 10,000–15,000 autonomous miles per week on public streets with 23 Lexus SUVs and 30 prototypes on the road [18]. On the other hand, their dependence on technology can also be a cause of concern and can probably lead to a spike in crashes at times when, for example; communication networks fail. The need for redundancy and complementary systems therefore becomes an important consideration in the vehicle-infrastructure design.

But according to a RAND Corporation report [19], the vehicles would need to have driven around 275 million miles fatality free before they can be verified as safe as human drivers. This suggests that an accelerated self-driving testing programme is required including testing for 'corner cases' – extreme conditions in which vehicles will operate. For example, situations where the sensors are not reliable in good weather, in urban canyons, or places where the map databases are out of date. Gesture recognition has also been reported as a serious problem, especially in real world settings.

If self-driving vehicles deliver on their promise to eliminate the vast majority of fatal traffic accidents, there is good reason to believe that tens of millions of traffic fatalities will be prevented around the world over the next two decades [20–22]. This is not merely theoretical. There is already some precedent for change of this magnitude in the realms of vehicle culture and automotive safety. For example, in 1970, about 60,000 people died in traffic accidents in the United States alone. A dramatic shift towards safety – including required seat belts and ubiquitous airbags – helped vastly improve a person's chance of surviving in the decades that followed [20].

11.6.1 Human interaction and transferring control between driver and vehicle

This is a growing and emerging field of research. It aims to look at human factors and behaviour, the role of human interaction with the vehicle to ensure safety, implications for driver licensing and the design of successful systems for the transfer of control from the autonomous vehicle to the human driver when the need arises. These are quite complex issues because of the differences between drivers, situational contexts and differences between the vehicles in terms of technology and the sophistication of the self-driving software used. Research in this area is expected to focus on how higher levels of automation could have negative impacts such as over-trust or under-trust, mode confusion, mental overload (or under-load) and skill degeneration. The research in this field is also expected to explore which tasks and driver behaviours would be considered acceptable by society while the automated vehicle is in control.

It is also expected that in the interim, there will be a need for research which looks at optimising the safety outcomes when a mix of self-driving and human-driven vehicles are operating on public road networks. As was mentioned before, the majority of crashes which the self-driving vehicles have been involved in so far have been due to human error (either drivers of other vehicles or as reported in a recent case due to a human error during the transfer of control from the autonomous mode to the driver). Future research should examine how collisions can be avoided in mixed traffic situations and the role of policy makers, vehicle manufacturers, and software developers in achieving safe outcomes.

There also seems to be a consensus in the literature that fully autonomous vehicles need to be at least as safe as human drivers, but there are no metrics being proposed to measure that. In some literature, it is suggested that this should include some metrics for testing and rating a vehicle [6]. A major delaying factor in allowing these vehicles on the road will be this standard for accreditation – without it insurers may be reluctant to offer insurance. Studies of human behaviour data whilst driving – blink rate and length, direction of eyeline, hand position on the wheel, etc. – would be key indicators of tiredness/distraction and could be detected using sensors. Test subjects would need to have these sensors in their vehicles for a long period of time. Once there is a good set of data from many test subjects, these can be compared against the statistics of an autonomous fleet and used to evaluate which is safer.

11.7 Potential impacts

The introduction of self-driving vehicles will have major socioeconomic impacts on our society – some good and some bad. Some of these impacts are described next.

11.7.1 Impacts on car ownership

A number of research studies reported in the literature show that car ownership is increasingly making less sense to many people, especially in urban areas [8,23]. These studies show that cars are expensive propositions. Beyond the cost of buying them, there are many costs of ownership, including fuel, maintenance and insurance. Autonomous technology will no doubt make new cars more expensive, but some of that expense can be mitigated by reducing the need for each family to own small fleets of them. This is already happening in some countries – in the United States, for example, the rates of motor-vehicle licensure are already plummeting among young Americans [23].

Other studies [24] have reported that consumers are finding it difficult to justify tying up capital in an under-utilised asset that stays idle for 20–22 h every day. Morgan Stanley's research shows that vehicles in the United States are driven just 4% of the year – a considerable expense considering that the average cost of car ownership is nearly $9,000 per year [24]. These studies have argued that autonomous on-demand vehicle sharing may provide a sensible option to a second car for many people and as the trend becomes more widespread, they may also begin to challenge the 'first' car.

Results from a recent study by the International Transport Forum [25] which modelled the impacts of shared self-driving vehicle fleets for the city of Lisbon in Portugal demonstrate the impacts on car ownership. The results showed that the city's mobility needs can be delivered with only 35% of vehicles during peak hours, when using shared self-driving vehicles complementing high capacity rail. Over 24 h, the city would need only 10% of the existing cars to meet its transportation needs. The study also found that a shared self-driving fleet that replaces cars and buses is also likely to remove the need for all on-street parking, freeing an area equivalent to 20% of the total kerb-to-kerb street space. Other studies have shown that dynamic ridesharing using self-driving vehicles will increase vehicle utilisation up to 8 h/day [26].

11.7.2 Impacts on vehicle sales

Although autonomous vehicles are not on the market yet, a recent study has suggested that self-driving vehicles could slash vehicle sales in the United States by nearly 50% [27].

A study from the University of Michigan Transportation Research Institute suggests that today's vehicle sales figures could plummet in the next decade, as autonomous vehicles start hitting the streets. The research team analysed data from the US National Household Travel Survey which looked at the habits of 150,147 households across the United States, including when each member of each household started and ended their daily trips. The study found a general lack of 'trip overlap' between drivers within a majority of households based on vehicle sharing. That meant families rarely used more than one vehicle at a time because their vehicles are often parked in very different locations when their trips take place. With an autonomous vehicle – one that could drive itself to a new destination – families could more easily share a vehicle.

With regard to the US National Household Travel Survey, the researchers found that sharing would have been possible in about 84% of all household trips. Furthermore, the researchers estimated that autonomous vehicles could cut demand for new vehicles by up to 43%, which would reduce the number of vehicles. That would reduce the number of vehicles in the average US household from 2.1 to 1.2 vehicles. However, they also noted that because of all the sharing (Chapter 10), American households could see a 75% increase in individual vehicle usage – from 11,661 to 20,406 annual miles per vehicle (this increase in mileage does not include additional miles that would be generated during each 'return-to-home' trip). As a result, vehicles might wear out faster which could ease the drop in demand.

With regard to the self-driving vehicle market share, a study by the McKinsey Global Institute [28] predicts that fully autonomous vehicles will be experiencing mass-market sales by 2030 or possibly earlier. The study estimates that 15% of global auto sales will consist of fully autonomous vehicles by 2030. The study also suggests that mass-market sales will begin around 2020–22. The study touched upon the impact of ride-sharing companies like Uber and Lyft, as well as any potential future ventures who are interested in the self-driving market. The study predicts that in the near-term, ridesharing companies will slow the annual growth of global auto sales, from the current average of 3.6% down to 2%. However, in the long term, the study found that growth will rebound, and that ridesharing companies will actually help grow sales by around 30% by 2030. By 2050, McKinsey predicts that one in three vehicles sold globally will be a shared vehicle.

11.7.3 Impacts on road users

Autonomous driving has the potential to reduce stress and increase mobility options for people unable or unwilling to drive. Some of these impacts are described next.

11.7.3.1 Passengers experience less stress and have more productive and leisure time

Some studies (Chapter 8) have reported that repeated and prolonged driving in congested conditions constitutes a form of chronic stress and a serious health risk in the long run [29]. In addition to reducing driver stress, advocates of self-driving vehicles argue that it would free passengers from the task of driving and would provide a personal mobility option to people unable or unwilling to drive. It comes as no surprise that ride sharing services – like Uber and Lyft, and car ownership-sharing services – like GoGet and Flexicar, are quickly gaining popularity as an alternative to vehicle ownership, and are increasingly paving the way for this market. A study by the McKinsey Global Institute [30] estimates that autonomous vehicles could free as much as 50 min a day for users, who will be able to spend their travel-time working, relaxing, or accessing entertainment. The study estimates that the time saved by commuters every day might add up globally to a mind-blowing one billion hours [30]. On the other hand, autonomous vehicles might have a negative impact on health and well-being (Chapter 9) if people abandon walking and cycling and replace these with autonomous rides particularly for shorter travel distances.

11.7.3.2 Mobility options for an untapped market including the aged and younger people

Shared autonomous mobility is also likely to change the number of passengers that pass through every vehicle – including a vast untapped market that does not drive today [23]. For example, the higher levels of autonomy, when the vehicle does not require a human driver, would enable transport and mobility for the blind, disabled or those too young to drive. The benefits for these groups would include independence, reduction in social isolation, and access to essential services.

These same benefits would return mobility to millions on the margins, including the elderly, the working poor and those who have lost their driving privileges, for example due to repeated speeding or infringement fines and those who may be considered a high road safety or insurance risk [31].

11.7.3.3 Impacts on pedestrians

There are increasing calls for vehicle technology developers and urban planners to design transport systems involving automated vehicles such that pedestrians intuitively understand how such vehicles operate. The presence of automated vehicles should in no way discourage pedestrians from completing their journey on foot but the vehicles should integrate seamlessly with other modes, particularly walking and cycling. There will be human factors work required to ensure automated vehicles communicate with pedestrians in a highly intuitive manner and ultimately standards will be developed to address this topic.

11.7.4 Impact on infrastructure and connected vehicle investments

Some technology providers (e.g. Google) claim that their technology does not require any special instrumentation of infrastructure for successfully operating their vehicles. Cameras in the vehicle leverage image recognition programs to help the vehicle read signs, traffic signs and other elements in the driving environment. Other automobile manufacturers are making efforts in the same direction and focusing efforts on modification and expansion of vehicle systems to assist the automated driving. However, the foreseen step-wise introduction of automated vehicles in traffic will face a transition period where the coexistence of conventional and highly automated vehicles will have to be managed in order to ensure uninterrupted levels of safety and efficiency. Instrumentation of road infrastructure will still play a major role in managing this transition period.

11.7.4.1 Impact on infrastructure investment

At the international level, there have been large investments in the Cooperative Intelligent Transport Systems (C-ITS) space. This includes vehicle-to-vehicle and vehicle-to-infrastructure communications and associated technologies. These investments will continue to be made over the next 15–20 years and will likely remain a priority until we reach full or near full autonomy market penetration rates. As long as there is a mix of human-driven and autonomous vehicles, C-ITS investments will remain important to ensure safety of the travelling public. It is premature at this stage to speculate what might happen once we reach full autonomy as the enabling technologies are developing very fast. With more autonomy over the next 15–20 years, more intelligence will be transferred to the vehicle which may reduce the need for infrastructure instrumentation.

11.7.5 Impact on intersection control and congestion on urban arterials

A team of researchers at the University of Texas in Austin are investigating how road authorities can manage autonomous vehicle traffic at intersections [32]. Their project, Autonomous Intersection Management or 'AIM', is developing a multi-agent framework for managing autonomous vehicles at intersections. The researchers argue that in the long term, it will make more sense to have intersection control mechanisms that offer more precise control, better sensors and quicker reaction times. The researchers claim that intersections of the future will not need traffic lights or stop signs. Instead, the autonomous vehicles' movement around intersections will be managed by a virtual traffic controller. The research team proposes an intersection control system using sensor technologies, a standardised communication protocol and the ability to deploy gradually over time to safely accommodate a mix of autonomous and human-driven vehicles in changing proportions.

For intersection control, the researchers are developing a system of 'reservations', in which a vehicle approaching the intersection 'calls ahead' via digital short-range communications to the 'intersection manager' to request a block of space and time to cross the intersection. The request transmits the vehicle's time of arrival, speed, size, acceleration capability and arrival and departure lanes. An intersection controller then simulates the vehicle's trajectory across the intersection, calculating the clear space required for its passage. It approves or denies the vehicle's request based on the space–time requirements of the processed trajectories of other vehicles, whether human-driven or autonomous. The intersection manager transmits a confirmation, rejection, or alternative slot to the approaching vehicle. Simulations of this system, with up to three lanes in each of four directions, showed dramatic reductions in waiting time compared to intersections with traditional controls. Their project proposes to take full advantage of the sensing and control capabilities of autonomous vehicles to reduce congestion. When traffic flow is optimised, autonomous vehicles have the potential to substantially increase roadway capacity while eliminating crashes caused by human error. The researchers are currently working on the next steps to determine real-world feasibility. Following evaluation of the protocols and algorithms for an individual intersection and a full-failure mode analysis, the researchers will simulate the feasibility of the system in networks of intersections. First, they will use small autonomous lab robots and later a full size autonomous vehicle. Ultimately, the researchers are working to deliver a prototype software implementation of the intersection manager and a detailed communication protocol. In a recent paper, the researchers demonstrate that their experimental results showed that the system can greatly decrease traffic delay when most vehicles are semi-autonomous, even when few (if any) are fully autonomous [33]. Their incremental deployment study shows that traffic delays keep decreasing as more vehicles employ features of autonomy.

11.7.6 Impact on mobility, parking, public spaces and congestion

Public acceptance of autonomous vehicles will rely to a large extent on how they perceive the benefits of these technologies, and how driving autonomy is likely to improve their quality of life without compromising safety. Given that there are not many large-scale deployments to date, this is an active area of research in which the impacts are being evaluated in advanced traffic simulation environments which rely on agent-based modelling techniques.

11.7.6.1 Modelling studies – agent-based frameworks

Most of the case studies reported in the literature have utilised different types of agent-based models for assessing the impacts of disruptive transport technologies including shared autonomous vehicles. These models offer a number of features which allow for modelling network performance using end-to-end trips made by travellers over multiple modes of transport, rather than single-mode trips made in a vehicle or walking. This approach also allows for modelling individual traveller behaviour including dynamic decision processing incorporating a dynamic mode-choice function of individual travellers. This provides new capabilities to allow a traveller in the model to make instantaneous choices between available modes as well as choices between available routes. For instance, consider a traveller in the model who initially intended to walk to a bus stop and take the bus to her destination. On arriving at the bus stop, the information display advises that the bus is running late as a result of traffic congestion. The traveller, depending on her delay tolerance threshold, may decide to request an autonomous vehicle using her smartphone app. Although existing micro-simulation tools can model dynamic route choice within a mode, the demand is specified by an Origin-Destination matrix of mode-specific trips making it difficult to model a person dynamically switching from one mode of transport to another. An agent-based model can represent dynamic mode switching by allowing each individual agent to choose a new mode of transport during its trip.

11.7.6.2 Impacts of autonomous mobility-on-demand (AMoD)

Shareable autonomous vehicles appear like a promising proposition for decreasing the overall number of private vehicles. This would in turn directly address the problems of oil dependency and pollution (particularly if they are all electric or fuel-cell), promote higher utilisation rates and reduce parking lot sprawls. To date, a number of studies have dealt with the implications of autonomous mobility-on-demand (AMoD) systems. The findings from these studies are summarised below.

Lisbon

The Lisbon study [25] examined the potential impacts that would result from the implementation of a shared and fully autonomous vehicle fleet. To perform this assessment, the researchers developed an agent-based model to simulate the behaviour of all entities in the system: travellers, as potential users of the shared mobility system; vehicles, which are dynamically routed on the road network to pick-up and drop-off clients, or to move to, from, and between stations; and dispatcher systems tasked with efficiently assigning vehicles to clients while meeting pre-defined service quality standards, for example with regard to waiting time and detour time. The analysis was based on a real urban context, the city of Lisbon in Portugal. The simulation used a representation of the street network, using origin and destination data derived from a fine-grained database of trips on the basis of a detailed travel survey. Trips were allocated to different modes: walking, shared self-driving vehicles or high-capacity public transport. A set of constraints were established (e.g. that all trips should take at most 5 min longer than today's vehicle trips, and assumed that all trips are done by shared vehicles and none by buses or private vehicles). The study also modelled a scenario which included high-capacity public transport (Metro in the case of Lisbon). The study modelled two different vehicle-sharing concepts, 'TaxiBots', a term the researchers coined for self-driving vehicles shared simultaneously by several passengers (i.e. ride sharing), and 'AutoVots', vehicles which pick-up and drop-off single passengers sequentially (vehicle sharing). For the different scenarios, the researchers measured the number of vehicles, kilometres travelled, impacts on congestion and impacts on parking space.

The results indicated that shared self-driving fleets can deliver the same mobility with significantly fewer vehicles. When serviced by ride-sharing TaxiBots and a good public transport high capacity underground system, 90% of vehicles could be removed from the city. Even in the scenario that least reduces the number of vehicles (AutoVots without underground), nearly half of all vehicles could be removed without impacting the level of service. Even at peak hours, only about one-third (35%) of today's vehicles would be needed on the roads (TaxiBots with underground), without reducing overall mobility. On-street parking could be totally removed with a fleet of shared self-driving vehicles, allowing in a medium-sized European city such as Lisbon, reallocating 1.5 million square metres to other public uses. This equates to almost 20% of the surface of kerb-to-kerb street area. In terms of environmental impact, only 2% more vehicles would be needed for a fleet of cleaner, electric, shared self-driving vehicles, to compensate for reduced range and battery charging time.

Stockholm

In the Stockholm study [34], the assessments included both a fleet consisting of currently in-use gasoline and diesel vehicles as well as electric vehicles. The results showed that an autonomous vehicle-based personal transport system has the potential to provide an on-demand door-to-door transport with a high level of service, using less than 10% of today's private vehicles and parking places. In order to provide an environmental benefit and lower congestion, the autonomous vehicle would require users to accept ride-sharing, allowing a maximum 30% increase of their travel time (15% on average) and a start-time window of 10 min. In a scenario where users were not inclined to accept a lower level of service, that is no ride-sharing and no delay, empty vehicle travel will lead to increased road traffic increasing environmental impacts and congestion. In a scenario which looked at electric vehicles, an autonomous vehicle-based system and electric vehicle technology seemed to provide a 'perfect' match that could contribute to a sustainable transport system in Stockholm.

Austin

The Austin case study [26] investigated the potential travel and environmental implications of autonomous shared mobility systems by simulating a 12-mile by 24-mile area in Austin, Texas. The multi-agent transport simulation (MATSim) software was used for conducting this experiment using 100,000 randomly drawn person-trips out of 4.5 million Austin's regional trips. The study reported that each autonomous shared vehicle would almost replace around nine conventional vehicles within the 24-mile by 12-mile area while providing the same level of service, but would generate approximately 8% more vehicle-mile travelled. Their study also confirmed that this system would decrease the emissions by not only replacing the heavier vehicles with higher emissions rates, but also by cutting down on the number of cold starts.

New York

The New York case study [35] introduced the Expand and Target algorithm which was integrated with three different scheduling strategies for dispatching autonomous vehicles. The study also implemented an agent-based simulation platform and empirically evaluated the proposed approaches using New York City taxi data. Experimental results demonstrated that the algorithms significantly improve passengers' experience by reducing the average passenger waiting time by around 30% and increasing the trip success rate by around 8% (a successful trip is one that meets the key performance indicators such as maximum allowed waiting time).

Melbourne

The Melbourne case study [36–39] used an agent-based simulation approach to investigate the impacts of deploying AMoD services on a small suburban network (Figure 11.2). A base case (BC) scenario was first defined, which replicated existing conditions in which all trips (2,136 trips) were completed using single occupant traditional vehicles during AM-Peak (07:00 a.m.–09:00 a.m.). In the second scenario, the traditional system was replaced by AMoD services in which 25% of people utilised privately owned AVs, and the other 75% used shared autonomous vehicles (SAVs) that featured autonomous vehicles with capacities for two, three or four people. In both cases, passengers would be picked up and dropped-off at their destinations by the autonomous vehicles. After dropping their passengers off, the privately owned AVs head back to their starting point (Home) and wait for further instructions from their owners to avoid parking charges. The SAVs, on the other hand, would typically be owned by a commercial fleet company which would direct the vehicles to nearby waiting areas where they also would wait for further instructions.

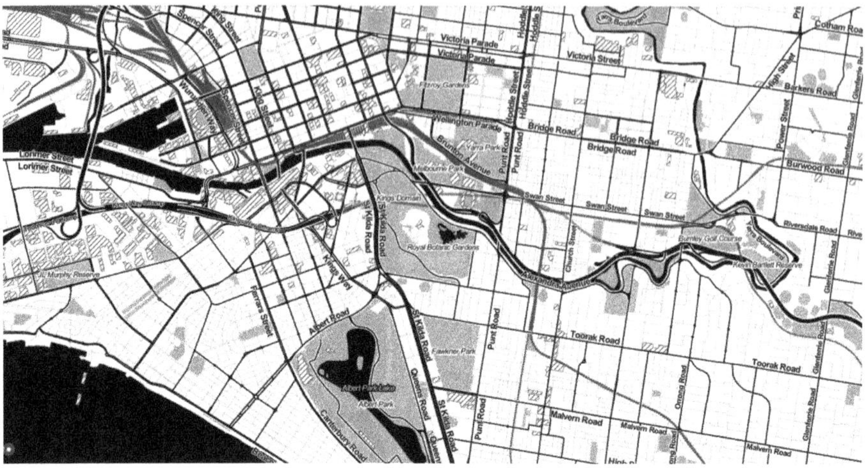

Figure 11.2 Pilot study area. Source: Dia et al. [36]

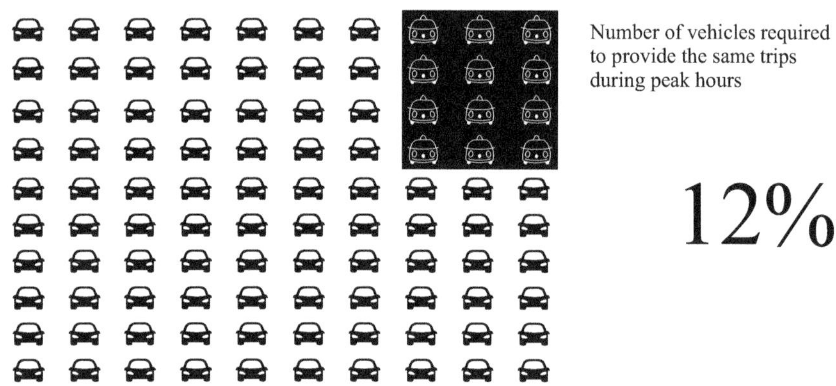

Figure 11.3 Impact of autonomous shared mobility on size of vehicle fleet – Melbourne pilot study. Source: Dia et al. [36]

Other future scenarios of AMoD were also simulated. In the first scenario (AMoD1), it was assumed that the on-demand vehicles were immediately available to passengers (maximum waiting times is zero). This constraint was tightened in the second scenario (AMoD2) by increasing the allowable passenger waiting times up to a maximum of 5 min.

The results (Table 11.1 and Figure 11.3) showed that using the AMoD system resulted in a significant reduction in both the number of vehicles required to meet the transport needs of the community (reduction of 43% in AMoD1, and 88% in AMoD2), and the required on-street parking space (reduction of 58% in AMoD1 and 83% in AMoD2).

However, the simulation also showed that this was achieved at the expense of an increase in the total VKT (increase of 29% in AMoD1 and 10% in AMoD2). In addition to assessing the mobility impacts of AMoD, the researchers also assessed how mode choice preferences impacted fleet operations. Investigations of shared mode choice preferences (vehicle-share versus ride-share) showed that shifting 40% of travellers to autonomous on-demand ride-sharing has the potential to deliver a 70% reduction in the total vehicle fleet size and 14% reduction in the total vehicle-kilometres-travelled compared to the BC Scenario. Further investigations were conducted to explore how the VKT under AMoD scenarios can be reduced if supported by public transport services. The results showed that shifting 30% to 70% of AMoD users to bus services will result in approximately 8% to 15% decrease in the vehicle-kilometres travelled by the autonomous vehicles.

Three more scenarios were also examined to investigate the impacts of vehicle-share and ride-share mode choice preferences. In Scenario 2 (Table 11.1), a homogeneous population of travellers was assumed where all travellers had identical mode-choice preferences and used only vehicle sharing with single occupant self-driving vehicles to reach their destinations. In reality, the value of travel time is generally distributed heterogeneously across individuals within a population, and according to time of day and trip purpose. It would therefore be expected that the preference towards vehicle sharing versus ride sharing will differ between travellers based on the increased travel time required to rideshare. Scenarios were developed to explore the effects of travel cost on mode choice behaviour of four categories of travellers. These include travellers with high value of travel time (HVoTT) who choose vehicle sharing systems (single-occupant autonomous vehicles with zero waiting time for passengers). This would necessitate that a sufficient number of vehicles are available all the time to serve these customers (at a premium cost). Travellers with a low value of travel time (LVoTT), on the other hand, would share their ride with other passengers using autonomous vehicles with capacities of 2, 3 and 4 people. For the ride-sharing system, it was assumed that travellers would need to wait until a vehicle is available. After dropping their passengers off, vehicles available for HVoTT customers will stay at the taxi rank to pick up other customers even if there is no current demand. For vehicles servicing LVoTT

Table 11.1 Comparative evaluation of base case and AMoD scenarios – Melbourne pilot study

Scenario name	Reduction in number of vehicles (%)	Increase in the total VKT (%)	Reduction in required parking space (%)
Scenario 1 (AMoD1) compared to base case (BCS)	43	29	58
Scenario 2 (AMoD2) compared to base case (BCS)	88	10	83

Table 11.2 Proportions of ride-share and car-share travellers in Scenarios 3–5

Scenario / Capacity	Car-sharing (%) Single occupant	Ride-sharing (%) Two passengers	Three passengers	Four passengers
Scenario 3	10	30	30	30
Scenario 4	20	20	30	30
Scenario 5	60		20	20

customers, these vehicles may relocate to other taxi ranks if there is a need. The same rebalancing system used in Scenario 2 was utilised for rebalancing the empty vehicles in these scenarios. To develop a better understanding of the impacts of mode choice preferences, three scenarios were simulated in which the proportion of ride sharing travellers varied between 40% and 90% as shown in Table 11.2.

The simulation results are reported in Figure 11.4. As expected, the results show marked reductions in the total number of vehicles, total VKT travelled and parking spaces requirements when more people choose to ride-share instead of car-share. For example, the results show that the total number of required vehicles increased by 132% (from 273 in AMoD3 to 632 in AMoD5) as a result of a 50% decrease in the proportion of ride-share travellers (from 90% to 40%). However, the number of vehicles in AMoD5 was still substantially lower (69% less) than the BCS.

Comparison of the results for AMoD5 and BCS also suggests that shifting 40% of the population to autonomous on-demand ride-sharing will result in a 70% decrease in the total vehicle fleet size (from 2059 to 632); 14% reduction in the total VKT (from 4,660 to 4,027) and 57% reduction in the required parking spaces.

11.7.7 Impact on public transport

The findings from the Lisbon simulation study suggested that the largest benefits for autonomous mobility-on-demand systems are achieved in the presence of a high-capacity public transport system. This was a significant finding which reinforces the need for on-going investment in both traditional and innovative types of public transport.

At the same time, there is currently significant research being undertaken on developing autonomous buses which can be used for shared on-demand mobility. Although difficult to predict how things will develop, there are opposite arguments which predict that disruptive technologies such as Uber's ride-sharing and car-pooling could actually be the biggest threat to public transport, and that the same could apply to autonomous vehicles [40].

These disruptive mobility solutions keep fares down by making the utilisation of vehicles very high (typically on hire for 50 of every 60 min). The prediction is that in some selected markets, these vehicles would soon start to carry parcels as well as passengers. Lyft in the United States, for example, recently launched a

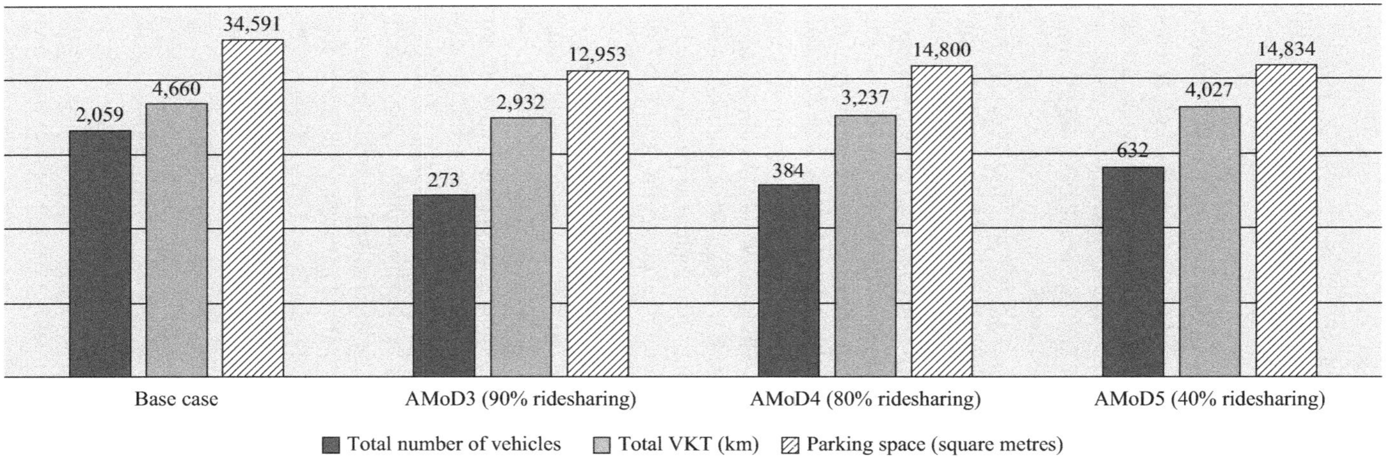

Figure 11.4 Impact of autonomous shared mobility on size of vehicle fleet – Melbourne pilot study. Source: Dia et al. [36]

program called 'Friends with Transit' that advocates partnerships with public transit agencies so that riders can use Lyft in conjunction with fixed-line transit. The company sees ride-hailing and public transport working in tandem, with ride-pooling (supported in the future by autonomous driving) filling in gaps in the service such as first-mile and last-mile problems.

11.7.7.1 Potential changes in the role of public transport

The new mobility options to be made available through autonomous driving may provide an important complement to existing public transport systems. The Lisbon simulation study, for example, showed that the best city-wide outcomes for deployment of autonomous on-demand mobility occur in conjunction with high capacity underground public transport.

Some related evidence may be drawn from some of today's cities where technology-based vehicle-sharing and ride-sharing (e.g. through use of Uber) is very popular [41]. Results from studies in San Francisco, for example, suggest that there has not been a dramatically negative effect on public transport usage thus far. As of December 2013, Uber was already providing 160,000 trips per week in the Bay Area, a number that has likely increased in the subsequent months. Those trips, however, do not appear to have reduced public transport ridership [41]. According to statistics from the American Public Transportation Association, ridership on buses and trains operated by all four of that region's major transit operators increased between 2013 and 2014. It is suggested that the prices being charged for these taxi-like services are too high to attract most people out of public transport for their daily trips. UberPool, for example, charges around 5 to 7 dollars a trip, which is still not what people can afford to get to work. At 14 dollars a day, bus and train ridership is not likely to be dramatically affected. Autonomous driving, however, may very well challenge this given that that prices may be much cheaper because there is no human driver labour involved (although vehicles maybe more expensive to buy and operate). It is still too early to draw conclusions on this, but research shows that in terms of city-wide impacts that in terms of city-wide impacts on mobility, public transport – particularly high-capacity for longer distances – will still have a strong role to play in urban mobility. Low-capacity public transport modes like buses, shuttles, mini-vans and school buses may however be impacted and replaced with autonomous vehicles particularly for those trips that are too long to walk and too short to drive. It should be noted, however, that eliminating bus lines altogether may disassemble the public transport grid which makes the network work in most places. Autonomous vehicles could fill some of that gap, but it is hard to see them replacing public transport bus routes entirely.

Other studies [41] have suggested that this may also alter the public sector's role in public transport, as governments may switch from subsidising service provision for all to providing vouchers for automated transport for low income people. This would require a very significant public role through subsidies if mobility for this segment of society, which does not have access to their own vehicles, is to be maintained. This would raise many challenges to governments around the world, for example, how would those subsidies work and would people have access to unlimited trips and travel distances?

Another area where public transport may be impacted is in the government-subsidised mobility services to people with disabilities. Paratransit services provided by many cities, especially in North America, already offer people with limited personal mobility a point-to-point alternative similar to that which could be offered by automated vehicles. In the United States, in particular, paratransit trips cost the public purse more than three times as much to provide as regular bus and rail services, but that is in part because of the low capacity of paratransit vehicles, high labour costs and their non-fixed-route services. Eligible customers in most regions are allowed to take as many trips as they'd like upon advanced reservation, generally at a per-ride fare on par with traditional transit.

It is also worth noting that any advances in technology that provide for automated vehicles could also result in automated public transport vehicles, potentially saving significantly on the cost of operations by eliminating the need for driver labour. Automated trains are already common for new subway and elevated rail systems, and some train lines in cities like Paris have been converted from driver into automated operation. Buses moving around the city with no drivers could be more frequent because of reduced labour costs, and certain bus lines could probably be operated profitably. In other words, automobile automation could have a genuine competitor in automated public transport. But automated buses and trains, much like automated taxis, do come at a cost: significantly fewer jobs in the transport business.

11.7.8 Impacts of autonomous vehicles on active transport

In many cities around the world, creating cycle routes 'for all ages and abilities' has become a focus for city councils and planners. In order to do this, the focus has been on the creation of physically separated bike lanes. The design is intended to protect cyclists from the physical danger of vehicles. But an all seeing, all obeying computer-driven vehicle could presumably behave much better. It could be programmed to travel at, say, 20 kph on designed 'low stress bicycle routes' and be programmed to give people cycling the right of way. It could also avoid certain streets altogether unless required to access a destination.

Instead of having to create separate spaces for different types of street-users because they are incompatible in some way, handing over control of vehicles to computers presents the opportunity for safe, low stress shared spaces. There are many planning advantages to this including narrower, more friendly feeling streets, making more efficient use of space. The public would have to trust the technology, and this will not happen overnight. It is difficult to know how the future will evolve, but it is worthwhile to start thinking through some of these scenarios so that society is better prepared as professionals and advocates when these questions and issues inevitably arise.

As was mentioned before, there is a concern that autonomous vehicles might impact on health and well-being if people abandon walking and cycling and replace these trips with autonomous rides, particularly for shorter travel distances.

11.7.9 Impact on jobs and labour

It is unclear at this stage how autonomous driving and the rise of robots may threaten the jobs of the future [42]. It has been suggested in some literature that the fallout from vehicle automation and self-driving capability will likely eliminate some 10 million jobs in the United States alone within 10–15 years of the rollout of self-driving vehicles. This speculation was based on a report by Pricewaterhouse-Coopers (PwC) [42] which predicted that the number of vehicles on the road will be reduced by 99%, and that the fleet will fall from 245 million to just 2.4 million vehicles. It was also based on statistics from the Bureau of Labour which listed that 884,000 people are currently employed in motor vehicles and parts manufacturing, and an additional 3.02 million in the dealer and maintenance network. According to the Bureau, truck, bus, delivery and taxi drivers account for nearly 6 million professional driving jobs in the United States. These 10 million jobs have been speculated to be in danger of elimination within 10–15 years. But the same literature also speculates that despite the job losses, eliminating the needs for car ownership will yield over $1 trillion in additional disposable income – and that is likely to usher in an era of unprecedented efficiency, innovation, and job creation.

It is certainly a possibility that as some jobs are eliminated, more will be created to deal with the innovations of a new era. There is also a counter possibility. As technology continues to accelerate, and vehicles become smarter, fewer labour may be necessary especially drivers. Some speculation in the industry is that AI may impact the wider economy beyond autonomous vehicles, and is already well on its way making 'good jobs' obsolete with many blue and white collar jobs poised to be replaced by robots and smart software over the next 20 years. This would clearly result in job inequality and economic insecurity for some segments of the labour market. This is ultimately a societal issue that must be debated and decided on publicly to ensure that the future will see broad-based prosperity levels as automation becomes more common and widespread.

11.7.10 Impact on the environment and pollutants emissions

Research shows that outdoor air pollution contributes to more than 2.5 million deaths worldwide each year [43]. In some countries, transport is estimated as the source of nearly 30% of all emissions responsible for global warming. In Australia, transport activity is one of the major sources of emissions related to the combustion of fossil fuels [44]. Passenger vehicle usage in urban areas was the largest transport source, contributing 8.5% of Australia's net emissions and accounting for around 39.7 $MtCO_2$. It is this category in particular, passenger vehicles, where potential reductions in emissions can be made with autonomous vehicles.

11.7.10.1 Impacts of fleets of shared autonomous vehicles

Modelling results from the Lisbon study found that the overall volume of car travel would likely increase with autonomous shared driving because the vehicles will need to re-position after they drop off passengers. However, autonomous vehicles could still be turned into a major positive in the fight against air pollution if they were all-electric and would have the potential to curb transport emissions and reduce our dependence on fossil fuels. Google has recently announced plans to roll out a handful of cordless charging prototype autonomous electric vehicles in California, and it is very likely that most autonomous vehicles in the future will also be electric or use hydrogen fuel-cell technology.

11.7.10.2 Impacts of fleets of autonomous electric taxis

A recent study by the Lawrence Berkeley National Laboratory in the United States found that self-driving electric taxis – battery powered and driven without human intervention – could reduce greenhouse gas (GHG) emissions by around 94% by 2030 [45]. The reductions in GHG emissions would be achieved by running the vehicles from the electricity grid, which by 2030 will use a greater proportion of renewable power. A more tailored array of vehicle sizes, including specialised vehicles for transporting just one or two people, would also contribute to reduced carbon emissions. In addition, human drivers are reported to be responsible for between 20% and 30% of inefficiencies in vehicles, so the shift to autonomy has the ability to use the vehicle in a more efficient manner. The study, published in the journal Nature Climate Change, compared four types of vehicles. This included traditional internal combustion engine vehicles (ICEV), hybrid-electric vehicles (HEV), hydrogen fuel cell vehicles (HFCV) and battery-powered electric vehicles. For each type, GHG emissions per mile travelled were predicted for 2014, and in 2030. The study found that, on average, 62% of vehicle miles travelled in the United States are for a single person, often travelling in a much larger five-seat vehicle. To overcome this inefficiency, the researchers suggested future autonomous vehicles will include smaller vehicles designed for just one, or two people, as well as larger vehicles. In the study, these hypothetical vehicles were based on the five-seat Nissan Leaf, but with a reduced width of 40%, accommodating a single seat. Electric-powered vehicles in 2030 are expected to benefit from reduced GHG emissions as the energy industry switches to more renewable power sources. The results from the scenarios conducted in the study suggest that a self-driving electricity-powered vehicle in 2030 will contribute to the release of 87%–94% less GHG emissions, as compared to a conventional petrol-driven vehicle today. This range is dependent on which of the three electricity generation models turns out to be correct. Although such savings would be dramatic, their impact is dependent on autonomous taxis playing a significant role in personal transit over the next couple of decades.

11.7.10.3 The energy and carbon impact of highly automated vehicles

In another study by researchers at the University of Leeds [46], the authors argue that vehicle automation may affect road vehicle energy consumption and GHG emissions in a host of ways, positive and negative, by causing changes in travel demand, vehicle design, vehicle operating profiles and choices of fuels. In their research, the authors identify specific mechanisms through which automation may affect travel and energy demand. The authors reviewed the literature for estimates of the energy impacts of each mechanism and, where the literature is lacking, developed their own estimates using engineering and economic analysis. They also considered how widely applicable each mechanism is and quantified the potential impacts of each mechanism on a common basis: the percentage change it is expected to cause in total GHG emissions from light-duty or heavy-duty vehicles in the United States. Their primary focus was travel-related energy consumption and emissions, as potential lifecycle impacts are generally smaller in magnitude. The authors explored the net effects of automation on emissions through several illustrative scenarios, finding that automation might plausibly reduce road transport GHG emissions and energy use by nearly half – or nearly double them – depending on which effects come to dominate. They also found that many potential energy-reduction benefits may be realised through partial automation, while the major energy/emission downside risks appeared more likely at full automation.

11.7.11 Impact on vehicle insurance and other industries

As autonomous driving technology advances, perhaps the most notable benefit is the promise of a striking reduction in accidents. But what impact will this have on the insurance industry? Some estimates put the potential reduction in accidents due to introduction of autonomous vehicles at higher than 90% [47]. If automation will make crashes far less likely, then why does one buy vehicle insurance? Some studies have speculated that premiums could be reduced by 75%, especially if drivers are no longer required to get coverage, and liability is shifted from drivers to product liability, manufacturers and technology companies. Under this scenario, insurers might move away from covering private customers from risk tied to 'human error' to covering manufacturers and mobility providers against technical failure.

11.7.12 Liability – who is responsible?

Developing a framework for determining who, or what, is at fault for everything from single-car to serious multicar incidents threatens to keep self-driving vehicles off roads even as the technology that makes them possible gets closer to reality. Is the occupant to blame? The automaker? The suppliers who provided relevant hardware or pieces of code? Last year, Volvo said it would accept liability for accidents involving their autonomous vehicles. Eager to make good on its promise to roll out autonomous vehicles within the decade, the company said it had found a relatively straightforward way to address this: It will accept 'full liability whenever one of its vehicles is in autonomous mode'. Google and Mercedes-Benz have also reportedly said they will accept full liability if their self-driving vehicles cause a collision while driving in autonomous mode. Whether Volvo's approach will work remains to be seen, especially that the primary threat to autonomous vehicles seems to be non-autonomous vehicles and their human drivers.

11.8 Opportunities for new business models

Autonomous driving is already shaping some new business models. For example, Uber is reportedly developing a self-driving vehicle, while Google is also reported to be looking at the shared mobility model. While Apple is gearing up to challenge Tesla in electric vehicles, Silicon Valley is also extending its reach into the auto industry. These developments show there is more to the connected and autonomous vehicle than technology and infotainment, and may also signal the beginning of inventions and creation of entire new shared economy businesses. These opportunities have potential to tap into new markets that could see smart mobility seamlessly integrated in people's lives.

11.8.1 Collaborative mobility

Probably the biggest anticipated impact of self-driving vehicles will be around the collaborative mobility space. To date, the transport sector is the most funded industry in the Collaborative Economy. This is manifested by one of the most promising trends within the disruptive mobility space: 'Mobility-as-a-Service' or MaaS. The key concept behind MaaS is to place the road users at the core of transport services, offering them mobility solutions based on their individual needs. This can be achieved by providing a single platform for combining all mobility options and presenting them to the customer in a simple and completely integrated manner. This means that easy access to the most appropriate transport mode or service will be included in a bundle of flexible travel service options for end users. MaaS has the potential to fundamentally change the behaviour of people and reduce reliance on car ownership by providing easy on-demand access to the mobility services they need. The trend is therefore gradually shifting, particularly in the context of smart cities, from the provision of urban transport networks, that is buses, trams and trains, to a focus on what people require, and how a more considered and integrated approach could produce far better outcomes.

Self-driving vehicles will have a dramatic impact on urban life especially when they begin to blur the distinction between private and public modes of transportation. By combining ride sharing with vehicle sharing, it may be possible to take every passenger to his or her destination at the time they need to be there, with much fewer vehicles.

11.9 Public attitudes to autonomous vehicles: an international perspective

One of the largest surveys of public opinion about autonomous and self-driving vehicles was conducted by the University of Michigan [48]. The survey examined public opinion regarding self-driving-vehicle technology in three major English-speaking countries – the United States, the United Kingdom and Australia.

The survey yielded useable responses from 1,533 persons 18 years and older. The main findings (applicable to each of the three countries) were as follows:

1. *The majority of respondents had previously heard of autonomous or self-driving vehicles, had a positive initial opinion of the technology, and had high expectations about the benefits of the technology.*
2. *The majority of respondents expressed high levels of concern about riding in self-driving vehicles, security issues related to self-driving vehicles, and self-driving vehicle not performing as well as actual drivers.*
3. *Respondents also expressed high levels of concern about vehicles without driver controls; self-driving vehicles moving while unoccupied; and self-driving commercial vehicles, busses and taxis.*
4. *The majority of respondents expressed a desire to have this technology in their vehicle. However, a majority was also unwilling to pay extra for the technology; those who were willing to pay offered similar amounts in each country.*
5. *Females expressed higher levels of concern with self-driving vehicles than did males. Similarly, females were more cautious about their expectations concerning benefits from using self-driving vehicles.*
6. *In comparison to the respondents in the United Kingdom and Australia, respondents in the United States expressed greater concern about riding in self-driving vehicles, data privacy, interacting with non-self-driving vehicles, self-driving vehicles not driving as well as human drivers in general, and riding in a self-driving vehicle with no driver controls available.*

The main implications of these results are that motorists and the general public in the three countries surveyed, feel positive about self-driving vehicles, have optimistic expectations of the benefits, and generally desire self-driving-vehicle technology when it becomes available (though some expressed high levels of concern about riding in vehicles equipped with this technology, and majority was not willing to pay extra).

11.10 Ethical challenges

Autonomy in general, and autonomous vehicles in particular, pose a number of ethical challenges. Among the various obstacles and roadblocks that must be solved before the public can embrace autonomous vehicles, the issue of ethics is particularly crucial to their success. The most important ethics hurdle is related to safety and the fact that driving and mobility can never be perfectly safe, which raises some difficult issues such as the nature of the dilemma and safe versus legal.

11.10.1 The nature of the dilemma

Imagine that in the not-too-distant future, you own a self-driving vehicle. One day, while you are driving along, an unfortunate set of events causes the vehicle to head towards a crowd of 10 people crossing the road. It cannot stop in time but it can avoid killing 10 people by steering into a wall. However, this collision would kill you, the owner and occupant. What should it do?

How should the vehicle be programmed to act in the event of an unavoidable accident? Should it minimise the loss of life, even if it means sacrificing the occupants, or should it protect the occupants at all costs? Should it choose between these extremes?

One way to approach this kind of problem is to act in a way that minimises the loss of life. By this way of thinking, killing one person is better than killing 10. But that approach may have other consequences. If fewer people buy self-driving vehicles because they are programmed to sacrifice their owners, then more people are likely to die because ordinary vehicles are involved in so many more accidents. This result can be counter-productive.

Researchers are seeking to find a way through this ethical dilemma by gauging public opinion [49]. Their idea is that the public is much more likely to go along with a scenario that aligns with their own views. So researchers in France posed these types of ethical dilemmas to a sample of several hundred people to find out what they thought. The participants were given scenarios in which one or more pedestrians could be saved if a vehicle were to swerve into a barrier, killing its occupant or a pedestrian. At the same time, the researchers varied some of the details such as the actual number of pedestrians that could be saved, whether the driver or an on-board computer made the decision to swerve, and whether the participants were asked to imagine themselves as the occupant or an anonymous person.

The results showed that in general, people are comfortable with the idea that self-driving vehicles should be programmed to minimise the death toll. This utilitarian approach is certainly laudable but the participants were willing to go only so far. They were not as confident that autonomous vehicles would be programmed that way in reality – and for a good reason: they actually wished others to cruise in utilitarian autonomous vehicles, more than they wanted to buy utilitarian autonomous vehicles themselves, according to the research team. And therein lies the paradox. People are in favour of vehicles that sacrifice the occupant to save other lives – as long they do not have to drive one themselves. The research team also points out that their work represents the first few steps into what is likely to be a complex moral maze. Other issues that will need to be factored into future thinking are the nature of uncertainty and the assignment of blame.

The researchers also argue that these issues raise many more important questions: 'Is it acceptable for an autonomous vehicle to avoid a motorcycle by swerving into a wall, considering that the probability of survival is greater for the passenger of the vehicle, than for the rider of the motorcycle? When an accident is unavoidable, should a self-driving vehicle be programmed to aim for the smallest object to protect its occupant? What if that object turns out to be a baby stroller? If a vehicle must choose between hitting a group of pedestrians and risking the life of its occupant, what is the moral choice? If a manufacturer offers different versions of its moral algorithm, and a buyer knowingly chooses one of them, is the buyer to blame for the harmful consequences of the algorithm's decisions?

Other researchers [50] believe that the situation is not as complicated. For example, researchers who study the legal and social implications of self-driving vehicles say plenty of ethical decisions are already made in automotive engineering. Inherent in airbags, for example, is the assumption that they are going to save a lot of lives and only kill a few. The researchers argue that given the number of fatal traffic accidents that involve human error today, it could be considered unethical to introduce self-driving technology too slowly. They also argue that the biggest ethical question is how quickly we move towards full automation given that we have a technology that potentially could save a lot of people but is going to be imperfect and is going to kill.

11.10.2 Safe versus legal

Should autonomous vehicles be programmed to break the law in cases in which safety is compromised? Take the double-yellow line problem. A road crew or other objects suddenly appear on a road and the autonomous vehicle does not have enough time to stop and has to make a decision: Obey the law against crossing the double-yellow line and hit the object, or break the law and spare the road crew? Or does it split the difference, veering at the last moment and nearly colliding with the cones. It is clear that the vehicle should cross to avoid the road crew. Less clear is how to go about programming a machine to break the law or to make still more complex ethical calls. Although researchers working in this field agree that these may not even be the right questions to ask, they argue that societies need to start thinking about traffic codes reflecting actual behaviour to avoid putting the software developer or vehicle manufacturer in a situation of deciding what is safe versus what is legal.

11.11 Regulations

Regulations will play an important role in the emergence and development of autonomous vehicle technologies. They are also likely to be the biggest hurdle for their deployment beyond the testing and trial periods.

Several jurisdictions around the world (including South Australia) have passed or are considering regulations that would enable the research and development testing of autonomous vehicles. Most of these regulations address safety related to operation of these vehicles on public roads. More jurisdictions are now moving beyond regulations for testing and focusing their efforts on implications for large-scale deployment particularly the new business models that are likely to emerge as a result of their deployment. Authorities will have to adapt and rethink their approaches to regulating these activities in order to avoid conflicts without stifling the innovative uses of these technologies.

Regulation under uncertainty is difficult. Effective responses would require an early and on-going dialogue between regulators and developers in which regulators would actively plan to minimise future risks and create legal frameworks that are flexible but robust. An important role for the regulators will be to limit the physical risks especially those that might be posed during the interim years when legacy fleets of vehicles would interact with autonomous vehicles.

A number of studies reported in the literature provide some guidance for policymakers around the types of regulation, regulatory approaches, and policy choices. The response from regulators would vary according to the frameworks within which they operate in their individual countries.

11.12 Summary and conclusions

The automated vehicles landscape shows a strong and growing momentum around different aspects of technological and business model developments.

The fast pace of scientific advances is continually providing opportunities for improving the technologies and making them more affordable. The convergence of autonomous driving and collaborative mobility marketplaces over the next decade will bring about a new wave of disruptions that will impact mobility and beyond.

When they arrive, autonomous vehicles will bring with them many opportunities. These include fundamental improvements in road safety and significant community benefits through reduced crashes; new business models such as MaaS integration with other transport options; and public transport augmentation (e.g. first and last kilometre services).

Before that happens, though, a number of key challenges will need to be addressed. These include: regulations (these need to be flexible yet robust); public acceptance (understanding of benefits and impacts); institutional issues (preparedness of governments to guide, govern and maximise benefits to the community); managing the transition during the period when automated and human-driven vehicles share the road (safety and integration with other users); human factors (managing the interface between machine and humans); and a better understanding of their impacts on congestion, mobility and accessibility.

Acknowledgements

The author would like to acknowledge the contributions of a number of higher degree students who made useful interventions while the chapter was in preparation. These include Farid Javanshour and Jack Hill from Swinburne University of Technology.

References

[1] Maynard, A. (2016, Jan). *The fourth industrial revolution: what does WEF's Klaus Schwab leave out?* [Online]. Available: https://theconversation.com/the-fourth-industrial-revolution-what-does-wefs-klaus-schwab-leave-out-53049

[2] Kageyama, Y. (2015, May 18). *CEO: Nissan will be ready with autonomous driving by 2020* [Online]. Available: http://phys.org/news/2015-05-ceo-nissan-ready-autonomous.html

[3] Mercedez-Benz (2015, Jan 5). *World premiere of the Mercedes-Benz F 015 Luxury in Motion research vehicle* [Online]. Available: https://www.youtube.com/watch?v=DYTV4d-Gn0s&feature=youtu.be
[4] Cathers, T. (2014, Jul 17). *When will you be able to buy a driverless car?* [Online]. Available: http://www.mojomotors.com/blog/when-will-you-be-able-to-buy-a-driverless-car/
[5] Kessler, A. M. (2015, May 4). *Hands-free cars set to drive into a legal void* [Online]. Available: http://www.afr.com/lifestyle/cars-bikes-and-boats/handsfree-cars-set-to-drive-into-a-legal-void-20150504-1mz9ww?stb=lkn
[6] International Transport Forum (2015). *Automated and autonomous driving regulation under uncertainty* [Online]. Available: http://www.internationaltransportforum.org/pub/pdf/15CPB_AutonomousDriving.pdf
[7] Horwitz, J. (2015, Dec 23). *Elon Musk says he now knows when the safe, fully autonomous self-driving car will arrive*. Available: http://qz.com
[8] Deloitte (2014). *Gen Y automotive consumer study – the changing nature of mobility* [Online]. Available: http://www2.deloitte.com/us/en/pages/manufacturing/articles/2014-gen-y-automotive-consumer-study.html
[9] Davies, A. (2015a, Oct 11). *Ford's skipping the trickiest thing about self-driving cars.* Available: http://wired.com
[10] Davies, A. (2015b, Nov 01). *The clever way Ford's self-driving cars navigate in snow.* Available: http://wired.com
[11] Segnet (2015). *A deep convolutional encoder–decoder architecture for robust semantic pixel-wise labelling.* Available: http://mi.eng.cam.ac.uk/projects/segnet/demo.php#demo
[12] Brueck, H. (2015, Dec 29). *The key to perfecting driverless cars could be hiding in your smartphone.* Available: http://fortune.com
[13] Protin Pictures (2015, Mar). *How does Google's driverless car work?* [Online]. Available: https://www.youtube.com/watch?v=ftouPdU1-Bo
[14] Toyota (2015, Nov 5). *Toyota will establish new artificial intelligence research and development company.* Available: http://pressroom.toyota.com/
[15] World Health Organisation (2015, May). *Road traffic injuries* [Online]. Available: http://www.who.int/mediacentre/factsheets/fs358/en/
[16] Smith, B. W. (2013, Dec 18). *Human error as a cause of vehicles crashes* [Online]. Available: http://cyberlaw.stanford.edu/blog/2013/12/human-error-cause-vehicle-crashes
[17] The Guardian (2015, May 12). *Google acknowledges its self-driving cars had 11 minor accidents* [Online]. Available: http://www.theguardian.com/technology/2015/may/12/google-acknowledges-its-self-driving-cars-had-11-minor-accidents
[18] Olanoff, D. (2015). *Google's November self-driving car report details learnings, a pull-over, and a rear-end collision.* Available: www.techcrunch.com
[19] RAND (2016). *Driving to safety: how many miles of driving would it take to demonstrate autonomous vehicle reliability.* Available: http://www.rand.org/pubs/research_reports/RR1478.html

[20] Lafrance, A. (2015a, Sep 29). *Self-driving cars could save 300,000 lives per decade in America.* Available: http://www.theatlantic.com
[21] Lafrance, A. (2015b, Jun 27). *How airbags are supposed to work.* Available: http://www.theatlantic.com
[22] Skeptical Raptor(2015, Oct 2). *It's simple math – vaccines saved 700,000 children's lives.* Available: www.skepticalraptor.com
[23] Neil, D. (2015, Dec 1). *Could self-driving cars spell the end of ownership?* Available: www.wsj.com
[24] McDuling, J. (2014, Sep 12). *The latest attack on America's car culture comes from Wall Street* [Online]. Available: http://qz.com/264781/the-latest-attack-on-americas-car-culture-comes-from-wall-street/
[25] International Transport Forum (2015). *Urban mobility system upgrade: how shared self-driving cars could change city traffic* [Online]. Available: http://www.internationaltransportforum.org/cpb/projects/urban-mobility.html
[26] Fagnant, D., Kockelman, K. and Bansal, P. (2015). Operations of a Shared Autonomous Vehicle Fleet for the Austin, Texas Market. *Proceedings of the 94th Annual TRB Meeting* [Online]. Available: http://www.caee.utexas.edu/prof/kockelman/public_html/TRB15SAVsinAustin.pdf
[27] Read, R. (2015, February). *Self-driving cars could slash auto sales by nearly 50%* [Online]. Available: http://www.thecarconnection.com/news/1096761_self-driving-cars-could-slash-auto-sales-by-nearly-50
[28] McKinsey (2016). Automotive revolution – perspective towards 2030. *How the convergence of disruptive technology-driven trends could transform the auto industry* [Online]. Available: https://www.mckinsey.de/sites/mck_files/files/automotive_revolution_perspective_towards_2030.pdf
[29] Abu Lebdeh, G. (2015, March 22). *Transport systems and public health* [Online]. Available: https://www.linkedin.com/pulse/transport-systems-public-health-ghassan-abu-lebdeh
[30] McKinsey (2015). T*en ways autonomous driving could redefine the automotive world* [Online]. Available: http://www.mckinsey.com/industries/automotive-and-assembly/our-insights/ten-ways-autonomous-driving-could-redefine-the-automotive-world
[31] Cox, J. (2015). *Tesla is 'Uniquely Positioned to Dominate' auto business* [Online]. Available: http://blogs.wsj.com/moneybeat/2015/08/17/tesla-is-uniquely-positioned-to-dominate-auto-business/
[32] AIM (2016). *Autonomous intersection management* [Online]. Available: http://www.cs.utexas.edu/~aim/
[33] Au, T.C., Zhang, S. and Stone, P. (2016). *Autonomous intersection management for semi-autonomous vehicles* [Online]. Available: http://www.cs.utexas.edu/~aim/papers/BookChapter14-au.pdf
[34] Rigole, P. 2014. *Study of a shared autonomous vehicles based mobility solution in Stockholm.* Master of Science Thesis, KTH University. Available: http://bit.ly/1Qig7Cu
[35] Shen, W. and Lopes, C. (2015). Managing autonomous mobility on demand systems for better passenger experience. *Principles and practice of*

multi-agent systems. Springer International Publishing. Available: http://bit.ly/1nC1WyN [accessed 11 January 2016].

[36] Dia, H., Javanshour, F. and Hill, J. (2016). Network impacts of autonomous shared mobility. *Proceedings of the Machine Learning for Large Scale Transportation Systems Workshop, The 22nd ACM SIGKDD International Conference on Knowledge Discovery and Data Mining*, San Francisco, United States, 13–17 August 2016.

[37] Dia, H. and Javanshour, F. (2016). Autonomous shared mobility-on-demand: Melbourne pilot simulation study. *Proceedings of the 19th EURO Working Group on Transportation Meeting,* EWGT 2016, 5–7 September, Istanbul, Turkey.

[38] Dia, H., Javanshour, F. and Hill, J. (2016). Modelling the impacts of autonomous shared mobility systems. *Proceedings of the 23rd ITS World Congress*, Melbourne, Australia, 10–14 October 2016.

[39] Dia, H., Hill, J. and Javanshour, F. (2016). Autonomous mobility on demand – review of recent literature and insights for shaping policy directions. *Proceedings of the 23rd ITS World Congress*, Melbourne, Australia, 10–14 October 2016.

[40] Passenger Transport (2016). *Uber is biggest threat to buses, says TfL chief* [Online]. Available: http://www.passengertransport.co.uk/2016/02/uber-is-biggest-threat-to-buses-says-tfl-chief/

[41] Freemark, Y. (2015). *Will autonomous cars change the role and value of public transportation?* [Online]. Available: http://www.thetransportpolitic.com/2015/06/23/will-autonomous-cars-change-the-role-and-value-of-public-transportation/

[42] Ford, M. (2015). *Rise of the robots: technology and the threat of a jobless future* [Online]. Available: http://www.amazon.com

[43] Kinver, M. (2013, Jul 15). *Air pollution kills millions each year, says study* [Online]. Available: http://www.bbc.com/news/science-environment-23315781

[44] NGGI (2011). *Australian national greenhouse gas accounts* [Online]. Available: http://www.environment.gov.au/climate-change

[45] Greenblatt, J. and Saxena, S. (2015). Autonomous taxis could greatly reduce greenhouse-gas emissions of US light-duty vehicles. *Nature Clim. Change*, vol. 5 (9), pp. 860–863. Nature Publishing Group. Available: http://dx.doi.org/10.1038/nclimate2685

[46] Wadud, Z., MacKenzie, D. and Leiby, P. (2016). Help or hindrance? The travel, energy and carbon impact of highly automated vehicles. *Transp. Res., A: Policy Pract,* vol. 86, pp. 1–18. ISSN 0965-8564

[47] Manyika, J. (2013, May). *Disruptive technologies: advances that will transform life, business, and the global economy, McKinsey* [Online]. Available: http://www.mckinsey.com/insights/business_technology/disruptive_technologies

[48] Schoettle, B. and Sivak, M. (2014). *A survey of public opinion about autonomous and self-driving vehicles in the U.S., the U.K. and Australia.* Available: https://deepblue.lib.umich.edu/bitstream/handle/2027.42/108384/103024.pdf

[49] Bonnefon, J.-F., Shariff, A. and Rahwan, I. (2016). The social dilemma of autonomous vehicles. *Science,* vol. 352 (6293), pp. 1573–1576

[50] Knight, W. (2015). How to help self-driving cars make ethical decisions. *MIT Technology Review.* Available: https://www.technologyreview.com/s/539731/how-to-help-self-driving-cars-make-ethical-decisions/

Glossary

AI	artificial intelligence
AIM	Autonomous Intersection Management
AMoD	autonomous mobility-on-demand
ANN	artificial neural network
BC	base case
C-ITS	cooperative intelligent transport systems
FIA	Federation Internationale de L'Automobile
GPS	global positioning system
HFCV	hydrogen fuel cell vehicles
HEV	hybrid-electric vehicles
HVoTT	high value of travel time
ICEV	internal combustion engine vehicles
ITS	intelligent transport systems
MaaS	mobility as a service
MATSim	multi-agent transport simulation
NHTSA	National Highway Traffic Safety Administration
LVoTT	low value of travel time
LiDAR	light detection and ranging
SAE	society of automotive engineers
SAV	shared autonomous vehicle
VISTA	Victorian Integrated Survey of Travel Activity
VKT	vehicle-kilometres travelled

Further Reading

[1] Barclays (2015b). *Disruptive mobility: AV deployment risks and possibilities.* Available: http://orfe.princeton.edu/~alaink/SmartDrivingCars/PDFs/Brian_Johnson_DisruptiveMobility.072015.pdf

[2] Fishman, T. (2012). *Digital-age transportation: the future of urban mobility.* Deloitte University Press. Available: http://dupress.deloitte.com/dup-us-en/industry/automotive/digital-age-transportation.html

[3] Litman, T. (2015). *Autonomous vehicle implementation predictions – implications for transport planning.* Victoria Transport Policy Institute. Available: http://www.vtpi.org/avip.pdf

[4] Michael, S. and Brandon, S. (2015). *Potential impact of self-driving vehicles on household vehicle demand and usage.* Sustainable Worldwide Transportation Program, University of Michigan. Available: www.umich.edu/~umtriswt

Chapter 12

Gamification and sustainable mobility: challenges and opportunities in a changing transportation landscape

Eleni I. Vlahogianni[1] *and Emmanouil N. Barmpounakis*[1]

Abstract

Motivating people to change their mobility behavior toward more sustainable ways to travel is one of the major challenges in transportation. Nowadays, with information communication and technology solutions and especially the smartphones, it is possible to deliver accurate mobility-related information to users to make smart travel decisions with the aim to reduce congestion and air pollution or improve safety. However, delivering mobility-related information is necessary, but not sufficient condition for initiating behavioral change. The last can only be achieved, if the communicated information targets to influence the attitudes, and motivators of road users. But, how we can persuade users to change their behavior?

This chapter will attempt to shed light to the drivers for success of gamified applications dedicated to mobility. Gamification as a tool to trigger behavioral change is analyzed and critically discussed. Recent applications are reviewed with emphasis on their effects to traffic operations, urban mobility, and environmental conditions in cities. Finally, the opportunities stemming from the advances in technology and communications and the potential of gamified mobility solutions for engaging citizens in sustainable mobility are discussed.

Keywords

Gamification; mobility; smart cities; urbanization

12.1 Introduction

Urbanization jointly considered with the current growth rates in car ownership form unsustainable conditions for cities with far reaching implications to mobility and

[1]Department of Transportation Planning and Engineering School of Civil Engineering, National Technical University of Athens, Athens, Greece

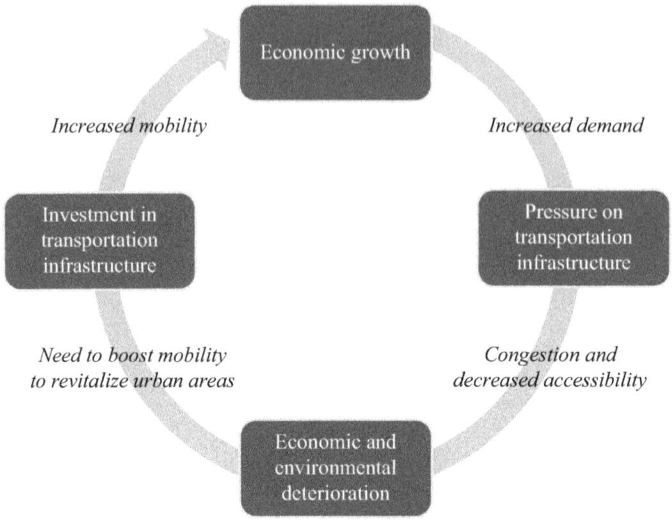

Figure 12.1 The cycle of urban mobility

environment [1,2]. According to UN, in 2014, 54% of the world's population was urban; this percentage is expected to reach 66% by 2050 [3]. In addition, the difficult economic landscape, which exists in many developed cities worldwide, is reflected to limited funding for large transportation works, as well as the inability to operate and maintain costly transportation management systems [4,5]. Nevertheless, even in financially robust environments, it may bring significant economic and environmental impacts (Chapter 2). As described in Figure 12.1, there is a relentless cycle related to mobility; services that enhance mobility in cities enable economic growth, which, in turn, is responsible for induced demand and increased motorization with significant effects to the transportation network. These are reflected to congested roads and decreased accessibility, which lead to deteriorated environmental and economic conditions. For transportation engineers and urban planners, the main approach all these years was to intervene to the transportation infrastructure, for example by increasing capacity or build large underground metro networks. However, tackling the economic turbulences in today's rapid changing and unstable fiscal conditions, innovative transportation solutions need to be developed aiming to boost mobility and accessibility of urban areas and achieve a region's economic efficiency.

In the pursuit of a trade-off between enhanced mobility and sustainability, transportation planners are forced to think of novel ideas on how to go beyond the car and help users find the optimum sustainable way to travel. Motivating people to change their mobility behavior and adopt a more sustainable way of traveling is one of the major challenges in transportation with far reaching effects to climate change and quality of life. The emergence of novel information communication and technology (ICT) solutions, as well as the extended use of smartphones, have formed a new mobility landscape where it is possible to deliver accurate mobility-related information to users to make smart-travel decisions with the aim to reduce congestion and air pollution or improve safety. This may have significant effect to the short-term evolution and dynamic characteristics of mobility patterns making well-established mobility management tools (tolls, rings, etc.) obsolete or inefficient [6].

Delivering mobility-related information is necessary, but not a sufficient condition for initiating behavioral changes: this can only be achieved, if the communicated information targets to influence the attitudes, and motivators of road users. Toward this direction, literature has to demonstrate several approaches ranging from classic advertising campaigns to promote eco-friendly mobility and transit use [7], to the use of simulation or simulators as a training/educational tool. Wang and Tang [8] presented the concept of artificial transportation systems (ATS), a mixture of real-word transportation systems with agent-based programming and modeling, large-scale distributed computing techniques, and new concepts and methods developed in complex systems. Rossetti *et al.* [9] discussed the integration of serious games into an ATS based on agent-based simulators, virtual reality simulators, and real-world transportation systems with a triple aim to (i) educate users to improve their skills and decisions (behavior assimilation), (ii) acquire information on the manner users make decisions (behavior elicitation), and (iii) provide incentives and related policies to enable users to perform certain actions (behavior persuasion). Miller [1] emphasized on the need to use ICT, geographic information system (GIS) and social media to deliver new transportation systems and services and cultivate users to share, cooperate and act collectively to solve operational, tactical and strategic mobility and accessibility problems.

To this end, transportation modelers and policy makers have put a significant amount of effort to develop systems and measures that target the awareness of transportation system users on the effects of their actions to the sustainability of the system, as well as the behavioral change of individual travelers. These approaches, usually identified by the term persuasive approaches, are based on social media, personalized messaging, social recognition, etc. An emerging concept related to persuasive systems that has recently attracted the attention of the transportation community is Gamification, or else the use of game-like concepts to nongame environments. Nowadays, gamification is seen as the third pillar of process innovation, along with big data analytics and crowdsourcing to gather data and solve problems, which reflect the essence of a smart/connected city [10]. The usefulness of gamification is established on the concept that all traffic and transportation activities may be seen as a game, where the user is to execute a series of actions and make a series of decisions in order to reach a certain goal [11].

This chapter will attempt to shed light to the drivers for success of gamified applications dedicated to mobility. Recent successful applications will be critically reviewed with emphasis on their effects to traffic operations, urban mobility, and environmental conditions in cities. The opportunities stemming from the advances in technology and communications and the potential of gamified mobility solutions for engaging citizens in sustainable mobility will be discussed.

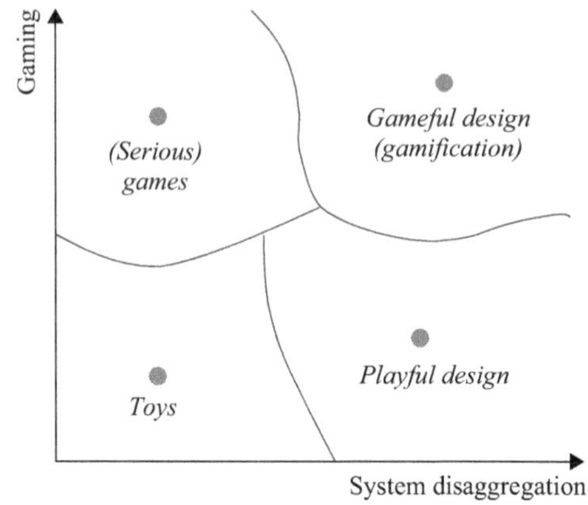

Figure 12.2 The space of game-related design

12.2 Gamification: what it is, how it works, and why it works

12.2.1 Definition and principles

Gamification is a useful tool to trigger behavior change. Although the term Gamification contains the word 'game', it does not refer to the use of actual games or Game Theory that some may think. It refers to the application of game-oriented design approaches and game-inspired mechanics (e.g., point scoring, leaderboards, methods to measure achievements) to originally nongame contexts, such as transportation and mobility [12,13]. Gamification clearly differentiates from games or toys via the conceptual two-dimensional space of gaming and system disaggregation [13] (Figure 12.2). Low degree of gaming (problem solving) and an approach, which focuses on the entire system, rather than its parts is the so-called toy, whereas a high degree of gaming, which is integrated to different tasks of a specific system forms the framework of a gameful design.

Robson *et al.* [14] underline three basic reasons why the interest on gamified applications has started to grown exponentially in all disciplines:

- The unpreceded advances in the gaming industry,
- The emergence of social media and web-based technologies, and
- The need to improve management through influencing behaviors.

In an effective gamified experience, a player should progress from novice to expert and, finally, to master. For each one of these categories of users, the motivations may differ; for example, novice users need onboarding, experts need fresh content, activities and challenges, and masters need exclusivity, recognition, and impact. To accomplish the above, a gamified experience is supported by the following principles [9,15]: (i) mechanics, (ii) dynamics, and (iii) esthetics (Figure 12.3).

These three principles form the interface between a gamified application designer and the user. Mechanics are the various actions, behaviors, and control

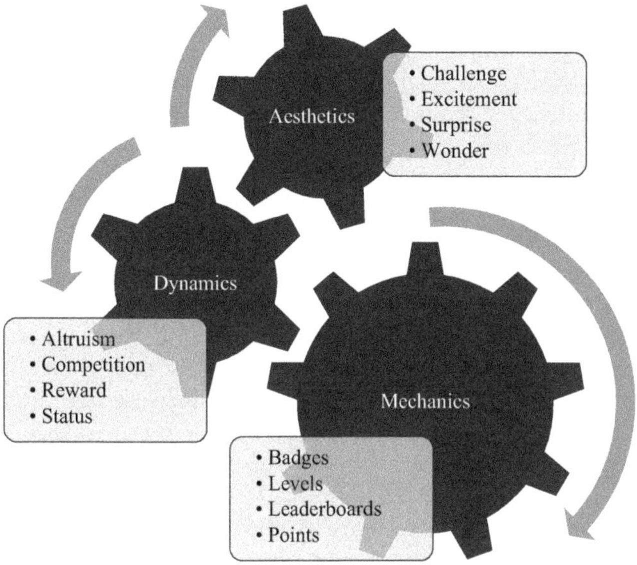

Figure 12.3 The three principles of an efficient gamified experience

mechanisms afforded to the player within a game context [16]. The mechanics involve the processes that drive action forward and may support a highly motivational user experience, for example badges, levels, leaderboards, points, etc. These gamification mechanics do not change from one player to other and remain constant throughout out the user's experience. Dynamics can be described as the players' interactions with the mechanics, for example altruism, competition, status, reward, etc. Players have a natural desire to be recognized, be in competition in one another or gain achievements, thus dynamics can further be described as the tools for satisfying desires. When the abovementioned components work well together, the user can better immerse in the game and fulfill the gamified experience. Finally, esthetics can be described as the tools that are used to evoke players' reactions and emotional responses, for example challenge, excitement, surprise, wonder, etc. Thus, they are of importance when it comes to continuing playing the game [16]. Evidently, too much information may act as a deterrent and overwhelm the player who is not yet skilled enough to process.

An effective gamification framework is based on a series of metrics to quantify the success on a predetermined goal, which is achieved through a set of tasks. Based on the metrics, users gain feedback about their performance and achievements. Users may also communicate with others, compare their performance, and compete. To improve performance, users may need to change their mobility patterns and undertake different behavior or perform alternative tasks. Typical examples of successful gamification frameworks can be seen in other aspects of everyday life, such as health, fitness, education, and finance [13]. In transportation, apart from mobility applications for travel behavior change, which will be analyzed in the following chapters, gamification is mainly used as an instrument for enhancing driving educational programs, supporting learning and adaptation of automotive user interfaces and vehicle functions [17], developing attractive in-vehicle infotainment systems for eco-driving, and driving safety [18,19].

12.2.2 Gamification as a mobility change instrument

Inducing changes in the individual mobility patterns depends on three main issues [20]: (i) awareness of mobility characteristics, decisions, and consequences of them, (ii) awareness of alternatives toward improving themselves, and (iii) monitoring self-improvement through multiobjective goals. Citizens are rarely aware of their daily mobility footprint, as well as the effects of their choices with respect to the efficiency of their travel patterns, as well as the sustainability of the transportation system [21]. Providing them with means to gather, visualize, and quantify their mobility characteristics is a critical step toward improving their mobility choices.

Nevertheless, even if the mobility profile is known, citizens cannot easily make the right decisions, as they are not familiar with the full spectrum of alternatives or they cannot process the multiple alternatives and decisions they have to make in order to achieve a more sustainable mobility footprint. In the decision-making process, users can be assisted by technology and communications systems. Nevertheless, there exists a certain misconception. Although new technologies and existing soft policy tools can address the need to change route and departure times to improve once travel, the issues of mode and trip choice have been found to be unaffected by existing soft policy instruments [22]. They seem to be more affected by instruments sourcing from psychological aspects of mobility choices, such as Travel Feedback Programs and Personalized Travel Planning tools [21–24]. These instruments use a person-to-person approach to demystify people choices and try to change them toward more sustainable mobility choices [22]. Jariyasunant *et al.* [23] go one-step forward by introducing automated methods to identify choices and study their effects to users that adopt them on the basis of signal processing, machine-learning techniques, and no human interaction.

The integration of information and technology instruments with gamification characteristics can enrich the users' motivation and engagement to activities and behaviors, which may improve their mobility profile. Gamification may be seen as a complementary soft tool to evoke a change in commuters' behavior in a way that they have the feeling of self-improvement, achieving goals, and social recognition. Moreover, gamification may trigger changes that are beneficial for the macroscopic operation of the entire transportation system. With the rapid penetration of ICT and intelligent transportation systems (ITS) (navigation systems, web addresses with traffic information, mobile applications dedicated to traffic, and transportation) travel patterns—especially daily ones—are becoming more and more asymmetrical (real-time changes of route to avoid congestion or incidents etc.) [22]. Asymmetries in mobility patterns significantly affect the predictability of traffic and the equilibrium of the transportation system. Users make constant route choices and mode choices with negative impacts on the effectiveness of certain static managerial policies and tools. Evidence also shows that using ICT and ITS to inform users may work in favor of choosing to travel with private car over other alternatives due to the users' selfish way of thinking [25]. Conceptually, gamified applications may work well in asymmetrical conditions, as they are based on cooperation and constant data sharing. Gamified applications can provide less selfish alternatives for decision-making based on up-to-date information. Users become aware of the effects of their behavior to the environment and transportation system and of how they can make informed decisions.

The less enthusiasts have emphasized the possible negative effects of gamified design in mobility applications. McCall *et al.* [26] and Hamari *et al.* [27] stress that game mechanics may lead to choices and behaviors other than intended. Weiser *et al.* [20] argue that such systems usually depend on the rationality of the users, which does not hold. Users are also not threatened by the intrusion of technologies and communications to their individual choices and are too broad to be able to address all individualized needs and choices. Seaborn and Fels [28] mention the criticism on the simplification of game concepts for the purpose of easy profit but argue that this is not applicable to gamification approaches that try to motivate users intrinsically.

12.2.3 Some gamified mobility applications

A typical example of gamification is Waze [29], the largest community-based traffic and navigation app, which creates localized communities of drivers in order to provide constantly updated information to its users. Instead of being a simple navigation app, users are supported by information provided by other users such as avoiding congested arterials, alerts for road hazards, getting gas, synchronize Calendar events and creating route plans based on current traffic conditions, etc. Another successful gamified application is Tripit [30], which helps companies achieve cost-effective travel management by optimizing travel booking behavior, making it possible to reduce both travel time and cost. Moovit supports over 800 cities across 60 countries and enables users to find the fastest and least crowded route when using public transport [31]. It also provides information to the community that many users may find useful and attractive, like real-time information, station crowdedness, incidents, etc. Changers—CO_2 fit, an app to automatically measures the commuters' distance traveled and calculate his/her carbon footprint depending on the mean of travel [32]. The core and most motivating feature of the app is that healthy and CO_2-saving mobility types (bike, foot, or public transport) will be rewarded with bonus points.

In Europe, several prominent examples aimed to trigger behavior change using gamified concepts. The Chromaroma project, a London's Oyster card powered travel game, rewards players for traveling better by encouraging nonrush hour travel, getting off early, and walking or riding bikes instead of traveling on buses and train [33]. The MOBI project targets modal shift through gamification promoting smart mobility to employees [34]. The gamified experience is mainly based on the competition element between teams of employees to make as many sustainable trips as possible. The SUNSET project, which uses the mobile Tripzoom-app [35], aims to improve the personal mobility by helping travelers getting better insight in his/her mobility pattern, the incentives for a travel behavior change, and how sharing information via social networks can create a social experience. In the project called SUPERHUB, the prototype app aimed to induce the user to change his/her travel behaviors by adopting more sustainable transport choices [36]. The user was invited to set weekly goals for mode of transport to use, track progress toward goals to support self-monitoring, and finally according to his/her profile the user would receive personalized messages that encouraged him/her toward sustainable transport choice.

Another European project named STREETLIFE used gamification mechanisms to promote more sustainable mobility means and habits [37]. The Peacox project targeted the development of mobile and web applications specifically designed to provide users with personalized multimodal tools in order to plan their journeys and reduce CO_2 emissions [38]. Some gamified practices are also implemented in the project MITOS, which aimed to provide advanced transportation services in the city of Santander, Spain [39]. The TEAM project aims to be an ecosystem for connected, collaborative, and green mobility with significant value proposition for both drivers and stakeholders [40]. Another project that focuses on urban public transport through gamified concepts is the Seamless Mobility Project, which integrates a broad array of applications (secure mobile payments methods, data aggregation from different components, etc.) to positively impact the interface between travelers and the transit system to provide responsive mobility services [41].

Apart from the abovementioned apps and projects, some other attempts target the change in drivers' behavior but with limited degree of gamification. Merugu *et al.* [42] describe an experimental project that took place in Bangalore, India, for influencing a population of commuters to travel at less congested times by awarding credits to users every day depending on their arrival time, instead of charging commuters who travel during congestion. A similar approach is followed by Pluntke and Balaji [43]. In TruCentive platform, users are divided into contributors and consumers in order to exchange parking availability information [44]. The authors integrate concepts from gamification and Game Theory to encourage users to be honest and provide accurate data to the rest of the community. UbiGreen application aims to reduce CO_2 emissions, while increasing green transportation [45]. Another application that aims to reduce congestion is nunav (first known as Greenway), which uses smart swarm algorithms to recommend the shortest paths, while trying to prevent the formation of congested arterials [46]. A collaborative sustainable mobility application for the city of Trento, Italy, called ViaggiaTrento is reported to have a significant potential in changing the traveling behavior of commuters [47]. A smartphone "endless runner" game called Grr–Grr–Bike aimed to educate commuters on safe bike riding, by avoiding obstacles or stopping at red lights [48]. All the above efforts are depicted in Table 12.1 with respect to the purpose for which they have been developed and the main gamification elements they entail.

Table 12.1 Gamified transportation applications

Application	Purpose	Main gamification elements
Waze [29]	Navigation	Avatars, social network, points, leaderboards, achievements, levels
Tripit [30]	Travel management	Points, badges, leaderboards
Moovit [31]	Public transport navigation	Points, levels
Changers—CO_2 fit [32]	Calculate carbon footprint	Leaderboards
Chromaroma Project [33]	Encouraging nonrush hour travel	Points, interactive map
MOBI Project [34]	Promoting smart mobility to employees	Points, virtual prizes, rewards, avatars
SUNSET Project (Tripzoom App) [35]	Improve personal mobility	Rewards, social networks
SUPERHUB [36]	Adopting sustainable transport choices	Goals, track progress
STREETLIFE [37]	Adopting sustainable transport choices	Points, leaderboards, badges, rewards
Peacox Project [38]	Personalized multimodal tools	Virtual environment, points, credits
MITOS Project [39]	Advanced Transportation Services in Santander, Spain	Points, rewards
TEAM Project [40]	Improve transportation safety and efficiency	Virtual coins, rewards, social networking
Seamless Mobility Project [41]	Integrating personal mobile devices and providing information	Points, credits
INSTANT Project [42]	Travel less congested hours	Credits, awards
INSINC Project [43]	Travel less congested hours	Rewards
TruCentive [44]	Exchange parking availability information	Credits, rewards
UbiGreen [45]	Reduction of CO_2 emissions	Points, visual designs
Nunav [46]	Navigation	–
ViaggiaTrento [47]	Adopting sustainable transport choices	–
Grr–Grr–Bike [48]	Educate commuters on safe bike riding	Educational elements

12.3 Quantifying the effects of gamified transportation applications

Gamified transportation mobility applications are the new hype in the field of "soft" policy measures to improve transportation system efficiency. Nevertheless, when it comes to tractable and quantified results on the long-term acceptance and effectiveness of such approaches, limited knowledge exists, at least in a scientific research form. Interdisciplinary review studies also support the scarcity of empirical evidence in relation to the positive/ negative effects of gamified applications [28]. The few studies that claim positive effects of gamification are related to the implementation context and the users' characteristics [27]. In a previous research, Gamification is considered as a subset of a larger effort to improve the user experience of interactive systems through gameful design, whereas the authors conclude that "only continued theoretical deliberations, playful, practical explorations, and rigorous evaluation with end-users will reveal whether or not gamification is here to stay" [28].

In the transportation context, there have been limited attempts to quantify the effect of gamified transportation applications. For example, the MOBI project entailed the development of a game is to encourage employees to travel to work more smartly (e.g., walking, cycling, public transport, and car sharing), as well as having fun competing against their friends and colleagues at the same time [34]. Results showed that the number of employees using sustainable modes of transport for their commute and business travel trips has been increased by 20%. During the first years of the campaign, 23,400 players improved their commuting habits directly in the game while the aim of primary energy savings was set to 684 CO_2 t. In the SUNSET project, although the majority of the participants did not report a change in their travel behavior, 23% of them that did respond to the challenge, indicated to have changed their behavior, which was seen in their different departure times [49]. Although some other effects on the environmental and well-being aspects are reported, the most important gains were related to the optimization of mobility of people and goods and improved quality of life in urban environments. Moreover, the positive effect of Changers—CO_2 fit app is mainly focused on the well-being of the users, whereas some corporate advantages of teambuilding and employee engagement are also reported.

In a different study, the use of game design elements in drivers' training were motivating and made the entire procedure more fun, whereas the competitive aspect of the game had a negative impact, as users aimed to complete tasks quickly without focusing on safe driving [50]. Then the authors identify the need for more indications of whether gamification is suitable for training better drivers. The results of the SUPERHUB project pointed out that three out of seven subjects were surprised by the emission information, reporting a consciousness rise on the matter, but only the same number of persons reported an effort in reducing their emissions [51]. During both the INSTANT [42] and the INSINC [43] projects, commuters were induced to shift from peak-to-off-peak travel times. Finally, the nunav smart application, which aims to prevent congestion via collaborative routing, reports that 832 h were saved in daily commutes in just 4 weeks, whereas a decrease of 30% in distance and a 10% reduction of vehicles is also reported [46].

The effects of gamified biking initiatives, both a small-scale experiment and an annual national campaign, have been documented in Wunsch *et al.* [52]. Both initiatives were found to have positive short-term effects in relation to the bike use. Approximately 25% in the experimental group and 36% participating in the campaign increased their amount of biking during the intervention period. The authors emphasize that, to induce long-term behavior changes, gamified biking initiatives have to be embedded into everyday life, to enable social interactions and provide mutual encouragement.

Evidence so far shows that all reported effects are based on limited sample of users, for example companies or small cities. The reason for this may be the fact that gamification is a rather new concept both for practitioners and researchers; therefore, the first applications were short scaled to avoid technical and other issues. In addition, in a small-scale case study, especially when it comes to experimental practices, it is easier for the researcher to receive direct feedback and have a better insight of the effects of the gamified concept. As far as the larger scale apps are concerned, their effect cannot still be quantified as it should be examined in a longer timeframe. As reported in Hamari *et al.* [27], only two of the studies that were reviewed lasted longer than 1 month. Other issues concerning the evaluation of gamified applications that have been reported in the literature is the lack of psychometric measurements (e.g., the effect of awards or other gamified concepts to the user) [53]. Evidently, the bigger the sample gets, the more complex the problem gets. Therefore, to be able to quantify the effect that gamification has on users' behavior, new methods of critical analyses and investigation need to be used, and new models to be estimated and evaluated, using concepts from a variety of different sectors. For example, due to advancement in technology and the recent burst in the social media area, media psychology has been one of the hot topics of psychology, which could be conceptually connected to the motivating framework gamification uses.

Along with the growing research on the specific subject, there is also a concurrent interest on emphasizing the need to be able to assess—at least qualitatively—the possible negative of undesired effects of these soft policies. According to te Brömmelstroet [54], there is limited information on the undesired effects of the so-called soft mobility policies and especially the gamification and emphasize on whether gamification principles are able to create creative, responsible, and empowered citizens, rather than conformist, egocentric, and powerless consumers that are just maximizing their own, financial, or gaming, budgets.

12.4 Deployment issues to consider

12.4.1 Drivers for success

The success of the gamified transportation applications is strongly dependent on the willingness of the travelers to use them and share their mobility habits. If the users are not willing or interested in gathering their mobility information, gamification will not be effective. Therefore, questions, such as "Who are the players and how to engage them? What information they need?", are critical. Incentives are important in mobility games. These include the following:

- Save time and money,
- Receive information and optimize trips,
- Achieve (social) recognition for being eco-friendly, healthy, etc.

In mobility games, a good transportation planner should think like a game designer. For a mobility game to be successful, users should have a sense of autonomy and control that are integral parts of its sustainability in long term. Long-term use is imperative to achieve travel behavior change, as the latter is a continuing process [55]. Communicating the correct information directly to the user without patronizing him, therefore, can have an additive impact to the willingness to change his/her behavior [53]. To this end, tailor-made information, which is not always in the form of competition, may increase longer term involvement [56]. This tailored-made information should be based on the user's individual needs. Other requirements for long-term involvement should target to promote bidirectional help, provide incentives, avoid additional stress and finally create trustworthy travel communities [57]. The need for high-quality information and personalized experience is also reported in Rehrl *et al.* [58].

In addition, although the psychological and social motivational needs are of great significance, when it comes to behavioral change, they alone only partially address the issue of motivation generation, users' abilities, and their triggers. Emphasis should also be given to the applied game mechanics that trigger behavioral change [20,59]. Special attention should be given to the uniqueness in mobility behavior, as well as the perceptions and beliefs of travelers. The more flexible the gamified design is—in terms of metrics and tasks involved—the greater the short and long run acceptance by the users and the more likely the behavioral change.

Moreover, it is important to consider the manner in which the user takes advantage of this information. For example, users may selfishly choose to use this information and cause additional problems to a transportation system [60]. Therefore, a gamified experience should be able to promote some of them to make sacrifices for the general good or advantages that will be obvious in the long run. Rewards of any type are important and should be tangible, unexpected, and performance contingent.

12.4.2 New management paradigms by leveraging ICT

Most of the tools discussed in the previous section are strongly related to the recent growth of the mobile phone industry. The latest advances in smartphones and smart applications gave the opportunity to transportation planners to approach the commuter in a more direct way and create a bidirectional communication [61]. This bidirectional relationship is fundamentally more capable of tackling everyday problems than any other imposed measure. The success story behind mobility related gamified applications is evident in the Waze paradigm [29]. The fact that Waze (with over 50 million users) was acquired by the already-strong mapping company Google over a $1 billion shows that the impact a successful gamified experience may have in current practices. Similarly, Moovit [31] with its over 30 million users and its expanding global presence shows that there is not only a need for better local transit, but also that gamification can engage commuters in contributing to its improvement. Such crowdsourcing applications may become an indispensable tool to enhance systems of traffic management and city administrations.

Nevertheless, managing urban mobility incorporates several conflicting goals. The integration of all interacting systems (from transportation to government services, business, healthcare, etc.) is an indispensable part of the success of developing holistic mobility solutions. In a holistic manner to think of transportation system, mobility behavior changes are linked to the possible alternatives and are constrained by cost, availability of a specific mode, familiarity, etc. If no feasible alternatives exist, gamification will have limited or no effect to transportation systems.

Further, gamified applications are mainly based on cooperation and information sharing. An emerging challenge, therefore, is to analyze the circumstances under which cooperative conditions are more efficient. Moreover, how can cooperation be sustainable in a system that is mainly considered as a competitive one? Commuters use the same infrastructure, same resources or even the same mean of transportation, whilst trying to move in the fastest, safest, and most economically sustainable manner possible. Drivers of different modes of transport may rarely develop competitive actions when moving in a heterogeneous environment [62]. As contradictory this may seem, theory suggests that cooperation can emerge even without friendship between individuals [63]. Therefore, the primary objective of a Gamified Transportation system should be to transform commuters from selfish agents that seek the short-term advantage to cooperative agents that are willing to wait for a greater long-term advantage in a fun and entertaining way.

Novel urban planning and mobility management concepts are rapidly evolving from the classic top-down thinking to a more user-centric perspective, which requires high intensity synergies between users, public authorities, and city transportation providers. Some prominent examples are shared mobility and private sector companies travel plans—for both business and commuting—as a means to mobility, reduce travel costs and efficiency in business [64,65]. In this framework, the ICT have a central role to play [53,66]: (i) gather and fuse the necessary data to monitor mobility activities, (ii) identify and predict mobility patterns, (iii) support the real-time decision-making, and (iv) evaluate the effects of users' behaviors to the mobility of cities. The role of ICT to mobility is a hot subject of discussion in transportation literature [67–70]. According to Banister and Stead [69], ICT will not reduce travel, but they may provide users with new options for activities that replace any gains from more efficient or eliminated trips.

Although ICT is a strong enabler of novel mobility solutions, their large-scale deployment comes with several challenges. First, the development of user-centric mobility solutions is fundamentally dependent on crowdsourcing data. The larger the datasets the greater the capabilities of gamified mobility solutions to address all types of behaviors and social norms. To this end, the concept of *open* and *linked* big data is to play a critical role to innovation, public engagement, and private–public sectors cooperation [71]. Management solutions based on gamification and ICT will be efficient, when the relevant data are gathered consistently over time, of high and standardized quality, so as they can be trusted for analyses, linked, to be easily distributed, and available in a standardized format, so as they can be easily processed. The integration of gamification, ICT, and openness of data will enable data reuse, which will foster innovation, and trigger new mobility policies for the public agencies to benefit.

12.4.3 Data privacy and security and social equity

The strong dependence of innovative mobility services on big data arises issues of privacy and security, as they are purely dependent on participatory sensing. In contrast to traditional data collection approaches, participatory sensing involves human probes with the aim to collect high volume, high granularity, and high spatial coverage data of any type (structured or unstructured). In crowdsourcing approaches, there is no explicit consent by the citizens who participate to the experiment. Moreover, no design method is followed, which may assure of the data validity and reliability. Yang *et al.* [72] summarize the privacy and security challenges in three categories: (i) security threats (stemming from data and from the actual tasks the application performs), (ii) reliability threats (in relation to the collected data, the computations, as well as the communication protocols) and (iii) availability threats (denial of service issues). There are various approaches followed in literature related to prevent violation of privacy, such as "privacy by design," data anonymization, data aggregation, "access control," and "need to know" [73,74]. Privacy and security in sharing data are a hot discussion and research topic and legislative and other initiatives are currently undertaken from governments, relevant agencies, and privacy sector companies [10].

Mobility equity concerns (Chapters 2 and 3) have been a significant problem in transportation management solutions [75]. Many soft transportation measures and policies are to affect mobility equity in both positively and negatively in an intended or unintended manner [76]. For example, road pricing, which has been extensively applied to many urban areas, have been extensively criticized as having unintended equity effects (Chapter 13). The use of ICT to develop and operate largely accepted gamified applications can merge the gap between user groups and reduce social exclusion, a key ingredient of sustainable transportation planning.

12.4.4 Advanced skills and novel educational paradigms

The changing nature of transportation problems often drives the need to test and evaluate new modeling paradigms robust to big data, imperfections, and real-time computations [61]. Ali *et al.* [77] report that efficient data aggregation, verification algorithms, traffic prediction, and efficient interface design should be in the main part of ICT research and ITS in general. Computational intelligence and machine learning for optimization and predictive analytics—in a synergistic environment with classical modeling tools and statistical inference—are the new hypes in mobility-related problem solving due to their ability to learn from vast and imperfect datasets and produce meaningful transportation information [78]. Moreover, traditional simulation tools based on classical macroscopic theories and four-step modeling are currently being replaced by more detailed simulation approaches such as the popular agent based models to integrate social, economic and environmental details in mobility models and users travel choices [6,9,79].

Inevitably, new educational paradigms will emerge to improve the skills of future transportation planners. Khisty and Kikuchi [80] highlighted several topics in which transportation students should develop deeper understanding, such as intermodalism and systemicity, soft systems methodologies, applied ethics, and communication. Zhu *et al.* [81] and Di *et al.* [82] proposed the use of simulation based tools to educate transportation engineers and improve students' engagement. Huang and Levinson [83] provided evidence on the effectiveness of teaching transportation planning with board games. However, when it comes to the indispensable skills to develop ITS and mobility-related applications in a smart city environment, the acquired skills and the training of transportation engineers are at a native stage. Chowdhury and Dey [84] review current practices and educational paradigms and emphasize on the need to train a new generation of mobility professionals to master traditional methods, but also cyber-physical systems (e.g., sensor technologies, data analytics, security and privacy, hardware–software integration, etc.).

12.5 Conclusion

Sustainable mobility is becoming more than ever a challenging concept. The emergences of ICT and Internet of Things (IoT) in a smart urban environment are strong enablers of novel ITS services, yet form a changing transportation management landscape. Novel policies are to emerge from a user-centric perspective, rather than the traditional top-down thinking, which requires high intensity synergies between users, public authorities, and city transportation providers. In these conditions, delivering mobility-related information is necessary, but not sufficient condition for initiating behavioral change to achieve sustainability. Gamification comes as a driver to deliver targeted and user-centric information to influence the attitudes, and motivators of road users.

Findings from the literature review show that integrating gamification concepts to current and emerging ICT solutions can enrich the users' motivation and engagement to activities and less selfish behaviors, which may improve their mobility profile. Moreover, gamified applications may cope with the asymmetries in travel patterns induced by the effects of delivering information on real-time changes on the transportation system, shift users to choices that are more efficient and macroscopically improve the operation of the transportation system. Moreover, it was evident that, although there exist many enthusiasts of gamified mobility applications, limited evidence can be found on quantified results regarding the effects of such applications. Further research should focus on the proper selection of the involving elements capable to trigger behavioral change and make such applications efficient and sustainable.

Finally, this chapter elaborated on a series of challenges stemming from the large-scale deployment of gameful individualized ITS solutions. Clearly, gamification to work needs to act complementary with other novel managerial policies and tools that depart from the classic top-down approach in a competitive transportation system with selfish users, to a more user-centric, equitable and cooperative transportation environment. It is, therefore, important to identify the conditions under which cooperative behaviors are more efficient, to have the proper ICT and algorithmic tools to gather, store, process, and model big data to transform them to meaningful personalized information and motivate mobility change. In this framework, providing a sense of security and autonomy to users who will share information is a key consideration from both a legislative and a technical aspect.

Evidently, the above conditions will challenge the role of transportation planners in both a theoretic and technical level. In a theoretic level, transportation researchers will have to expand their skills from the classical transportation modeling to advance data mining approaches and highly detailed simulations schemes. From the technical aspect, transportation engineers will face challenges, such as cloud storage and computing, coding in new language forms for fast computations, technologies for gathering and disseminating information, etc. In this demanding landscape, transportation planners should not aim to replace other disciplines experts, but to be able to understand and efficiently use the full potential of ICT applications and the accompanying modeling tools. This new professional and research landscape should be supported by the deployment of new educational paradigms to train the next generation of researchers and professionals.

References

[1] Miller, H. J. (2013) "Beyond sharing: Cultivating cooperative transportation systems through geographic information science." *Journal of Transport Geography*, 31, 296–308.

[2] Hao, H., Geng, Y., & Sarkis, J. (2016). "Carbon footprint of global passenger cars: Scenarios through 2050." Energy, 101, 121–131.

[3] United Nations, Department of Economic and Social Affairs, Population Division (2014). *World Urbanization Prospects: The 2014 Revision, Highlights* (ST/ESA/SER.A/352).

[4] Pnevmatikou, A., Milioti C., & Karlaftis M. (June 2014). "Economic Recession and the impact of ICT on Integrated Mobility during Disruptive Events." *Conference on Engineering and Applied Sciences Optimization (OPT-i), Greece.*

[5] Milioti, C. P., Pnevmatikou, A., & Karlaftis, M. G. (2014). "Mobility Patterns, User Behavior and Financial Crisis: An Exploratory and Revealed Preference Analysis." *In European Transport Conference 2014.*

[6] Kokkinogenis, Z., Monteiro, N., Rossetti, R. J., Bazzan, A. L., & Campos, P. (2014). Policy and incentive designs evaluation: A social-oriented framework for Artificial Transportation Systems. *In Intelligent Transportation Systems (ITSC), 2014 IEEE 17th International Conference* (pp. 151–156). IEEE.

[7] Hiselius, L. W., & Rosqvist, L. S. (2016). "Mobility Management campaigns as part of the transition towards changing social norms on sustainable travel behavior." *Journal of Cleaner Production*, 123, 34–41.

[8] Wang, F.-Y., & Tang, S.M. (2004). "Concepts and frameworks of artificial transportation systems." *Complex Systems and Complexity Science*, 1(2), 52–59.

[9] Rossetti, R. J., Almeida, J. E., Kokkinogenis, Z., & Gonçalves, J. (2013). "Playing transportation seriously: Applications of serious games to artificial transportation systems." *IEEE Intelligent Systems*, 28(4), 107–112.

[10] Matthew C., Alexander E., Cynthia M., *et al.* (October 14, 2014). *The Smart/Connected City and Its Implications for Connected Transportation*. Available: www.its.dot.gov/index.htm White Paper. FHWA-JPO-14-148 VNTSC-FHWA-14-10 U.S. Department of Transportation Intelligent Transportation Systems Joint Program Office 1200 New Jersey Avenue, SE Washington, DC 20590.

[11] Montola, M. (2005, December). "Exploring the edge of the magic circle: Defining pervasive games." *In Proceedings of DAC* (Vol. 1966, p. 103).

[12] Zichermann, G., & Cunningham, C. (2011). *Gamification by Design: Implementing Game Mechanics in Web and Mobile Apps*. O'Reilly Media, Inc.

[13] Deterding, S., Dixon, D., Khaled, R., & Nacke, L. (2011). "From game design elements to gamefulness: Defining gamification." *Proceedings of the 15th International Academic MindTrek Conference: Envisioning Future Media Environments*. ACM.

[14] Robson, K., Plangger, K., Kietzmann, J. H., McCarthy, I. & Pitt, L. (2015). "Is it all a game? Understanding the principles of gamification." *Business Horizons*, 58(4), 411–420.

[15] Hunicke, R., LeBlanc, M., & Zubek, R. (2004). MDA: A formal approach to game design and game research. *Paper presented at the Proceedings of the AAAI Workshop on Challenges in Game AI*, San Jose, CA.

[16] Xu, Y. (2011). *Literature Review on Web Application Gamification and Analytics*. Honolulu, HI, Collaborative Software Development Lab, Department of Information and Computer Sciences, University of Hawai'i, CSDL Technical Report. 11–05.

[17] Diewald, S., Lindemann, P., Möller, A., Stockinger, T., Koelle, M., & Kranz, M. (2014, November). "Gamified training for vehicular user interfaces—Effects on drivers' behavior." *In 2014 International Conference on Connected Vehicles and Expo (ICCVE)* (pp. 250–257). IEEE.

[18] Diewald, S., Möller, A., Roalter, L., Stockinger, T., & Kranz, M. (2013, October). "Gameful design in the automotive domain: review, outlook and challenges." *In Proceedings of the Fifth International Conference on Automotive User Interfaces and Interactive Vehicular Applications* (pp. 262–265). ACM.

[19] Vaezipour, A., Rakotonirainy, A., & Haworth, N. (2016, July). "Design of a gamified interface to improve fuel efficiency and safe driving."

In International Conference of Design, User Experience, and Usability (pp. 322–332). Springer International Publishing.

[20] Weiser, P., Bucher, D., Cellina, F., & De Luca, V. (2015). "A Taxonomy of Motivational Affordances for Meaningful Gamified and Persuasive Technologies." ICT for Sustainability (ICT4S).

[21] Cairns, S., Sloman, L., Newson, C., Anable, J., Kirkbride, A., & Goodwin, P. (2004). Smarter choices–changing the way we travel, final report of the research project "The influence of soft factor interventions on travel demand". Published by the Department for Transport on the *Sustainable Travel* section of www.dft.gov.uk.

[22] Salomon, I., & Singer, M. E. (2014). "'Informal travel': A new conceptualization of travel patterns?" *Transport Reviews*, 34(5), 562–582.

[23] Jariyasunant, J., Abou-Zeid, M., Carrel, A., *et al.* (2015). "Quantified traveler: Travel feedback meets the cloud to change behavior." *Journal of Intelligent Transportation Systems*, 19(2), 109–124.

[24] Taniguchi, A., Suzuki, H., & Fujii, S. (2007). "Mobility management in Japan: Its development and meta-analysis of travel feedback programs." *Transportation Research Record: Journal of the Transportation Research Board*, 2021(1), 100–109.

[25] Abdel-Aty, M. A., & Abdalla, M. F. (2006). "Examination of multiple mode/route-choice paradigms under ATIS." *IEEE Transactions on Intelligent Transportation Systems*, 7, 332–348.

[26] MCCall, R. Koenig, V., and Kracheel, M. (2013) "Using gamification and metaphor to design a mobility platform for commuters." *International Journal of Mobile Human-Computer Interaction*, 5(1): 1–15.

[27] Hamari, J., Jonna, K., & Harri, S. (2014) "Does gamification work?—A literature review of empirical studies on gamification." *System Sciences (HICSS), 2014 47th Hawaii International Conference on* IEEE.

[28] Seaborn, K., & Deborah I. F. (2015) "Gamification in theory and action: A survey." *International Journal of Human-Computer Studies*, 74: 14–31.

[29] "Free Community-based Mapping, Traffic & Navigation App", *Waze.com*, 2016. [Online]. Available: https://www.waze.com. [Accessed: 15 Apr 2016].

[30] "TripIt – Travel Itinerary – Trip Planner", *TripIt*, 2016. [Online]. Available: https://www.tripit.com. [Accessed: 15 Apr 2016].

[31] "Moovit: Public Transport | Your City Your Local Transit App", *Moovit*, 2016. [Online]. Available: http://moovitapp.com. [Accessed: 15 Apr 2016].

[32] "Home | Changers", *Changers*, 2016. [Online]. Available: https://changers.com. [Accessed: 15 Apr 2016].

[33] "Chromaroma | Mudlark", *Wearemudlark.com*, 2016. [Online]. Available: http://wearemudlark.com/projects/chromaroma/. [Accessed: 15 Apr 2016].

[34] Buningh, S., Martijnse-Hartikka, R., & Christiaens, J. (2014). "Mobi-Modal Shift Through Gamification," *in Transport Research Arena (TRA) Fifth Conference: Transport Solutions from Research to Deployment*.

[35] Broll, Gregor, Cao, H., Ebben, P., Holleis, P., Jacobs, K., Koolwaaij, J., *et al.* (2012). "Tripzoom: An app to improve your mobility behavior."

Proceedings of the 11th International Conference on Mobile and Ubiquitous Multimedia. ACM.

[36] Wells, S., Kotkanen, H., Schlafli, M., *et al.* (2014). "Towards an Applied Gamification Model for Tracking, Managing, & Encouraging Sustainable Travel Behaviours," *EAI Endorsed Transactions on Ambient Systems.*

[37] Kazhamiakin, R., Marconi, A., Perillo, M., Pistore, M., Valetto, G., Piras, *et al.* (2015). "Using gamification to incentivize sustainable urban mobility." *Smart Cities Conference (ISC2)*, 2015 IEEE First International. IEEE.

[38] Schrammel, J., Busch, M., & Tscheligi, M. (2013). "Peacox–persuasive advisor for CO_2-reducing cross-modal trip planning." *In First International Conference on Behavior Change Support Systems* (pp. 3–5).

[39] Diamantaki, K., Rizopoulos, C., Tsetsos, V., Theona, I., Charitos, D., & Kaimakamis, N. (2013). "Integrating game elements for increasing engagement and enhancing user experience in a smart city context." *Intelligent Environments (Workshops)* (pp. 160–171).

[40] Belloti, F., Berta, R., De Gloria, A., *et al.* (2015). A smart mobility serious game concept and business development study. *GALA Conference 2015 Workshop on Games for Mobility and Intelligent Transportation Systems (G4MITS)*, Rome, Italy, 9–11 December 2015.

[41] Costa, P. M., Fontes, T., Nunes, A. A., Ferreira, M. C., Costa, V., Dias, T. G., *et al.* (2016). "Application of collaborative information exchange in urban public transport: The seamless mobility solution." *Transportation Research Procedia*, 14, 1201–1210.

[42] Merugu, D., Prabhakar, B., & Rama, N. (2009), "An Incentive Mechanism for Decongesting the Roads: A Pilot Program in Bangalore," in *Proceedings of ACM NetEcon Workshop*, 2009.

[43] Pluntke, C., & Balaji, P. (2013). INSINC: "A Platform for Managing Peak Demand in Public Transit." *JOURNEYS, Land Transport Authority Academy of Singapore*, Singapore.

[44] Hoh, B., Yan, T., Ganesan, D., Tracton, K., Iwuchukwu, T., & Lee, J. (2012). TruCentive: A game-theoretic incentive platform for trustworthy mobile crowdsourcing parking services. In *Intelligent Transportation Systems (ITSC), 2012 15th International IEEE Conference.*

[45] Froehlich, J., Dillahunt, T., Klasnja, P., Mankoff, J., Consolvo, S., Harrison, B., *et al.* (2009). UbiGreen: Investigating a mobile tool for tracking and supporting green transportation habits. *Proceedings of the SIGCHI Conference on Human Factors in Computing Systems*. ACM.

[46] "Graphmasters | No more Traffic Jams." *Nunav.net*, 2016. [Online]. Available: http://www.nunav.net. [Accessed: 15 Apr 2016].

[47] Bordin, S., Menendez, M., & De Angeli, A. (2014). "ViaggiaTrento: An application for collaborative sustainable mobility." *ICST Transactions on Ambient Systems*, 4, e5.

[48] Nash, A., Purgathofer P., and Kayali, F. (2013) "Using Online Games in Transport: Grr–Grr–Bike Case Study," Transportation Research Board 93rd Annual Meeting Compendium of Papers, 14-3805, Washington, DC, USA.

[49] *SUNSET Project Final Report* (2014) (D9.2). Available: http://www.sunset-project.eu/pdf/SUNSET_Project_Final_Report.pdf.

[50] Diewald, S., Lindemann, P., Möller, A., Stockinger, T., Koelle, M., & Kranz, M. (2014). Gamified training for vehicular user interfaces—effects on drivers' behavior. *Connected Vehicles and Expo (ICCVE), 2014 International Conference*. IEEE.

[51] Jylhä, A., Nurmi, P., Sirén, M., Hemminki, S., & Jacucci, G. (2013). Matkahupi: A persuasive mobile application for sustainable mobility. *Proceedings of the 2013 ACM Conference on Pervasive and Ubiquitous Computing Adjunct Publication*. ACM.

[52] Wunsch, M., Millonig, A., Seer, S., Schechtner, K., Stibe, A., & Chin, R. C. (2016). Challenged to Bike: Assessing the Potential Impact of Gamified Cycling Initiatives. In *Transportation Research Board 95th Annual Meeting* (No. 16-6249).

[53] Weiser, P., Scheider, S., Bucher, D., Kiefer, P., & Raubal, M. (2015). "Towards sustainable mobility behavior: Research challenges for location-aware information and communication technology." *GeoInformatica*, 20(2), 1–27.

[54] te Brömmelstroet, M. (2014). "Sometimes you want people to make the right choices for the right reasons: Potential perversity and jeopardy of behavioural change campaigns in the mobility domain." *Journal of Transport Geography*, 39, 141–144.

[55] Gabrielli, S., Forbes, P., Jylhä, A., *et al.* (2014). "Design challenges in motivating change for sustainable urban mobility." *Computers in Human Behavior*, 41, 416–423.

[56] Noar, S. M., Benac, C. N., & Harris, M. S. (2007). "Does tailoring matter? Meta-analytic review of tailored print health behavior change interventions." *Psychological Bulletin*, 133(4), 673.

[57] Dickinson, J. E., Cherrett, T., Hibbert, J. F., Winstanley, C., Shingleton, D., Davies, N., *et al.* (2015). "Fundamental challenges in designing a collaborative travel app." *Transport Policy*, 44, 28–36.

[58] Rehrl, K., Stefan, B., & Hans-Joachim, M. (2007). "Assisting multimodal travelers: Design and prototypical implementation of a personal travel companion." *IEEE Transactions on Intelligent Transportation Systems*, 8(1), 31–42.

[59] Millonig, A., and Konstantin, M. (2014). "Playful mobility choices: Motivating informed mobility decision making by applying game mechanics." *ICST Transactions on Ambient Systems*, 4, e3.

[60] Roughgarden, T., & Éva, T. (2002). "How bad is selfish routing?" *Journal of the ACM (JACM)*, 49(2), 236–259.

[61] Vlahogianni, E. I. (2015). Computational intelligence and optimization for transportation big data: Challenges and opportunities. In *Engineering and Applied Sciences Optimization* (pp. 107–128). Springer International Publishing, Cham, Switzerland.

[62] Barmpounakis, E., Vlahogianni, E., & Golias, I. (2016). "Intelligent transportation systems and powered two wheelers traffic." *IEEE Transactions on Intelligent Transportation Systems*, 17(4), 908–916.

[63] Axelrod, R. (1984). The Evolution of Cooperation. *Basic Books*, Cambridge, MA.

[64] Shaheen, S., Chan, N., Bansal, A., & Cohen, A. (2015). Shared Mobility: A Sustainability & Technologies Workshop: Definitions, Industry Developments, and Early Understanding, Caltrans, Transportation Sustainability Research Center UC Berkeley, November, San Francisco California.

[65] Robèrt, M. (2017). "Engaging private actors in transport planning to achieve future emission targets—upscaling the Climate and Economic Research in Organisations (CERO) process to regional perspectives." *Journal of Cleaner Production*, in press. Volume 140, Part 1, Pages 324–332, DOI:http://dx.doi.org/10.1016/j.jclepro.2015.05.025.

[66] Vlahogianni, E. I., Karlaftis, M. G., & Golias, J. C. (2014). "Short-term traffic forecasting: Where we are and where we're going." *Transportation Research Part C: Emerging Technologies*, 43, 3–19.

[67] Townsend, A. M. (2000). "Life in the real-time city: Mobile telephones and urban metabolism." *Journal of Urban Technology*, 7(2), 85–104. doi: 10.1080/713684114.

[68] Golob, T. F., & Regan, A. C. (2001). "Impacts of information technology on personal travel and commercial vehicle operations: Research challenges and opportunities." *Transportation Research Part C: Emerging Technologies*, 9(2), 87–121. doi: 10.1016/s0968-090x(00)00042-5.

[69] Banister, D., & Stead, D. (2004). "Impact of information and communications technology on transport." *Transport Reviews*, 24(5), 611–632. doi: 10.1080/0144164042000206060.

[70] Cohen-Blankshtain, G., & Rotem-Mindali, O. (2016). "Key research themes on ICT and sustainable urban mobility." *International Journal of Sustainable Transportation*, 10(1), 9–17.

[71] Mondschein, A. (2014). *Re-Programming Mobility Literature Review*. Rudin Center for Transportation Policy and Management, New York.

[72] Yang, K., Zhang, K., Ren, J., & Shen, X. (2015). "Security and privacy in mobile crowdsourcing networks: Challenges and opportunities." *IEEE Communications Magazine*, 53(8), 75–81.

[73] Bargh, M. S., Meijer, R., Choenni, S., & Conradie, P. (2014, October). Privacy protection in data sharing: Towards feedback based solutions. In *Proceedings of the Eighth International Conference on Theory and Practice of Electronic Governance* (pp. 28–36). ACM.

[74] Choenni, S., Bargh, M. S., Roepan, C., & Meijer, R. F. (2016). Privacy and security in smart data collection by citizens. In *Smarter as the New Urban Agenda* (pp. 349–366). Springer International Publishing.

[75] Shirmohammadli, A., Louen, C., & Vallée, D. (2016). "Exploring mobility equity in a society undergoing changes in travel behavior: A case study of Aachen, Germany." *Transport Policy*, 46, 32–39.

[76] Rietveld, P., Rouwendal, J., & van der Vlist, A. (2007). Equity issues in the evaluation of transport policies and transport infrastructure projects. *Policy Analysis of Transport Networks*, pp. 19–36, Routledge, UK.

[77] Ali, K., Al-Yaseen, D., Ejaz, A., Javed, T., & Hassanein, H.S. (2012). CrowdITS: Crowdsourcing in intelligent transportation systems. In: *Wireless Communications and Networking Conference (WCNC)*, 1–4 April 2012, IEEE, pp. 3307–3311.
[78] Karlaftis, M. G., & Vlahogianni, E. I. (2011). "Statistics versus neural networks in transportation research: Differences, similarities and some insights." *Transportation Research Part C: Emerging Technologies*, 19(3), 387–399.
[79] Lei Z., David M. L., & Shanjiang Z. (2008). "Agent-based model of price competition, capacity choice, and product differentiation on congested networks." *Journal of Transport Economics and Policy (JTEP)*, 42(3), 435–461.
[80] Khisty, C., & Kikuchi, S. (2003). "Urban transportation planning education revisited: Reading the dials and steering the ship." *Transportation Research Record: Journal of the Transportation Research Board*, 1848, 57–63.
[81] Zhu, S., Xie, F., & Levinson, D. (2010). "Enhancing transportation education through online simulation using an agent-based demand and assignment model." *Journal of Professional Issues in Engineering Education and Practice*, 137(1), 38–45.
[82] Di, X., Liu, H. X., & Levinson, D. M. (2015). "Multiagent route choice game for transportation engineering." *Transportation Research Record: Journal of the Transportation Research Board*, 2480, 55–63.
[83] Huang, A., & Levinson, D. (2012). "To game or not to game: Teaching transportation planning with board games." *Transportation Research Record: Journal of the Transportation Research Board*, 2307, 141–149.
[84] Chowdhury, M., & Dey, K. (2016). "Intelligent transportation systems—A frontier for breaking boundaries of traditional academic engineering disciplines [education]." *Intelligent Transportation Systems Magazine*, IEEE, 8(1), 4–8.

Glossary

ATS artificial transportation systems
GIS geographic information system
ICT information communication and technology
IoT Internet of Things
ITS intelligent transportation systems

Further Reading

[1] Frith, J. (2013). "Turning life into a game: foursquare, gamification, and personal mobility." *Mobile Media & Communication* 1(2), 248–262.
[2] Koster, R. (2013). *Theory of Fun for Game Design*. O'Reilly Media, Inc, Sebastopol, California, USA.

[3] Marczewski, A. (2013). *Gamification: A Simple Introduction*. Andrzej Marczewski, UK.
[4] McCall, R., Louveton, N., Kracheel, M., Avanesov, T., & Koenig, V. (2013). "Gamification as a methodology: a multi-part development process. Workshop on Designing Gamification." *ACM SIGCHI Conference on Human Factors in Computing Systems (CHI 2013)*, April 27th–May 2nd, Paris, France.
[5] McGonigal, J. (2011). *Reality is Broken: Why Games Make Us Better and How They Can Change the World*. Penguin, London, UK.
[6] Pink, D. H. (2011). *Drive: The Surprising Truth about What Motivates Us*. Penguin, London, UK.
[7] Werbach, K., & Dan, H. (2012). *For the Win: How Game Thinking Can Revolutionize Your Business*. Wharton Digital Press, University of Pennsylvania, Philadelphia, Penssylvania, USA.

Chapter 13

Digital innovations and smart mobility: mapping the value beyond the hype

Hussein Dia[1]

Abstract

The continuous waves in digital innovations over the past few years, and the sweeping changes anticipated by disruptive mobility, have inspired visions of a very different future as well as a good deal of hype. To distinguish between the hype and reality, this chapter discusses the role of disruptive technologies within low carbon mobility networks and describes how digital innovations are changing the urban mobility landscape, enabling real-time measurement and analysis of urban mobility, and promoting sustainable smart cities. The chapter maps the value of data-driven innovations in mobility and draws on practical applications, case studies and modelling results to demonstrate the benefits. The chapter also describes the major behavioural and technological changes, and the new mobility business models that would be required to achieve the desired targets and outcomes. Finally, the chapter provides a reflection of the value derived from typical smart mobility 'use cases'. It then identifies the policy principles which are central to the success of smart mobility, and how it will give our cities an opportunity to modernise transport systems and help drive economic growth and create jobs for the twenty-first century.

Keywords

Smart mobility; smart cities; disruptive technologies; intelligent transport systems; digital innovations

13.1 Introduction

As was discussed in Chapter 2, our cities are playing an increasingly active role in the global economy [1]. Today, just 100 cities account for 30% of the world's

[1]Centre for Sustainable Infrastructure, Swinburne University of Technology, Melbourne, Australia

economy. New York City and London, together, represent 40% of the global market capitalisation. In 2025, 600 cities are projected to generate 58% of the global gross domestic product (GDP) and accommodate 25% of the world's population. It is also expected that 136 new cities, driven by faster growth in GDP per capita, will make it into the top 600 by 2025, all from the developing world, 100 of them from China alone [2]. The twenty-first century appears more likely to be dominated by these global cities, which will become the magnets of economy and engines of globalisation [1–3].

Chapter 2 also described the multitude of challenges facing our cities including congestion, emissions, ageing infrastructure and limited budgets to maintain and upgrade urban transport systems [4–14]. Digital innovations are expected to have a role in addressing these challenges. Already, substantial progress has been achieved using intelligent transport systems (ITS) over the past 20 years. This includes a range of technologies that allowed operators to maximise the efficiency of the infrastructure.

Cities of the future will also include systems that utilise field sensors to detect and respond quickly to operational disruptions. Vital infrastructure downtimes will be cut using sensors that monitor the health of critical infrastructure, collect data on system functioning, alert operators inside an integrated urban control centre to the need for predictive maintenance, and identify potential breakdowns before they occur. Smarter vehicles, trains and public transport systems are increasingly sensing their surrounding environments and enhancing safety in situations where driver error is most common. Technology will also gradually influence car ownerships models, promote more ride-sharing and car-sharing in our cities and increasingly start to blur the line between on-road private and public transport systems.

13.2 The opportunities

The convergence of physical and digital worlds is creating unprecedented opportunities to enhance the travel experience for millions of people and businesses every day. Disruptive and emerging forces – including on-demand shared mobility, big data analytics and autonomous vehicles – are expected to change the mobility landscape and provide travellers with more choices to meet their travel needs while reducing reliance on building additional infrastructure. The coming together of these trends is providing new opportunities to 'sense the city' and unlock operational innovations and access to high-quality urban mobility. Through data mining, artificial intelligence and predictive analytics, smart mobility systems can help city managers monitor the performance of vital infrastructure, identify key areas where city services are lagging, and inform decision makers on how to manage city growth and make our cities more liveable [7].

Cities are essentially made up of a complex network of systems that are increasingly being instrumented and interconnected, providing an opportunity for better infrastructure management [15,16]. An 'Internet of Things' comprising sensors, monitors, video surveillance and radio frequency identification (RFID) tags, all communicating with each other to enhance infrastructure capability and resilience, and capturing volumes of data [16].

Decision makers and leaders who run our complex cities are increasingly recognising the role of smart technologies in improving the efficiency of existing infrastructure and sweating of assets through better utilisation of available infrastructure [17]. These systems can significantly improve operations, reliability, safety, and meet consumer demand for better services with relatively small levels of investment [15].

13.2.1 Smart cities: technology-driven urban infrastructure

The infrastructure marketplace is rapidly changing and is becoming more dependent on technology to manage assets, collect and analyse data, generate and collect revenue, and provide decision makers with real-time information on system performance and operational problems.

The smart infrastructure paradigm includes applications of information technology, data mining, sensors, smart algorithms and predictive analytics to improve the performance of infrastructure systems and asset management in the buildings, energy, health, water and communications fields as shown in Figure 13.1.

The concept of a "smart city" has gained considerable popularity in industry and academia over the past few years. The context in which this concept is used, however, is multi-faceted and is not always consistent. A large number of definitions have been proposed in the literature for what constitutes a 'smart city'. These vary and depend on whether the focus is placed on the "hard" domains such as infrastructure, buildings, energy grids etc. with information technology comprising the digital platform enabling better measurement and monitoring of these systems, or whether the term is applies to "soft" domains such as policy innovations, education, social inclusion where the applications of information technology are not usually decisive [18].

In this chapter, a smart city is defined as "a city that connects the social, physical, economic, and information infrastructure to create a vibrant urban

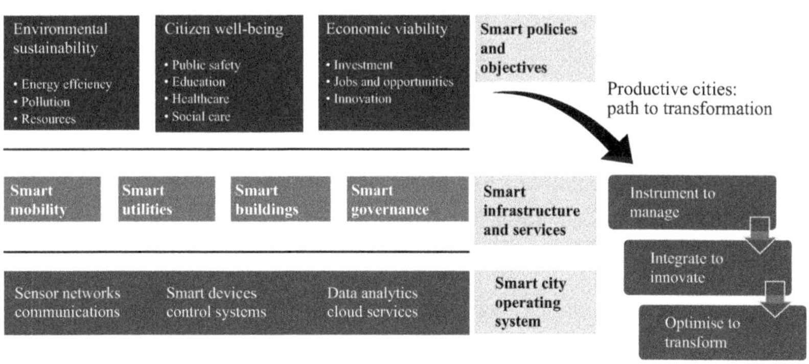

Figure 13.1 Smart cities model. Source: Dia [18]

environment that enhances access to services, places and economic opportunities, and improves the quality of life for its citizens."

Figure 13.1 illustrates that technology should be viewed and used only as an enabler to achieve a city's desired objectives. These objectives include, but not limited to, environmental sustainability (energy efficiency and improved air quality), citizen well-being (public safety, education and social care) and economic viability through providing access to jobs, services and opportunities. To achieve the desired transformation, cities would need to harness the increasingly available amounts of data from infrastructure and crowd-sourced information. To develop insights into this information, the data need to be integrated, fused and mined to establish patterns and trends which can be used to optimise city services and transform cities to meet the desired policies and objectives.

13.2.2 Smart mobility

The impacts of digital innovation have already started to be felt in cities around the world. Technology has brought with it smart phones, connected devices and infrastructure, highly automated vehicles and access to large amounts of real-time traffic data. For the first time in decades, road users have access to increasingly more sophisticated applications to help them plan their journeys and meet the demands of their travel. This fundamental shift offers consumers a real choice based on real time and in some cases predictive information that allows users to plan ahead and anticipate delays and disruptions to their commute. New technologies also offer consumers mode choices based on comparative pricing and current network status. As transport operators change their business models, and new start-ups and technology providers enter the urban mobility landscape, new business models will continue to transform the use of information, payment, integration and automation [19].

Providing access to high-quality urban mobility services requires a variety of planning and operational innovations, as well as better understanding of travel behaviour, operational processes and the factors which affect these issues. The last 5 years, in particular, have seen rapid developments in technology and a marked shift in the thinking towards the provision of mobility solutions to meet people's demand for travel in urban areas (see Table 2.1 in Chapter 2).

Smart mobility essentially includes systems that are used to provide seamless, efficient and flexible travel across all modes of transport. This is illustrated in Figure 13.2 which shows a number of elements which make up smart mobility including instrumented smart infrastructure, ITS, operational and strategic modelling. In addition, it also includes some of the emerging disruptive mobility solutions including mobility-as-a-service (MaaS) and the anticipated autonomous shared mobility-on-demand services.

In practice, this would translate into smarter connected vehicles, trains and public transport systems which would increasingly sense their surrounding environments and enhance safety in situations where driver error is most common [6].

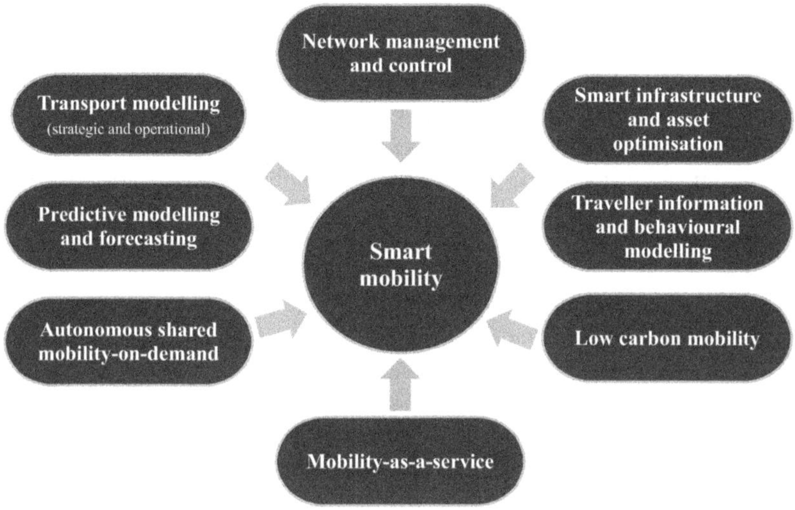

Figure 13.2 Smart mobility model [18]

For example, on-board public transport, a range of GPS, position fixing, video surveillance and communications equipment are increasingly providing more accurate and reliable multi-modal real-time passenger information, resulting in better informed travellers and ensuring a smoother, safer and more reliable experience for customers [7]. Back-office systems that leverage sensors, web, mobile and GPS technologies are increasingly utilising smarter algorithms, data mining and predictive modelling tools to reduce delays to passengers by optimising schedules and capacities in real time [7]. Near railroad level crossings, a range of train-to-infrastructure and train-to-vehicle technologies are improving passenger safety by better detection of fast approaching vehicles and providing warnings to avoid collisions [20]. Electric vehicle charging infrastructure will also increasingly being integrated into smart grid networks, providing consumers with access to sustainable and equitable forms of connected mobility [21]. A combination of technologies and sensors will also improve safety and security by permitting operators to remotely disable or enable a public transport service in the event of a security threat (e.g. an unauthorised driver).

13.2.2.1 Disruptive technologies

Disruptive forces are increasingly changing the mobility landscape and providing consumers with more choices to meet their travel needs while reducing reliance on building additional infrastructure. Although some of these technologies are still a few years away (e.g. self-driving vehicles), they have already started to shape a vision for a mobility transformation driven by a number of converging forces including vehicle electrification, automated self-driving, mobile computing, on-demand shared mobility services [18], mapping and predictive analytics.

The coming together of these powerful trends, and the ability to monitor and control things in the physical world electronically, have inspired a surge of innovation and enthusiasm about the future of urban mobility. Some of the most notable disruptions that are either being introduced or planned for cities include the collaborative on-demand shared mobility services and autonomous vehicles. These are aimed at improving the passenger experience, access to jobs, services and opportunities, reducing both capital and operations costs and enhancing the safety, resilience and efficiency of city services.

13.2.2.2 Autonomous vehicles

Vehicle automation is part of a much larger disruption in automation and connectivity. Chapter 11 provides a comprehensive coverage of the technologies and the anticipated impacts on mobility [22–26]. Vehicle autonomy will increasingly play a bigger role in the mix of travel options available to travellers in the future, although their benefits will not be fully realised until the opportunities are well developed for fully autonomous vehicles. Only then, when the human driver is no longer included in the costing of trips, will they reap their potential benefits. As was discussed in Chapter 11, their advent may also bring some negative impacts if they are not planned for as part of a holistic approach towards low carbon mobility.

13.2.2.3 Collaborative shared mobility

To date, the transport sector is the most funded industry in the Collaborative Economy [27]. This is manifested by one of the most promising trends within the disruptive mobility space, known as 'Mobility-as-a-Service' as shown in Table 13.1 [28]. The key concept behind MaaS is to place the road users at the core of transport services, offering them mobility solutions based on their individual needs (Chapter 10).

This means that easy access to the most appropriate transport mode or service will be included in a bundle of flexible travel options for end users. MaaS has the potential to fundamentally change the behaviour of people and reduce reliance on car ownership by providing easy on-demand access to the mobility services they need [29]. The trend is therefore gradually shifting, particularly in the context of smart cities, from the provision of urban transport networks, that is buses, trams and trains, to a focus on what people require, and how a more considered and integrated approach could produce far better outcomes [30]. The key advantages of these new services include reliability, predictability, convenience and ease of accessibility. Most services also offer easy and secure payment options using cashless mobile transactions. For a given trip, consumers can select from different types of services based on trip distance, waiting and travel times and levels of service [31].

Self-driving vehicles will have a dramatic impact on urban life especially when they begin to blur the distinction between private and public modes of transport. By combining ride sharing with car sharing, it may be possible to meet people's need for travel with much fewer vehicles. As more sensors and intelligence saturates our cities, it will become possible to collect more accurate real-time information, seamlessly, on every dimension of urban life. Consider HubCab, for example,

Table 13.1 New urban mobility services

	Traditional mobility solutions	**New mobility services**	
Individual-based mobility	Private car ownership	Car sharing: peer-to-peer	A peer-to-peer platform where individuals can rent out their private vehicles when not in use (e.g. Turo)
	Taxi	E-hailing	Process of ordering a car or taxi via on-demand app. App matches rider with driver and handles payment (e.g. Uber, Lyft)
	Rental cars	Car sharing: fleet operator	On-demand short-term car rentals with the vehicle owned and managed by a fleet operator (e.g. GoGet, Car2Go, ZipCar, Getaround)
Group-based mobility	Car pooling	Shared e-hailing	Allows riders going in the same direction to share the car, thereby splitting the fare and lowering the cost (e.g. UberPool, LyftLine)
	Public transport	On-demand private shuttles	App and technology enabled shuttle service. Cheaper than a taxi but more convenient than public transit (e.g. Bridj)

Source: Adapted from [28].

which is a web-based interactive visualisation tool that looks at how New York's 170 million annual taxi trips connect the city [32]. Gradually, it is becoming easier to see precisely where, how and at what times different parts of a city become stitched together as hubs of mobility. By using these technologies, decision makers can begin to unravel the complexity of travel patterns and identify how to reduce the social and environmental costs embedded in transport systems. HubCab, as an example, was developed to target taxi services as a way to understand the linkages between travel habits and the places people travel to and from most often. The same principles can be applied to optimise the performance of autonomous on-demand mobility systems.

Access versus ownership and the collaborative economy

The disruptive mobility trends have the potential to fundamentally change the relationship between the consumer and automobile. The rise of the collaborative or sharing economy, popularised over the past few years by companies such as Airbnb, Zipcar and Uber, looks set for rapid growth over the next decade [33]. Access to mobility rather than to car ownership will enable customers to be more selective in choosing from the door-to-door mobility services offered by 'mobility operators' for intercity, suburban as well as 'last kilometre' travel solutions. Today, consumers use their cars as 'all purpose' vehicles regardless of whether they are commuting alone to work or enjoying a leisurely drive with the family to the beach. There are already early signs that the importance of private car ownership is declining, and shared/collaborative mobility is increasing. With mobility offered as a service, consumers will have the flexibility to choose the best solution for a specific trip purpose, on-demand using their smart phones. According to a recent report [33], the shift to diverse mobility solutions is likely to result in 10% of new cars sold in 2030 to be a shared vehicle. This will reduce private-use vehicle sales but at the same time increase utilisation of shared vehicles. This will lead to a faster replacement rate of shared vehicles which may offset some of the reduction in private-use vehicles. The report also estimates that more than 30% of kilometres driven in new cars sold could be from shared mobility.

Opportunities through technologies

The shift towards embracing collaborative mobility is driven by the recognition of the potential opportunities, and risks, generated as a result of the interaction between existing and emerging disruptive technologies [33]. The use of collaborative mobility is seen as a way of harnessing new developments, like big data analytics and the Internet of Things, to mitigate against the effects of increased congestion and pollution resulting from continued urbanisation [22,34].

With population growth expected to add an additional billion people to the world's inhabitants within the next decade, and with 60% of these people expected to live in cities, transport operators are faced with a situation where traditional methods will no longer be sufficient to restrain congestion [29]. Under these conditions, increasingly large amounts of effort are being diverted into alternate means of providing mobility [28]. In addition to the efforts by transport authorities, who are already struggling to keep pace with existing congestion levels, the predicted migration of the earth's population towards its cities will cause congestion to spread to regions that have never before experienced traffic issues [29]. The following section reviews some of the opportunities and trends and identifies ways that they have potential to influence the drive towards collaborative mobility.

Digitalisation and the Internet of Things

Since the conception of collaborative mobility, attempts to organise large-scale programmes have been hampered by the difficulty of establishing an effective central administration [35]. The underlying premise behind any collaborative mobility enterprise is to increase transportation efficiency by increasing vehicle utilisation. This cannot be easily achieved without strong centralised control of the vehicle fleet [35]. Early examples of failed attempts included the Procotip (Montpelier, France, 1971) and Witkar (Amsterdam, 1973) shared car schemes. Both of these schemes made compact small vehicles available for use in and around the congested inner city areas using a payment token system. Both these schemes failed because of inefficiencies in vehicle distribution caused by a lack of control over the vehicle fleet, resulting in poor asset utilisation and excessive downtime between each trip [35].

At the same time, the advent of the 'Internet of Things' – whereby people and physical objects are constantly interconnected to the Internet – has also exacerbated effective control of such undertaking. Operators are now provided with unprecedented levels of data, including real-time analysis on everything from vehicle positions to weather patterns. With these data available, the challenge facing transport decision makers is no longer about gathering information, but how to use the information available for decision making in a meaningful and timely manner. In Finland, the Kutsuplus bus network received constant updates on traffic conditions, allowing drivers to provide updated estimates of arrival times to their passengers, often accurate to within 30 s [36–41]. Around the world, software developers have developed tools to combine information on every transport option available in a given city. Users simply enter their origin and destination, and the software would provide them with a route optimised for cost, travel time, comfort or environmental efficiency [41].

The MaaS and shared mobility trends are breaking down the boundaries between different transport modes. This is largely due to the fact that technology is creating an intermediate level between the different means of transport and their users, which is made possible by the fusion of data into a new data layer. This is introducing new challenges to the regulatory and policy frameworks of transport. For the users, the focus will no longer be on the transport mode, but rather on mobility which will increasingly become an information service with physical transportation products, rather than a transportation product with services [41]. In addition to ensuring public safety and access to services, regulators and policy makers must also oversee other aspects such as interconnection, interoperability, capacity management, standards and security. To achieve this, regulators must focus on the establishment of new comprehensive regulatory frameworks which would enable the use of information technology especially across the different transport modes. In particular, this will imply the need to develop an outcome-focused regulatory framework which (1) puts the users at the centre of the new mobility system, (2) encourages innovation and promotes safety, security, social equity and environmental sustainability. In the short term, MaaS will also provide public agencies with the opportunity to bring innovation to their own transport services. In the long term, public agencies may need to rethink their role and consider opportunities for public–private partnerships and service agreements with private mobility providers [31]. This new perspective will allow policy makers to focus on the regulation of the new data layers as well as the interface between the data layers and the physical transportation services.

In many cities in the United States, public agencies are increasingly considering MaaS as an opportunity to provide more mobility options and also address the first-and-last-kilometre problem, particularly during late hours of the night and in low density areas. Few cities are already collaborating with mobility providers such as Bridj and Uber to supplement and strengthen their existing public transport offerings and are viewing the technology and mobility providers as partners rather than competitors. Some examples from Tampa, Dallas, Atlanta, Memphis and San Francisco demonstrate how this can be achievable and cost-effective [31].

Another challenge to policy makers is protection of privacy. Those who have access to the data (and especially to the way the data is conveyed to the end-users) control the information and have immense power. The abuse of such data and information can result in market distortions, security risks and diminished privacy protection [41].

The arrival of shared autonomous mobility-on-demand systems is also likely to introduce major challenges to policy makers [42–47]. Regulations will play a key role in the emergence and development of autonomous vehicles and the new business models around them such as autonomous MaaS. They are also likely to be the biggest hurdle for the deployment and public acceptance of these services. The regulators are already adapting and rethinking their approaches on this issue to avoid stifling the innovative uses of these technologies. Effective responses require an early and on-going dialogue between regulators, developers and the public in which regulators would create legal frameworks that are flexible but robust. An important role for the regulators will be to limit physical risks especially those that might be posed during interim years when legacy fleets of cars would interact with the autonomous vehicles that are offering the MaaS solutions.

13.3 The benefits: more for less

Adoption of technology-based customer-centric approaches have the potential to introduce substantial improvements in customer satisfaction and create a shift in attitude to cost and value. A smarter city will mean better access to sustainable forms of transport; electricity and drinking water that can be counted on; and energy-efficient buildings resulting in enhanced standards and quality of life for today's increasingly empowered citizens and consumers. In addition to improving asset utilisation, these systems will also improve safety and security, enhance business and operational outcomes, boost productivity and economic growth, and deliver environmental benefits.

Given the maturity levels and affordability of smart technologies, these benefits can be achieved at a fraction of the cost of investment in new infrastructure. In a study for the Australian Market, Access Economics [48] reviewed the potential economic benefits from the adoption of smart technologies in transport, electricity, irrigation, health and broadband communications. The report examined how smart systems will allow the use of vast amounts of data collected in all areas of city activity far more effectively, providing the potential to radically alter our economy and society for the better. Their research demonstrated that smart technologies would have significant benefits including a 1.5% increase in GDP and increase in the net present value (NPV) of GDP by $35–80 billion over the first 10 years. In another report prepared by the Climate Group [49] on behalf of the Global e-Sustainability Initiative, it was estimated that a 15% reduction in emissions can be realised in 2020 through smart technologies that achieve energy and resource efficiency using adaptive and proactive technologies.

Globally, the benefits of investment in smart infrastructure is highlighted by research from the McKinsey Global Institute (MGI) which looked at how modernising our infrastructure to drive economic growth will require costly future investment in new projects [50]. The MGI study, which looked at the projected global infrastructure investment over the next 17 years, estimated that just keeping pace with projected global GDP growth will require $57 trillion in infrastructure investment between now and 2030. Nearly 60% more than the $36 trillion spent over the past 18 years and more than the estimated value of today's infrastructure.

Given widespread fiscal constraints, even assembling the minimum investment required to meet growth predictions is going to be a challenge. So rather than investment in new projects, governments should look to address some of their infrastructure needs by getting more out of existing capacity by adoption of smart infrastructure technologies. Boosting asset utilisation and expanding the use of demand-management measures can produce a decisive difference if scaled up globally and can result in savings of up to $400 billion a year according to the MGI research.

The literature on smart transport technologies is abundant with case studies which demonstrate the benefits of ITS [50]. In addition to reducing reliance on building additional capacity, smart infrastructure technologies also offer a much lower capital and investment cost. For example, the cost of the transport technology solution on the United Kingdom's M42 motorway was $150 million and took 2 years to implement; widening the road to produce the same outcome would have taken 10 years and cost $800 million [50]. Other studies have also suggested that using smart transport technologies for roads, rail, airports and ports can double or triple asset utilisation. The average cost–benefit ratios for each category of road improvement project are displayed in Figure 13.3.

13.3.1 Case studies – mapping the value beyond the hype

The ability to reduce congestion, cut travel times, improve trip efficiency, save petrol and reduce pollution through the use of collaborative mobility has long been

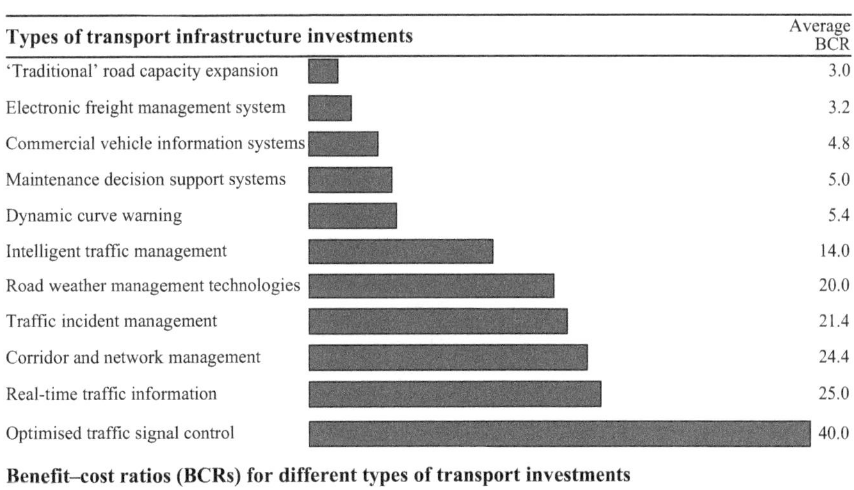

Figure 13.3 Comparison of cost–benefit ratios for different road investment strategies. Source: Dia (adapted from [50])

understood [35]. The addition of high funding levels and the potential to create an entire new consumer market within the transport sector offers a tempting reward for anyone who can unlock the potential of collaborative mobility [33]. With the capacity for profit, combined with environmental and social benefits, a wide range of international initiatives have been launched, with varying degrees of success, in an attempt to harness its potential [51]. The following section will analyse some of the key collaborative mobility projects that have been undertaken around the world. It will focus on identifying success and failure factors and will try to identify lessons that may be learnt from each case.

13.3.1.1 Early beginnings

Although recent advances and breakthroughs in technology have helped the push towards 'mobility as a service', the potential benefits have long been recognised [35]. Two examples of early experiments of collaborative mobility are presented next.

Purdue University – Mobility Enterprise (1983)

In January 1983, Purdue University began its Mobility Enterprise, which provided families who joined the scheme with a MAV (minimum attribute vehicle or 'small car') and access to a shared pool of larger vehicles. The aim of the project was to use the availability of shared cars to enable drivers to choose a fit-for-purpose vehicle for each trip, rather than having to use an all-purpose vehicle for every occasion. It was expected that this would increase the efficiency of the vehicle-kilometres travelled and reduce fuel consumption. The experiment ran for two and a half years and incorporated an average of 12 households. To encourage people to join their experiment, the Mobility Enterprise had to be competitive when compared with private car ownership. The results from this initiative showed a significant reduction in fuel consumption, but no change in the driving habits of those involved. The researchers found that they were not able to significantly influence the driving behaviour of those involved. However, the project was able to demonstrate that fuel consumption could be reduced through use of fit-for-purpose mobility and that a tangible monetary saving can be used to motivate change [35].

San Francisco – STAR Project (1983–85)

The STAR (short term auto rental) project was undertaken in the San Francisco area between 1983 and 1985. The project identified an apartment complex of 9,000 residents that was well serviced by public transport, and established a car rental dealership within the precinct which provided competitively-priced short-term rental agreements. The aim of the project was to demonstrate that it was possible to live in an American City without owning a motor vehicle. The study found that since the public transport services were well utilised during the commute to and from work, many of the trips made by private vehicle were discretionary in nature. For these discretionary trips, the presence of a rental dealership would allow drivers the choice of a fit-for-purpose vehicle, providing fuel savings and increased vehicle utilisation. The project was a partial success, with those who made infrequent trips electing to forgo private vehicle ownership (and in some cases saving as much as $1,000 a year). But the project was limited in that it was only utilised for discretionary, non-work related trips.

13.3.1.2 New beginnings

In recent years, the convergence of technology and infrastructure has renewed interest in shared mobility. Some of the emerging trends in this space are covered next.

Car-sharing services

As most private vehicles are only utilised less than 10% of the time, shared car services (Chapter 10) aim to increase vehicle utilisation and efficiency by maintaining a fleet of vehicles that can be accessed on either a subscription or hire basis [28]. Global car sharing provider Zip Car estimates that due to the increased utilisation of the vehicles, each of the shared vehicles can replace as many as 15 ordinary vehicles, though it should be noted that most Zip Car users drive between 60% and 80% less than the average motorist and this allows each shared car to provide for a greater number of people than would otherwise be the case [30]. Surveys of the users of Autolib, the French car-sharing operator revealed that of those who did not own a private vehicle, 70% listed Autolib as the reason that they had been able to move beyond private car ownership [30]. In recent years, some shared car providers have been developing new additions to their services and forming new partnerships to further realise the potential of collaborative mobility [52]. In an attempt to increase the attractiveness of car sharing in inner-city environments, companies are attempting to provide services that facilitate short term, impulse use of shared cars. BMW's Drive now programme uses flexible hire services to accommodate short distance trips. Proving initially to be successful, the scheme has now been established in five German cities as well as London and San Francisco [53]. Ford's City Driving on-demand scheme follows a similar approach and utilises a pay-by-the-minute hire scheme with the option for one way trips [53]. In Germany, the car-sharing operator Flinkster reached an agreement with Ford to have access to Ford's existing network of car dealerships as depots for the car-share scheme. In doing so, Ford and Flinkster have created a nation-wide car-share network without the capital for purchasing land and storage facilities [53].

Two other schemes currently being trialled by Ford are for non-corporately controlled car sharing. The 'Car Swap' register is an internal experiment where a registry has been compiled of Fords employees in the Dearborn area in the United States of America. The register includes people who are interested in car sharing their vehicles. If someone who is part of the scheme needs a vehicle with some particular feature, a search of the registry would provide a list of people who can be contacted to acquire the use of that vehicle. Ford's second user-managed scheme 'Share Car' is currently in the development stage, with a team in Bangalore partnering with Zoomcar to establish the legal framework that would allow a group of people to jointly share ownership of a private vehicle [53]. A final scheme from the Ford motor group is the Remote Positioning project being undertaken in Atlanta, USA. The project is aiming to perfect the use of drone vehicles that can be remotely controlled from a centralised facility, as shared vehicles. The use of these vehicles would allow a shared car operator to both deliver a vehicle to the customer's door and retrieve used vehicles from any point without the cost and time involved in physically ferrying valet drivers backwards and forth between jobs [53].

E-hailing

The largest, most heavily-funded and most well-known on-demand mobility providers are the e-hailing services such as Uber, Lyft and Didi [30]. Using private drivers who supply their own vehicles (to minimise the capital cost associated with traditional fleet operations), these companies operate an on-demand taxi-like service in most major cities around the world [30]. In particular, Uber is rapidly destabilising the traditional taxi industry [31,38] with recent reports showing that together with competitor ridesharing companies, they have captured more than 50% of the taxi market share for business trips. Uber also has a ridesharing variant of its service known as UberPool [54] which operates in a number of cities including London. Statistics released by Uber claimed that, after 6 months of UberPool being introduced to London, the system has saved 700,000 vehicle miles (1,120,000 km) or 52,000 l of fuel [55]. This saving translates into a reduction of 124 t of CO_2 emissions [54]. Uber also claimed that in some cities, over half of the trips are now being made using UberPool [55].

Public transport innovations

Although most cities are continuing to push ahead with the expansion of existing conventional public transport networks (Beijing alone constructed over 230 mi (380 km) of subway between 2008 and 2015), the world is increasing seeking ways to diversify the current transport options [28]. Active transport (walking or cycling) has recently been gaining acknowledgement as a legitimate transport option, with bikeshare schemes flourishing around the world [28]. In Paris, these two trends have been combined, with the city including its bikeshare service under the umbrella of its public transport network [28]. Probably the most innovative approach to public transport in recent times is the on-demand buses such as Kutsuplus and Bridj described next.

Kutsuplus – on-demand public transport

The city of Helsinki in Finland has been praised for its 'ambitious' plans and target to overcome the need for private car ownership by 2025. The Kutsuplus mobility experiment initiated is seen by many as the flagship project leading this attempt [40]. Hailed as the first true on-demand public transport service, the Helsinki Regional Transport Service ran a technology-driven minibus service which utilised advanced algorithms to assign vehicles based on passenger demand on a real-time basis [51]. The initiative entered active service in 2012 with a fleet of three dedicated minibuses. Kutsuplus allowed commuters to specify an origin and a destination point (within a defined service area), and the algorithm then identified a minibus travelling in that direction and instructed its driver to pick up the new passenger [39].

The Kutsuplus experiment was launched with two initial goals: (1) test the technological feasibility of using computer-based routing algorithms to overcome the difficulty of maintaining effective control over mobility; and (2) measure public support and willingness to pay for on-demand public transport [51]. The service received strong public support with positive feedback and continual growth in ridership figures [51]. A survey of user satisfaction recorded an overall satisfaction rating of 4.7 out of 5, 10% higher than the satisfaction rating received by Helsinki's conventional public transport networks [39]. An important finding from this experiment, which is relevant to other collaborative mobility operators, is that people were willing to accept longer travel times during a journey more readily than they would accept longer waiting times at the start of a journey. This finding helped the operators to revise the algorithm used for bus routing to minimise waiting times on pickup which resulted in a considerable increase in the number of users [39].

Although the cost per trip was reported to be lower than that seen in conventional bus services [39], the Kutsuplus bus service was deemed a financial strain on the public purse and was discontinued at the end of 2015 [37]. This was contrary to public expectations of seeing the service expanded into the future [51]. To increase ridership, the system needed to provide a more attractive service covering a larger service area, longer operating hours and shorter waiting times at pick-up. Although some improvements in service quality were achieved, larger scale increases in the services required the purchase of extra vehicles [39]. Data collected during the trial showed that every time the service was expanded, the increase in passenger numbers improved the efficiency of the entire system because a higher density of users allowed the computer algorithm to plot a more optimised route with less meandering to pick up lone passengers [39]. To that end, it was reported that a 31% increase in the system's capacity would result in an average of a 60% increase in revenue [39].

Other examples of on-demand public transport

The almost success story of Helsinki's on-demand minibus service has drawn interest from around the world [37]. Kutsuplus demonstrated to both governments and individuals that on-demand collaborative mobility was challenging but feasible from a technology perspective [39]. Since Kutsuplus' launch, cities around the world have begun to adopt similar on-demand shared mobility solutions. Many of these schemes are directly modelled after Kutsuplus, like the Via Transportation and Chariot systems [29]. Other initiatives, however, have used Kutsuplus as a base framework but have then tailored the system to overcome identified weaknesses and meet the needs of the operator [37]. An example is the Bridj service which includes a number of cities and projects in the United States including Kansas City (Missouri) where Bridj partnered with the city to provide a partial on-demand service. The city-owned Bridj busses follow a rough route in accordance with an established timetable (like in a traditional bus service) but will respond to customer request for pick up along that route [37].

In both London and New York, Ford has launched Dynamic Social Shuttle, its own on-demand minibus scheme with a focus on minimising response time.

Ford aims to use this project to gain data on shared mobility patterns and better understand the social dynamics that are driving shared mobility [53].

The Split project in Washington, DC, uses the same software as the Kutsuplus mini busses to optimise vehicle routes and travel times. However, it has succeeded in overcoming the growth difficulties that crippled its predecessor by using private vehicles for its fleet. This Uber-style approach where drivers provide their own vehicles reduces the financial barrier to growth which prevented Kutsuplus from achieving the scale of operations that would have allowed it to function effectively [53].

In Bangalore, rather than using dynamic pricing to affect the cost of bus fares throughout the day, the bus operators issue a raffle ticket to every person who travels on a bus outside peak hours, with cash prizes for the lucky winners. The scheme doubled the pre-peak ridership and reduced peak hour travel times by 24% at little cost to the bus operator [30]. As a point of comparison, Singapore's attempt to influence pre-peak ridership on their rail lines achieved a 7% shift in passenger numbers by making all travel before 8 a.m. free [30].

Crowd-sourcing and data fusion

One of the key challenges facing mobility providers and data analysts is how to combine and fuse data streams from separate, often incompatible sources and combine them into a meaningful, user-friendly and cohesive set of information [56]. The computer algorithm used to direct the Kutsuplus busses was one of the first to combine user requests with spatial positioning and real-time traffic reports to control route planning and predict accurate estimates of arrival times [39]. The success of this computer algorithm has paved the way for a new era of informed journey planning [37]. By combining multiple information sources into one interface, tools like London's 'CityMapper' application (now expanded to cover 10 EU cities) can increase the efficiency of people's mobility by allowing them to view all of their options in real time and optimise their journeys [56]. Other initiatives such as Moovit and Moovel are also being used around the world with each system offering some unique features [30]. For example, the Moovit App augments its data collection by utilising customer feedback to increase the accuracy of its route recommendations [30]. The WAZE App, while not a journey planning app, advises drivers of traffic congestion and disruption in real time, guiding their vehicles around hotspots to reduce travel time and save fuel [28].

13.4 The concerns

Although there is considerable support for the benefits of embracing the possibilities that smart mobility has opened through the availability of big data and the internet of things, international governments and service providers have proved slow to adopt these initiatives [7]. Although some of this delay has undoubtedly been caused by conservative decision makers reluctant to trial new technologies, this pragmatic view can in the long term hinder rather than help with adoption and uptake [6]. As with all new emerging technologies, there also exist community concerns that must be addressed in order to accelerate the wide-scale acceptance of the technology into society [17].

Research into the application of smart mobility has identified three main areas of concern that need to be addressed. These include legislative support, software resilience and personal privacy [16].

13.4.1 Legislative support

It is often argued that the current legislative frameworks provide insufficient guidance in dealing with the potential of where smart mobility and smart cities could lead. This lack of legislative support is understandable as the rapid advancements in technology that have been made in recent years have left the traditional legal policies out of date and insufficient to deal with the realities of big data and the interconnectedness of things at a city-wide level [56]. Although some law-making bodies are beginning to investigate and legislate (for or against) smart mobility (e.g., Europe is considering laws making communications software with inbuilt automatic distress functions mandatory in all new vehicles), the issue is still largely in limbo [11,57]. While the uncertainty regarding the legality of new policies and technologies remains, it will be difficult to formulate an effective plan for the future [3]. For example, this uncertainty is highlighted by the development of legislative governance surrounding self-driving cars in America. Out of fifteen states to debate laws allowing the trial of autonomous vehicle's, nine states (60%) rejected the proposal [57].

13.4.2 Software resilience

Although smart mobility is designed to improve the resilience of transport infrastructure by responding to disruptions in real time, its success (or in fact its entire operation) is based upon the quality decision making software and the quality of the digital network that supports that software [15]. Some researchers have expressed some valid concerns that the concept of smart mobility is based upon brittle foundations and that as more and more items are added to the existing internet of things (from which all smart mobility initiatives draw their decision making data), the likelihood of a software malfunction or incompatibility error compromising some or all of this web is real [16]. They argue that such service failures (power blackouts, loss of mobile networks, poorly handled version updates) have been a traditional part of life, but the interconnectedness of the modern world means that failure of one point could (if the system has not been properly designed) have flow-on effects for the rest of the smart city structure [16]. Although the obvious remedy for this concern is the careful, methodical, construction of a city's data collection web, the prohibitive cost of establishing duplicate infrastructure will mean that software developers will be working with the existing systems of dispersed data collection methodologies and all of the data compatibility issues that it implies [58].

Added to the concerns of unnoticed errors is the problem of cyber-attack, where an external party chooses to cause the software to fail [59]. In an age of growing tension and mistrust, this is a very real and present danger [16]. The issue that must be resolved with any smart system is simply: When this system fails, how much of the smart city will it take with it [16]?

13.4.3 Privacy

In the age of the internet of things, personal data are being created, transmitted, tracked and recorded at an unprecedented rate [58]. Take mobile phone data as an example. Even without the benefit of triangulation, a single tower can place the location of an individual's mobile phone to within a few kilometres. With triangulation, this radius can be decreased to as little as 50 m. If the individual's phone is GPS equipped, then the bearer's position can be accurately identified to within less than 5 m [56]. On its own, this would only be a mild concern. But in an interconnected world, everything from credit cards, public transport tickets, Wi-Fi strength from a passing café etc. leaves a recordable trace [58]. It is this level of detailed, real time, location-based data that are both the greatest strength and greatest cause for concern of a smart mobility system [16]. The main reason is that it is very hard to collect location-based data in an anonymous fashion. If a data set contains an individual's movement patterns for the last 3 years (where they live and work, which road they use, their times of departure), will it really matter if the data set does not use their name [56]?

In the past this information, whilst still recorded, was difficult to accumulate as it was obtained by different parties for different purposes, often using incompatible methods of data collection and storage [16]. However, this has begun to change with many governments recognising that to access the full benefits of smart mobility in a smart city requires a previously unknown level of data interconnectedness [17]. This level of interdependence, many now believe, can only be realised by the creation of a single-decision making entity with direct access to all of the data in and around its area of responsibility, for example Rio's City Control Centre in Brazil [16]. It is the creation of these urban intelligence centres that allows for the greatest real time, big data benefits and the greatest potential loss of individual privacy [16]. For this reason, it is the responsibility of the operators and decision makers to operate them with transparency [7]. For people to receive the full benefits of what a smart city has to offer, they need to be able to feel that their data is safe [56].

13.5 Policy lessons

A number of studies which examined the promise of smart mobility have repeatedly identified three nations as global leaders: South Korea, Japan and Singapore [60]. One of the most notable aspects of mobility in these countries is how widespread their smart technologies were, and the impact they had on the quality of life of people in terms of ease of travel, reduced congestion, and improved safety and reliability of transport services. These studies provide a unique insight into these countries' remarkable journeys in smart mobility and the factors which contributed to their success (Figure 13.4). Although non-policy factors such as geographical constraints, cultural and political issues played some role, it was ultimately their national ITS policies which made them successful in smart mobility innovations. The recurring policy themes in these leading nations are summarised below. These policy principles apply both to nations embarking on new smart infrastructure projects, as well as developed economies looking to make the most of existing assets.

National vision. South Korea, Japan and Singapore have all demonstrated a national level commitment to smart mobility. From the outset, their governments articulated and owned a clear vision for ITS and linked it to national Information Technology policies and long-term strategies for improving road safety and the quality of life for their citizens. Governments in these countries also demonstrated strong leadership in convening relevant stakeholders and spearheading implementation.

Commitment to funding. As a percentage of GDP, South Korea and Japan each invested more than twice as much in smart mobility than the United States. Not surprisingly, this level of annual spending (around 0.016% of GDP) allowed

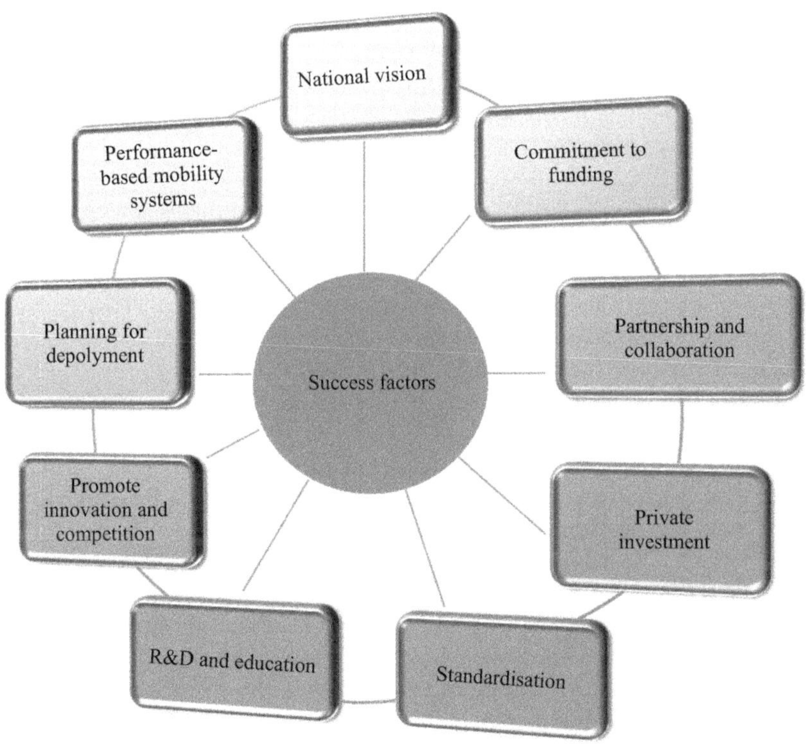

Figure 13.4 Policy principles for deployment of smart mobility. Source: [18]

South Korea to provide 100% coverage of ITS on all expressways (around 4,000 km), and 20% ITS coverage on national roads (2,500 km out of 13,000 km).

Partnership and collaboration. The public and private sectors in these leading nations played an important role in co-developing platforms that enabled government, industry, academic and professional associations to collaborate on developments at both local and national levels.

Private investment. The three leading countries were all successful at forging public private partnerships within their nations and viewed their investments as creating a platform through which the private sector can develop value-added products and services.

Standardisation. Leading countries developed national ITS architectures which provided the basis for interoperable ITS applications and assisted in the delivery of consistent, cohesive and cost-efficient services to citizens. For example, establishing common standards for electronic toll collection (ETC) in Japan encouraged high market penetration and uptake of on-board devices in more than 70% of vehicles. In Singapore, the single national standard of ETC also facilitated the implementation of a city-wide congestion charging scheme from early 1998.

R&D and education. The three leading nations recognised from early stages that ITS will not reach critical mass unless they commit to funding large-scale research and demonstration projects. For example, the 2025 ITS vision, announced by the Japanese Cabinet in June 2007, articulated policies on R&D and set a goal that: 'By 2025, ITS will have been constructed that integrate vehicles, pedestrians, roads, and communities; and that have made traffic smoother, and almost eliminated all fatal traffic accidents'.

Innovation and competition. This policy principle recognises the role of the private sector in developing and making available to governments and citizens innovations and technologies to improve their lives. Examples include alignment between the transport and telecommunications industries, where ITS was recognised as being inseparable from wireless technologies (for cooperative mobility applications) and high-speed networks (for video transmission).

Planning for deployment. A big portion of the funding available to these countries was allocated to supporting ITS technology development, test-beds and proof of concept demonstrations as a precursor to wide-spread deployment. This approach also helped to inform and educate their citizens about the tangible benefits of smart technologies in transport.

Performance-based mobility systems. Nations leading in ITS recognised the need to move from a political or jurisdiction-based system of allocating transport investment to one that uses performance and cost–benefit analysis as the basis for investment decisions. ITS promotes this principle by providing quality data needed to make sound performance-based investment decisions. These data can also be used by the private sector to provide value-added services. These nations also recognised smart mobility and ITS as a 'force multiplier'. Decision makers in these nations were informed to recognise the importance and high benefit–cost ratios provided by ITS.

13.6 Conclusions

The influence of digital innovations is changing the road user and traveller experience, and setting high expectations that will shape the future demand for transport services.

Cities around the world are anticipated to benefit from the use of smart technologies, but they must first overcome a number of challenges to improve infrastructure resilience and reliability. The deployment of these technologies, complemented by appropriate governance and regulatory changes, will deliver substantial benefits through improved city management systems, better informed consumers and enhanced connectivity between vital infrastructure systems. Although decision makers and leaders who run the world's cities are increasingly recognising the role of smart technologies in 'sweating of assets', deployment at a global scale is still in its infancy.

As this chapter has shown, the benefits of investing in smart systems are compelling, particularly given the improvements that could be made in terms of providing innovative solutions to lift our economic efficiency and living standards. To spur change programmes and capture potential savings, cities must move beyond a project-by-project view and upgrade systems for planning, operating, and delivering smart infrastructure. This sort of investment will give our cities an opportunity to modernise their infrastructure and help drive economic growth and create jobs for the twenty-first century.

References

[1] Dobbs, R, Remes, J, Manyika, J, Roxburgh, C, Smit, S & Schaer, F 2012, *Urban world: cities and the rise of the consuming class*, McKinsey & Company, The McKinsey Global Institute (MGI), Seoul.

[2] Dobbs, R, Smit, S, Remes, J, Manyika, J, Roxburgh, C & Restrepo, A 2011, *Urban world: mapping the economic power of cities*, McKinsey & Company, The McKinsey Global Institute (MGI), Seoul.

[3] McKinsey & Company 2016, *Automotive revolution – perspective towards 2030*, McKinsey & Company, The McKinsey Global Institute (MGI), Seoul.

[4] *United Nations World Urbanization Prospects 2014, revision*. Available at: https://esa.un.org/unpd/wup/Publications/Files/WUP2014-Report.pdf; p. xxi and p. 48.

[5] Wilson, M 2012, By 2050, 70% of the World's Population Will Be Urban. Is That A Good Thing? *Fast Company*. Available at: http://www.fastcodesign.com/1669244/by-2050-70-of-the-worlds-population-will-be-urban-is-that-a-good-thing [Accessed 17 May 2016].

[6] Winston, C & Mannering, F 2014, 'Implementing technology to improve public highway performance: a leapfrog technology from the private sector is going to be necessary', *Economics of Transportation*, vol. 3, no. 2, pp. 158–165.

[7] Neumann, C 2015, *Big data versus big congestion: Using information to improve transport*, McKinsey & Company. Available at: http://www.mckinsey.com/industries/infrastructure/our-insights/big-data-versus-big-congestion-using-information-to-improve-transport [Accessed 13 May 2016].

[8] World Health Organisation 2016(a), *United Nations Road Safety Collaboration, United Nations*. Available at: http://www.who.int/roadsafety/en/ [Accessed 13 May 2016].

[9] FIA Foundation 2016, *Road Safety Fund, FIA Foundation*. Available at: http://www.fiafoundation.org/our-work/road-safety-fund. [Accessed 13 May 2016].

[10] World Health Organisation 2016(b), *Road traffic injuries*, United Nations, World Health Organisation, Fact Sheets, fs358.

[11] BITRE 2015, *Traffic and congestion cost trends for Australian capital cities*. Department of Infrastructure and Regional Development. Available at: https://bitre.gov.au/publications/2015/files/is_074.pdf.

[12] European Commission 2011, *Roadmap to a Single European Transport Area, European Commission: Mobility & Transport*. Available at: http://ec.europa.eu/transport/strategies/facts-and-figures/transport-matters/index_en.htm [Accessed 13 May 2016].

[13] ITS America 2011, *What is Intelligent Transportation?* ITS America, ITSA WikiSpace.

[14] International Transport Forum 2010, Transport Outlook 2010, 'The 2010 International Transport Forum: Transport and Innovation: Unleashing the Potential', *Organisation for Economic Co-operation and Development*, Leipzig, Germany, May 26–May 28, 2010, pp. 1–28.

[15] Batty, M 2013, 'Big data, smart cities and city planning', *Dialogues in Human Geography*, vol. 3, no. 3, pp. 274–279.

[16] Kitchin, R 2013, 'The Real-Time City. Big Data and Smart Urbanism', *GeoJournal*, vol. 79, no. 1, pp. 1–14.

[17] Hobbs, A & Hanley, S 2014, *Big and Open Data in Transport, Parliamentary Office of Science & Technology*, POSTnotes Post-PN-472.

[18] Dia, H 2015, *Workshop on Smart Mobility*, Swinburne University of Technology, Booklet.

[19] Deloitte 2015. *Transportation in the Digital Age: Disruptive Trends for Smart Mobility*. Available at: http://www2.deloitte.com/uk/en/pages/business-and-professional-services/articles/transport-in-the-digital-age.html [Accessed 5 September 2016].

[20] Rail Level Crossing Group 2010, *National Railway Level Crossing Safety Strategy*, Australian Transport Council.

[21] Yi, P & Kandukuri, Y 2012, 'Optimum Location Identification of Plug-In Electric Vehicle Charging Stations Based on Graphic Weighting', *The Twelfth COTA International Conference of Transportation Professionals, American Society of Civil Engineers*, Beijing, China, August 3–August 6, 2012, pp. 3435–3440.

[22] Burt, M, Cuddy, M & Razo, M 2014, *Big Data's Implications for Transport Operations: An Exploration,* U.S. Department of Transportation, VNTSC-FHWA-14-13.

[23] Nvidia 2016. *Nvidia automotive.* Available at: http://www.nvidia.com/object/drive-automotive-technology.html [Accessed 2 June 2016].

[24] International Transport Forum 2015, *Automated and Autonomous Driving: Regulation Under Uncertainty.* Available at: http://www.itf-oecd.org/automated-and-autonomous-driving-regulation-under-uncertainty [Accessed 2 June 2016].

[25] Dia, H 2016, *Who or what is behind the wheel? The regulatory challenges of driverless cars. The Conversation.* Available at: https://theconversation.com/who-or-what-is-behind-the-wheel-the-regulatory-challenges-of-driverless-cars-55434 [Accessed 2 June 2016].

[26] Cunningham, W 2016, *Nvidia's computer for self-driving cars as powerful as 150 MacBook Pros.* Available at: http://www.cnet.com/au/news/nvidias-computer-for-self-driving-cars-as-powerful-as-150-macbook-pros/. [Accessed 2 June 2016].

[27] Owyang, J 2015, *Ten ways mobility-as-a-service changes your lifestyle.* Available at: http://www.web-strategist.com/blog/ [Accessed 2 June 2016].

[28] Bouton, S, Knupfer, SM, Mihov, I & Swartz, S, 2015, *Urban mobility at a tipping point. McKinsey & Company.* Available at: http://www.mckinsey.com/business-functions/sustainability-and-resource-productivity/our-insights/urban-mobility-at-a-tipping-point [Accessed 10 January 2016].

[29] Kamargiani, M, Matyas, M, Li, W & Schafer, A 2015, *Feasibility Study for 'Mobility as a Service' concept in London.* University College London. Available at: https://www.bartlett.ucl.ac.uk/energy/docs/fs-maas-compress-final. [Accessed 2 June 2016].

[30] Burrows, A 2015, *Journeys of the Future – Introducing Mobility as a Service. Atkins Global.* Available at: www.atkinsglobal.com.

[31] Center for Automated Research 2016, *The impact of new mobility services on the Automotive Industry.* Available at: http://www.cargroup.org/?module=Publications&event=View&pubID=138 [Accessed 2 August 2016].

[32] Santi, P, Resta, G, Szell, M, Sobosevsky, S, Strogatz, S, & Ratti, C 2013, *Quantifying the benefits of vehicle pooling with shareability networks.* Viewed 2 June 2016. Available at: http://arxiv.org/abs/1310.2963.

[33] Kaas, H-W, Mohr, D, Gao, P, Müller, N, Wee, D, Hensley, R, *et al.* 2016, *Automotive revolution – perspective towards 2030.* McKinsey & Company (January 1st). Available at: http://www.mckinsey.com/~/media/mckinsey/industries/high tech/our insights/disruptive trends that will transform.

[34] Hu, X, Liao, Z, Wang, J & He, J 2015, Shared parking policy analysis based on game theory. In *15th COTA International Conference of Beijing: American society of civil engineers (ASCE)*, pp. 3841–3853. Available at: http://ascelibrary.org/doi/abs/10.1061/9780784479292.354 [Accessed 3 August 2016].

[35] Doherty, MJ, Sparrow, FT & Sinha, KC, 1987, 'Public use of autos: mobility enterprise project,' *Journal of Transportation Engineering,* vol. 113, no. 1, pp. 84–94.

[36] Hietanen, S 2015, *Mobility as a service – European model of digital era transport, ITS Finland.* Available at: http://merjakyllonen.fi/merja/wp-content/uploads/2015/10/Hietanen-ITS-Finland.pdf.

[37] Kaufman, R 2016, *Chasing the next Uber – next city: which city will be the first to crack on-demand mobility? NEXTCITY.* Available at: https://nextcity.org/features/view/helsinki-kutsuplus-on-demand-transportation-mobility-next-uber.

[38] Liu, J, Lin, D & Li, C 2010, Designing and implementing a service oriented traveler information system. In *Traffic and transportation studies 2010.* Reston, VA: American Society of Civil Engineers, pp. 741–749. Available at: http://ascelibrary.org/doi/abs/10.1061/41123%28383%2971 [Accessed 3 August 2016].

[39] Rissanen, K 2016, *Helsinki regional transport Authority (HSL),* Available at: https://www.hsl.fi/en/helsinki-regional-transport-authority.

[40] Greenfield, A 2014, Helsinki's ambitious plan to make car ownership pointless in 10 years. *The Guardian.* Available at: https://www.theguardian.com/cities/2014/jul/10/helsinki-shared-public-transport-plan-car-ownership-pointless [Accessed 2 August 2016].

[41] Finger, M, Bert, N & Kupfer, D 2015, *Mobility-as-a-service: from the Helsinki experiment to a European model?* Available at: http://fsr.eui.eu/Documents/WorkshopPaper/Transport/2015/150309MaaSObserver.pdf [Accessed 2 August 2016].

[42] Dia, H 2015(b), *Driverless cars will change the way we think of car ownership. The Conversation.* Available at: https://theconversation.com/driverless-cars-will-change-the-way-we-think-of-car-ownership-50125 [Accessed 2 June 2016].

[43] Karl, C 2015, 'The role of autonomous vehicles in smart mobility', *Workshop on Smart Mobility,* Swinburne University of Technology, Melbourne, Australia, 17 November, 2015, pp. 1–315.

[44] Deloitte 2014, *The 2014 Gen Y automotive consumer study – the changing nature of mobility.* Available at: http://www2.deloitte.com/us/en/pages/manufacturing/articles/2014-gen-y-automotive-consumer-study.html [Accessed 2 June 2016].

[45] International Transport Forum 2015, *Urban mobility system upgrade: how shared self-driving cars could change city traffic.* Available at: http://www.internationaltransportforum.org/cpb/projects/urban-mobility.html [Accessed 2 June 2016].

[46] Zakharenko, R 2015, *Self-driving cars will change cities, Social Science Research Network.* Available at: http://papers.ssrn.com/sol3/papers.cfm?abstract_id=2683312 [Accessed 13 May 2016].

[47] Dia, H, Javanshour, F & Hill, J 2016, Network impacts of autonomous shared mobility. *Proceedings of the Machine Learning for Large Scale*

Transportation Systems Workshop, The 22nd ACM SIGKDD International Conference on Knowledge Discovery and Data Mining, San Francisco, United States, 13–17 August 2016.

[48] Access Economics 2009, *The economic benefits of intelligent technologies. Report by Access Economics Pty Limited for IBM.* Available at: https://www-03.ibm.com/press/au/en/pressrelease/27568.wss.

[49] Climate Group 2008. SMART 2020: Enabling the low carbon economy in the information age. *A report by The Climate Group on behalf of the Global eSustainability Initiative (GeSI).* Available at: http://www.smart2020.org/_assets/files/03_Smart2020Report_lo_res.pdf.

[50] Dobbs, R, Pohl, H, Lin, D, *et al.* 2013, *Infrastructure productivity: how to save $1 trillion a year,* McKinsey & Company, The McKinsey Global Institute (MGI), Seoul.

[51] Sulopuisto, O 2016, *Why Helsinki's innovative on-demand bus service failed. Citiscope.* Available at: http://citiscope.org/story/2016/why-helsinkis-innovative-demand-bus-service-failed [Accessed 10 July 2016].

[52] Wollschlaer, D, Foden, M, Cave, R & Stent, M 2015, Digital disruption and the future of the automotive industry. *IBM Centre for applied insights.* Available at: https://www.ibm.com/multimedia/portal/H752407R29967B14/IBMCAI-Digital-disruption-in-automotive.pdf.

[53] Ford 2015, *Ford smart mobility map. Ford smart mobility.* Available at: https://media.ford.com/content/dam/fordmedia/NorthAmerica/US/2015/03/02/MobilityExperiments.pdf [Accessed 2 August 2016].

[54] Titcomb, J 2016, *UberPool has been used more than 1 million times in London.* The Telegraph. Available at: http://www.telegraph.co.uk/technology/2016/06/07/uberpool-has-been-used-more-than-1-million-times-in-london/.

[55] Barber, L 2016, *Uber reveals London ridesharing figures for UberPool. CITYA.M.* Available at: http://www.cityam.com/242760/uber-reveals-london-ride-sharing-figures-for-uberpool [Accessed 6 July 2016].

[56] Crist, P, Greer, E, Ratti, C, *et al.* 2015, *Big data and transport: understanding and assessing options,* International Transport Forum Data Base, Paris.

[57] Benjamin, S 2015, 'Self-driving vehicles: how-to-guide', *Workshop on Smart Mobility,* Swinburne University of Technology, Melbourne, Australia, 17 November, 2015, pp. 1–34.

[58] Katal A, Wazid, M & Goudar, R 2013, 'Big data: issues, challenges, tools and Good practices', *2013 Sixth International Conference on Contemporary Computing (IC3),* Institute of Electrical and Electronics Engineers, Noida, August 8–August 10, 2013, pp. 404–409.

[59] Mims, C 2013, Coming Soon: *The Cybercrime of Things, The Atlantic.* Available at: http://www.theatlantic.com/technology/archive/2013/08/coming-soon-the-cybercrime-of-things/278409/ [Accessed 17 May 2016]

[60] Information Technology and Innovation Foundation 2010, *Explaining International IT Application Leadership: Intelligent Transportation Systems.* Available at: https://itif.org/media/explaining-international-it-application-leadership-intelligent-transportation-systems#video.

Glossary

AI	artificial intelligence
AV	autonomous vehicle
BTRE	Bureau of Transport and Regional Economics
ETC	electronic toll collection
GDP	gross domestic product
GPS	global positioning system
ITS	intelligent transport systems
MaaS	mobility as a service
MAV	minimum attribute vehicle
MGI	McKinsey Global Institute
NPV	net present value
RFID	radio frequency identification
STAR	short term auto rental

Further Reading

[1] ACOLA (2015). 'Delivering sustainable urban mobility.' *Australian Council of Learned Academies*.

[2] Barclays (2015a). *Disruptive mobility: A scenario for 2040. Barclays research insights.* Available at: https://www.investmentbank.barclays.com/content/dam/barclayspublic/docs/investment-bank/global-insights/barclays-disruptive-mobility-pdf-120115-459kb.pdf.

[3] Barclays (2015b). *Disruptive mobility: AV deployment risks and possibilities.* Available at: http://orfe.princeton.edu/~alaink/SmartDrivingCars/PDFs/Brian_Johnson_DisruptiveMobility.072015.pdf.

[4] Fishman, T (2012). *Digital-age transportation: The future of urban mobility.* Deloitte University Press. Available at: http://dupress.deloitte.com/dup-us-en/industry/automotive/digital-age-transportation.html.

[5] Global e-Sustainability Initiative (2008). 'Smart 2020: Enabling the low carbon economy in the information age.' GeSI, A report by Climate Group on behalf of GeSI.

[6] Litman, T (2015). *Autonomous Vehicle Implementation Predictions – Implications for Transport Planning.* Victoria Transport Policy Institute. Available at: http://www.vtpi.org/avip.pdf.

[7] Michael Sivak and Brandon Schoettle (2015), *Potential Impact of Self-Driving Vehicles on Household Vehicle Demand and Usage,* Sustainable Worldwide Transportation Program, University of Michigan. Available at: www.umich.edu/~umtriswt.

Chapter 14

Summary and future directions

Hussein Dia[1]

Abstract

This chapter provides a summary of the key themes arising from the individual chapters. It also outlines a framework for rethinking urban mobility and identifies the broad strategies for triggering policy changes. The chapter also provides a research agenda for future work in this area guided by the findings and recommendations from the individual chapter contributions.

14.1 Introduction

This book includes contributions from a number of multi-disciplinary research fields including urban planning, transport planning and strategy, traffic management and transport technology. Together, they represent a cohesive investigation into the landscape of urban mobility and the policy principles and practical solutions that will drive the development of sustainable urban mobility and improve accessibility to jobs and opportunities in our cities. The book also analyses the relationships between urban form and mobility and calls for a future with more integration of land-use and transport in compact cities. Although each city is different and policy responses vary, case studies were presented to show how engagement with a broad range of stakeholders, development of clear objectives and policy responses, on-going monitoring and evaluation, and communication of progress helps to achieve the outcomes of sustainable transport and reduce greenhouse gas emissions.

The previous chapters discussed the breadth and depth of a large number of contemporary challenges and best practices in the provision of sustainable urban transport. The book started by providing a comprehensive picture of the challenges facing urban mobility and identified a framework for transitioning the conventional 'predict-and-provide' approaches towards a framework for planning sustainable transport solutions (Chapter 2). It then detailed the policy pathways available to decision makers and reinforced these through case studies which demonstrated a range of benefits that accrued to communities which adopted and implemented these

[1]Centre for Sustainable Infrastructure, Swinburne University of Technology, Melbourne, Australia

policies (Chapter 3). By providing this background, the book laid the foundations for understanding the existing situation, emerging trends and the potential pathways for triggering change. The remainder of the book included contributions which tackled the specific challenges facing urban transport and provided solutions, pathways and best practices to address them. These included reducing automobile dependence (Chapter 4), integrated land-use and transport planning (Chapter 5), prospects for decarbonising suburban mobility (Chapter 6), public transport provisions (Chapter 7), motorised transport and impacts on public health (Chapter 8), policy direction for creating built environments which support active transport, health and well-being (Chapter 9), mobility through the sharing economy (Chapter 10), future vision for shared autonomous mobility-on-demand solutions (Chapter 11), gamification as a tool to trigger sustainable behaviours in urban mobility (Chapter 12) and digital innovations and disruptive mobility solutions – current and future (Chapter 13).

Overall, the book acknowledges that urban mobility and accessibility are key for promoting sustainable urban development. It also acknowledges their importance for enhancing the economies of urbanisation. Throughout the individual chapters and contributions, it also analyses how the urban function is improved through sustainable mobility solutions. However, the book also acknowledges that gaps exist towards achieving this vision, and that these gaps exist today largely due to the separation of land-use and transport, as well as the inadequate integration among transport modes. The book emphasises the urgent need to reframe urban mobility policies and practices in order to address these shortcomings.

This closing chapter ties the analysis together, emphasises the key messages advocated in the chapters, and summarises the policies and pathways that can be implemented to bring about the desired outcomes of sustainable urban mobility.

14.2 A framework for rethinking urban mobility

This book calls for a conceptual leap and renewed thinking in addressing urban mobility. A summary of the key elements for reframing how urban transport is planned and designed in cities, and how best to provide sustainable urban transport services is provided next. These elements have been recognised in previous studies which also explored similar pathways to reducing dependence on motorised transport (e.g. [1–5]). Here, they are updated and reinforced.

14.2.1 A systems approach

Cities are a complex network of interconnected systems. The challenges facing transport in our cities – rapid urban growth, dependence on motorised transport, reduced access to services and activities, etc. – are structural in nature and must be framed as part of a holistic approach to improve urban form in cities. Solutions and interventions which recognise this complexity will have a strong potential for charting a course towards sustainable urban mobility.

14.2.2 Transport as a 'derived demand' and 'valued activity'

In planning and designing urban mobility solutions, it is essential to recognise travel as a 'derived demand' but also a 'valued activity'. Travel originates from the need for people to access places, jobs, opportunities, services and activities. The purpose of most travel is to earn income, purchase goods, attend schools, etc. The transport infrastructure and the vehicles, cars, trains, buses and bikes that move on it are simply the means to achieve these ends [6]. As such, travel in itself has traditionally been seen as a cost and a wasteful activity, the duration of which should be minimised. But in the information age, the notion that all travel is a derived demand may become weaker. Today, passengers on trains and buses, and travellers using ride-sharing or other new modes of transport, are increasingly having better access to mobile devices and tablets which they use to occupy their time with a range of activities, including work. Travel time becomes a valued activity and would no longer be considered wasted time.

14.2.3 Reframing transport

Reframing the primary objective of urban transport as one for improving access to jobs and opportunities gives priority to policies and strategies which promote transit-oriented developments, improved public transport services, active transport infrastructure and less reliance on policies that encourage private vehicle usage. The concept of accessibility should apply to all segments of society to ensure that the poor and disadvantaged have adequate access to goods and services within the city.

14.3 Policies and strategies

Ten key principles from the 'Avoid, Shift, Share, Improve' framework were identified in Chapter 3 [7]. These included (1) Planning dense and human scale cities, (2) Developing transit-oriented cities, (3) Optimising the road network and its use, (4) Improving public transport, (5) Encouraging walking and cycling, (6) Controlling vehicle use, (7) Parking management, (8) Promoting clean vehicles, (9) Stakeholder consultation and engagement and (10) Creating pathways and adapting regulations to comprehensive deployment.

In this chapter, the broad policies and strategies are grouped into six categories to demonstrate their strategic linkage and their cumulative potential for triggering policy and operational change [1].

14.3.1 Strengthening the linkage between land-use and transport

The connection between land-use and transport needs to be re-built and strengthened to achieve sustainable urban mobility. An integrated approach to land-use and transport shifts the focus of planning from placement of structures and designation of land-use to that of enabling the realisation of people's needs and everyday functions in the most efficient and sustainable manner. Within this approach, the key challenge is therefore not merely to overcome the separate handling of transport and land-use planning. Rather, it is to foster an integration of multi-modal mobility within a holistic and sustainable land-use system [1].

14.3.2 Rethinking urban planning and transport engineering designs

As was discussed in Chapter 5, a strong link exists between transport supply and demand, and urban form. Mixed-land-use developments reduce the need for travel and promote active transport. Quality transport connections between functional places and facilities improve access and increases functionality of each place, leading to a reduction in the distances and number of trips between origins and destinations. This can be achieved through creative planning and urban designs, combined with innovative infrastructure and transport engineering designs. For example, compact configurations complemented with transport-oriented developments reduce private cars while still making it viable for cities to invest in different modes of public transport.

14.3.3 Realigning transport infrastructure investment

Achieving low carbon mobility requires prioritisation in the choices of infrastructure investments. The current imbalance in funding and investments between private and public modes of transport needs to be corrected [1,2]. This applies equally to developed and emerging countries. It is not sufficient to pursue policies that 'balance' investments between different modes of transport. More initial funding should be allocated to developing and expanding non-motorised and high-capacity public transport infrastructure. The option of value capture to complement public funding should be examined to generate sustainable funding streams. Besides being a politically appealing option, this funding model also reinforces the link between land-use and transport.

14.3.4 Integrating urban transport facilities and service operations

Efficient land-use patterns (e.g. compact, mixed and walkable) allow for less reliance on expensive mobility systems in general. Properly designed transport systems also contribute to business expansion, increased economic output and employment generation. Efficiency must underpin management, operational and system design practices throughout the urban transport sector. In the case of high-capacity public transport systems, this can take the form of redeploying buses and equipment to high-ridership markets that produce the highest fare revenues. Integrated transport and land-use planning development must also be emphasised in national urban development policies and plans.

14.3.5 Urban governance frameworks

The development of a fully integrated and sustainable multi-modal urban mobility system requires a robust urban governance structure. Innovations and policies geared towards sustainable mobility require strong institutional and governance structures to oversee their successful implementation. Political will, sound leadership, transparency and accountability are essential in building public trust. Also vital to the entire process are the planning institutions, as these are capable of creating compelling visions of urban futures. There is also a need to inject efficiencies, accountability and transparency into the urban transport decision-making process. This requires the development and institutionalisation of planning processes and evaluation approaches that are based on objective measures of performance and tied to well-articulated goals and hoped for outcomes. This promotes both transparency and accountability.

14.3.6 Regulatory frameworks

The current regulatory frameworks for management of road space and urban form have perpetuated urban transport design that favoured motorised transport. Transforming our cities towards sustainable urban mobility requires major reforms in the legal and regulatory framework relating to urban transport management. The interventions highlighted in this book call for changes in the management of space, the urban form, the engineering of transport, as well as in the institutional and financing arrangements related to urban development.

14.4 Research agenda to inform low carbon mobility policies – Australian perspective

Central to the success of low carbon mobility solutions is rigorous research that meets the twenty-first-century global challenges including energy provision, climate change and health. Cities and urban areas are expected to achieve big gains through well-planned policy implementations and smart thinking on urban transport. To deliver substantial improvements in mobility and reductions in emissions, major policy, behavioural and technological changes would be required to achieve desired outcomes and targets for low carbon living.

The establishment of a research agenda on low carbon mobility presents an opportunity to enhance support for research and innovation by building on existing success in transport research and intellectual capital. A thriving research agenda represents a major investment in the future of low carbon mobility and sustainable transport and would also provide a distinctive trajectory and direction for students, researchers and industry practitioners. This section provides a research framework that identifies high-level research needs to enable successful deployment and integration of low carbon mobility solutions in our cities. The framework articulates a vision for low carbon mobility, maps the landscape of research themes and identifies the research gaps that need to be bridged to enable successful deployment.

The research framework presented in this chapter draws on the current status of low carbon mobility research, and discussions with academics, researchers and industry stakeholders. This approach allowed for an objective analysis of the current challenges and opportunities, areas where there is general agreement that fundamental research is lacking, and highlighted new research opportunities available through new technologies. Soliciting the insights and judgements of researchers and industry specialists also helped to map the areas where the stakeholders believed insufficient work has been done, and where more focused research is needed. It also allowed for identifying those areas that are likely to become important in the future and where research is needed to support their development.

In compiling these priorities, the insights of researchers and specialists were solicited during workshops that were held in Adelaide and Melbourne, Australia, as part of the involvement of a few universities in low carbon mobility research [8,9]. The workshops included participants from academia and industry who presented on a wide range of topics which covered demand modelling and urban planning studies, through to new methodologies for urban mobility data collection, modelling, and establishment of living laboratories to evaluate impacts of new interventions. The workshops resulted in the mapping of a broad research framework that included inputs from the stakeholders who attended and participated in the workshops. The context of this framework was suburban travel, but the high-level research needs are equally applicable to low carbon mobility in urban environments (Figure 14.1).

The overarching vision for the low carbon mobility research framework is to reduce greenhouse gas emissions from passenger car usage. The proposed framework includes research programmes which would investigate travel demand in the age of connected mobility; new methods to provide travel supply including autonomous shared mobility and on-demand car and ride-sharing; and investigations of long-term impacts on mobility, energy, urban form, sustainability and quality of life. The research agenda also includes strategic research programs in urban governance and opportunities for improving urban transport in the developing world. The mapping in Figure 14.1 demonstrates the wide range of ideas that have potential to improve mobility, including options for electrification of road travel, teleworking and smart work centres. The key theme in this mapping is the need to provide reliable and cost-effective alternative mobility options that reduce reliance on private vehicle travel. The research framework includes studies which aim to understand:

Current situation and trends. This includes an assessment of the current situation and best practices in deployment of urban mobility and the new options for reducing the carbon footprint of urban transport with a particular focus on the opportunities available through disruptive technologies, shared car ownership and on-demand access to public transport and new mobility options. It also includes identification of emerging or anticipated future trends in urban mobility with a particular focus on the role of autonomous on-demand shared mobility solutions. Although it is important that comprehensive environmental scans are conducted, and a synthesis of published research is undertaken to identify gaps, it is also recognised that using the literature reviews as an extrapolation method has some limitations and will provide a limited insight into the trajectory of research in smart mobility. In these studies, it becomes important to solicit the insights of stakeholders through interviews, workshops, surveys and questionnaires on issues affecting their short- to long-term transport operations and planning.

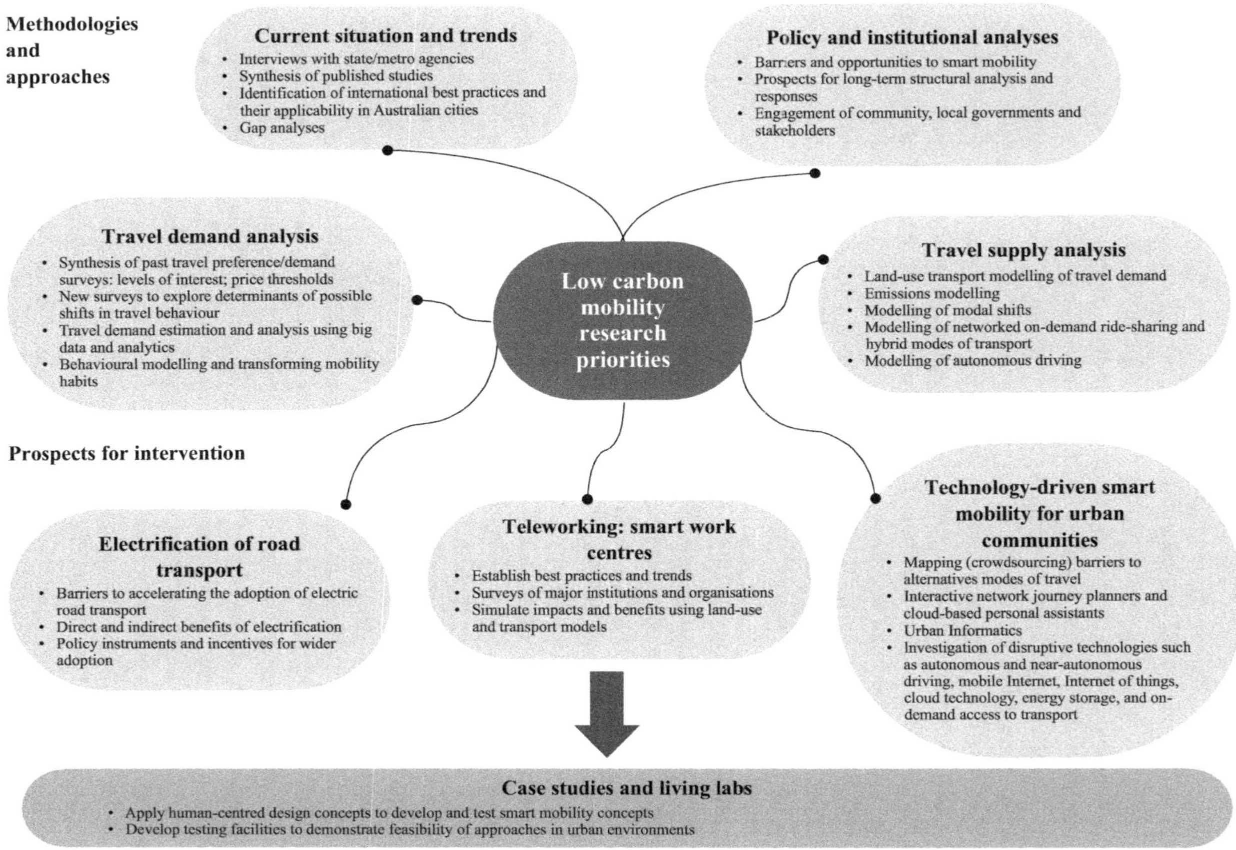

Figure 14.1 A research framework for low carbon mobility

Policy and institutional analyses. This includes assessment of the barriers and opportunities arising from low carbon mobility solutions. The analyses should include an evaluation of the current regulatory frameworks, pathways to developing outcome-focused research, regulations under uncertainty, public expectations and acceptance of technology-driven mobility solutions, the future role of government agencies in managing the transport network under scenarios of shared autonomous mobility, and enhancing engagement with the community, local governments and other stakeholders. An important focus under this category of studies is also the development of new urban governance systems to guide the creation of urban mobility innovations, corporate social responsibility, Internet-enabled multi-stakeholder communication, and new modes of community engagement to plan pathways for deployment of ubiquitous low carbon mobility solutions. These efforts would be directed towards developing and evaluating new models and processes by which social, design and technological innovation can be more effectively delivered into the planning of smart mobility.

Travel demand analysis. This includes analysis of existing household travel surveys to define travel behaviour (e.g. trip generation, destination choice and mode choice for households), the conduct of new studies to understand the drivers for travel demand in urban areas in the age of shared and connected mobility, and surveys to establish the determinants of shifts in travel behaviour from private vehicles to new mobility solutions. It also includes Stated Preference experiments to explore opportunities for changes in travel behaviour (trip frequency, destination and mode choices), focussing on areas of high car dependency/forced car usage/areas to explore determinants of shifts in travel behaviour. Other studies that would need to be completed under this framework include travel demand estimation and analysis using predictive analytics, machine learning tools and data mining of new sources of information from mobile use, smartcard automated fare collection and crowd-sourced data to uncover behavioural patterns using sophisticated modelling and analytics tools. Travel demand studies would also need to include projects that examine the land-use-transport interactions and their influence on the demand for travel modes under variable scenarios of supply and service patterns, especially those related to shared mobility and future autonomous vehicle applications such as first-and-last-kilometre travel solutions.

Travel supply analysis. This should include studies into the modelling of land-use-transport interaction, demand for travel and mode splits under different scenarios of infrastructure supply, including walking and cycling infrastructure, electric vehicle charging networks and networked public transport services. The modelling tools should allow for evaluating the impacts at a regional level (macro-models); precinct level (meso-models) and operational level (e.g. micro and increasingly sophisticated nano behavioural models). The application of these tools will allow for investigations of modal shifts, and the impacts of smart mobility intervention measures including their benefits in reducing greenhouse gas emissions, reducing social and economic costs of current levels of car-dependence, and their efficiency and cost-effectiveness in providing customers with alternative travel options. The travel supply analysis research should also look into transport planning and modelling studies focusing on pathways to increasing customer usage of different modes of travel. The modelling tools can also serve as decision support systems for investment in low carbon mobility and would be valuable for assessment of the feasibility and cost-effectiveness of the proposed interventions; assess a portfolio of investment options to encourage greater adoption of new mobility solutions that optimise the reduction in cities' carbon footprint and high areas and social cohorts that might be most receptive to adopting low carbon mobility solutions.

Prospects for intervention. A large number of interventions are potentially available for decision makers including smart transport interventions, teleworking and electrification of road transport. There is already considerable research and momentum around network management, teleworking [10] and Intelligent Transport Systems (ITS) interventions, but probably the area where important research is needed going forward is in disruptive mobility which includes collaborative shared mobility (Chapter 10), autonomous vehicles (Chapter 11) and digital innovations (Chapters 12 and 13). In these areas, more research is needed to address the following issues particularly about autonomous vehicles:

- Will they reduce or increase congestion?
- How will they impact total vehicle-kilometres of travel?
- Will they increase or decrease urban sprawl?
- How will they impact urban form?
- Will they induce more demand for travel?
- What impact will they have on parking?
- Will they reduce or increase emissions?
- How will they impact car ownership?

Case studies and living laboratories. Research into low carbon mobility solutions and interventions should also include opportunities for creation of case studies and pilots in living labs. For example, a number of universities in the United States have established off-road test-beds for testing autonomous driving. These are especially relevant to assessment of road safety impacts. Equally important are trials on public roads to provide travellers and consumers with opportunities to witness the new mobility solutions, evaluate their experiences and get a better understanding of the potential impacts and opportunities while also helping public authorities to learn more about the solutions and work towards their deployment.

14.5 Practical research routes to inform urban mobility policies

The high-level research framework described in this chapter is aimed at identifying some practical routes that can be used to steer transport policy on low carbon mobility. As the list of 'next big thing' grows longer with the fast pace of breakthroughs and scientific advances, it is important that policy making provides a guiding vision to ensure that technology interventions, in particular, are well applied and focused on user needs, and that they address genuine and practical problems that promote sustainable cities.

Some of the key research routes that have been identified in previous research [11] are relevant and valid for low carbon mobility policies. These include the following.

14.5.1 Looking beyond the immediate benefits and establishing the long-term impacts of new technologies

The sweeping changes anticipated by disruptive mobility have at times inspired visions of a very different future, as well as a good deal of hype. To distinguish between the hype and reality, rigorous and extensive research must go beyond the immediate obvious impacts. An example is autonomous vehicles which seem to have captured people's imaginations over the past few years. Although they are very likely to introduce some big benefits such as reducing traffic injuries and fatalities and free up people's time from the driving task, it is still not well understood how they will impact congestion and the total vehicle-kilometre-travelled per capita, and whether they are likely to replace or augment public transport, and their potential impacts on health and well-being. There are also concerns about over-extensive urban sprawl and also impacts on energy and land-use. Recent research [12] suggests that the current vehicle fleet could be reduced by up to 90 per cent in suburban areas when a shared network of driverless vehicles are introduced. It is not clear, though, whether they will induce new demand for travel given that travel time will become shorter because of fewer vehicles on the road, and because people would feel that travel time is no longer unproductive because they are free from the task of driving. It is therefore important to undertake research that looks past the immediate benefits and investigate how these technologies are likely to impact urban living, demand for travel, and social cohesion in the long term.

14.5.2 Develop rigorous but flexible evaluation frameworks and tools

Given that some of the disruptive mobility solutions have not been tested and deployed yet, their impacts are probably best evaluated on test-beds or in simulations. The key advantage of using test-beds is the ease of replicating reality and how consumers and drivers engage with and react to the technology. The limitation, however, is that they are usually small scale and can't capture wider long-term impacts on larger city networks. Agent-based simulations, properly calibrated and validated, provide a cost-effective solution to complement the test-bed evaluations, but these also have some limitations particularly in replicating behavioural issues. Further developments are also needed for decision-making tools. The most commonly used method to evaluate transport projects today is the cost-benefit analysis (CBA). Although this is effective in determining monetary value over a period of time, CBA does not consider other important factors in terms of how it will affect other facets of people's lives. This includes who exactly will benefit from the project, as well as who does not. Multi-criteria analysis is an improved evaluation tool compared to CBA as it analyses a broader set of criteria or policy objectives. However, this method is subjective, and inconsistencies may arise. Other methods that offer improved evaluation techniques must be developed. These techniques should complement the CBA with risk assessment, as well as consideration of qualitative impacts on the general public.

14.5.3 Adapt governance systems and develop agile and outcome-focused regulations

Regulations will play a key role in the emergence and development of disruptive mobility solutions. They are also likely to be the biggest hurdle for their deployment. The regulators must adapt and rethink their approaches to avoid stifling the innovative uses of these technologies. Effective responses require an early and on-going dialogue between regulators, developers and the public in which regulators would create legal frameworks that are flexible but robust. For example, with self-driving vehicles, an important role for the regulators will be to limit physical risks especially those that might be posed during interim years when legacy fleets of vehicles would interact with autonomous vehicles. Given the fast pace of technological developments, regulators would need to re-think their roles and focus on achieving agreed outcomes rather than enforcement of codes and standards.

14.5.4 Facilitate and encourage active transport and public transport innovations

Some of the current mobility disruptions are already impacting the taxi industry, and there is some concern that it may in the long-run impact public transport, particularly bus services. There are a number of innovative public transport initiatives around the world such as the Bridji bus service in the United States which has been deployed in a successful way. Adopting innovative solutions for bus transport, in particular, is crucial for moving towards a more sustainable future and will play an important role in bridging the gaps for first-and-last kilometre travel. Public transport must be efficient, attractive and of good quality to appeal to the general public, and its development must be part of a holistic solution that integrates public and active transport modes. Bike-sharing has become popular in many cities around the world and has become ‘trendy’ rather than simply ‘sustainable’. Smart technologies and strategies that can support and increase the attractiveness and appeal of this mode of travel would also contribute to more sustainable mobility solutions in our cities.

14.6 Concluding remarks

The need and urgency for steering our cities on a course towards low carbon mobility is important not only for reducing the costs associated with urban transport, but also for increasing the overall benefits to society [2]. Widespread deployment of sustainable mobility systems would cut across the intersection of the most urgent challenges confronting the global community today [1]. Measures that seek to reduce greenhouse gas emissions or improve the quality of life for urban dwellers cannot succeed without addressing the challenges of sustainable mobility, nor without redressing the prevailing distortions in accessibility to our urban centres.

The contributions in this book demonstrate that sustainable urban mobility is an essential ingredient to many of the indicators by which we measure our quality of life: our health, happiness, prosperity, connectedness and security. They also suggest that it is possible to promote interventions that can effectively enhance the mobility and accessibility of our cities by providing policy makers with the instruments and strategies to positively influence the well-being of city populations.

References

[1] Givoni, M and Banister, D (2013). Mobility, transport and carbon. In Givoni, M and Banister, D (Eds), *Moving Towards Low Carbon Mobility,* Edward Elgar Publishing, Cheltenham, UK, pp. 1–12.

[2] Banister, D (2013). City transport in a post carbon society. In Givoni, M and Banister, D (Eds), *Moving Towards Low Carbon Mobility*, Edward Elgar Publishing, Cheltenham, UK, pp. 255–266.

[3] Bishop, J (2013). Technology. In Givoni, M and Banister, D (Eds), *Moving Towards Low Carbon Mobility*, Edward Elgar Publishing, Cheltenham, UK, pp. 92–110.

[4] Givoni, M (2013). Alternative pathways to low carbon mobility. In Givoni, M and Banister, D (Eds), *Moving Towards Low Carbon Mobility*, Edward Elgar Publishing, Cheltenham, UK, pp. 209–230.

[5] Hickman, R (2013). Urbanization and future mobility. In Givoni, M and Banister, D (Eds), *Moving Towards Low Carbon Mobility*, Edward Elgar Publishing, Cheltenham, UK, pp. 60–74.

[6] UN-Habitat (2012). *Planning and design for sustainable urban mobility: global report on human settlements 2013*. Available at: http://unhabitat.org/planning-and-design-for-sustainable-urban-mobility-global-report-on-human-settlements-2013/ [Accessed 17 September 2016].

[7] Bongardt, D (2016). *10 principles SUT*. Available at: https://prezi.com/7ufnp8crzc1l/10-principles-sut/ [Accessed 18 September 2016].

[8] Philip, M and Taylor, M (2014). Research synthesis report: A research agenda for low carbon mobility. *CRC for low carbon living*. Available at: http://www.lowcarbonlivingcrc.com.au/resources/crc-publications/crclcl-project-reports/rp2009-research-synthesis-report-research-agenda. [Accessed 17 August 2016].

[9] Dia, H (2014). *Workshop on greening suburban travel*. Swinburne University of Technology. Available at: http://www.lowcarbonlivingcrc.com.au/research/program-2-low-carbon-precincts/rp2021-greening-suburban-transport [Accessed 17 August 2016].

[10] Aguiléra, A, Guillot, C and Rallet, A (2012). Mobile ICTs and physical mobility: Review and research agenda. *Transportation Research Part A: Policy and Practice*, 46(4), pp. 664–672.

[11] Bruun, E and Givoni, M (2015). Six research routes to steer transport policy. *Nature*, 523(7558), pp. 2–4.

[12] Dia, H, Javanshour, F and Hill, J (2016). Network impacts of autonomous shared mobility. In *Machine learning for large scale transportation systems workshop*. San Francisco. Available at: http://media.wix.com/ugd/ea3995_f398d3792c764647ae791aab726a8ad4.pdf [Accessed 17 August 2016].

Glossary

CBA cost-benefit analysis
MCA multi-criteria analysis

Further Reading

[1] Aditjandra, P, Mulley, C and Nelson, J (2013). The influence of neighbourhood design on travel behaviour: Empirical evidence from North East England. *Transport Policy*, 26, pp. 54–65.
[2] Akyelken, N (2013). Finance and investment in transport. In Givoni, M and Banister, D (Eds), *Moving Towards Low Carbon Mobility*. Edward Elgar Publishing, Cheltenham, UK, pp. 129–147.
[3] Banister, D (2011). Cities, mobility and climate change. *Journal of Transport Geography*, 19(6), pp. 1538–1546.
[4] Banister, D, Anderton, K, Bonilla, D, Givoni, M and Schwanen, T (2011). Transportation and the environment. *Annual Review of Environment and Resources*, 36, pp. 247–270.
[5] Banister, D, Stead, D, Steen, P, *et al.* (2000). *European Transport Policy and Sustainable Mobility*. Spon Press, London.
[6] Bishop, J (2013). Technology. In Givoni, M and Banister, D (Eds), *Moving Towards Low Carbon Mobility*. Edward Elgar Publishing, Cheltenham, UK, pp. 92–110.
[7] Hayashi, Y (2013). Drastic visioning and backcasting to leapfrog to low carbon transport in growing Asia. Keynote presentation. In *Workshop on a Research Agenda for Low Carbon Transport*. Adelaide, October. CRC for Low Carbon Living, Sydney.

Index